Plumbing Technology

Justin Duncan

BNP Business News Publishing Company
Troy • Michigan

Library of Congress Cataloging in Publication Data

Duncan, Justin
Plumbing technology /Justin Duncan.
p. cm.
Includes index.
ISBN 0-912524-76-6
1. Plumbing I. Title
TH6123.D86 1994 93-11615
696'.1--dc20 CIP

Editor: Joanna Turpin
Art Director: Mark Leibold
Copy Editor: Carolyn Thompson

Printed in the United States of America
7 6 5 4 3 2 1

Dedication

This book is dedicated to my children, Ruth and David.

Acknowledgments

Special thanks to my editor and publisher, Joanna Turpin, whose active cooperation and effort made this book possible.

Table of Contents

Preface

The teaching of the plumbing industry and its engineering differs from the normal, rigid formulas and procedures common to other building industries and engineering categories, and the differences manifest themselves as follows:

- The need to understand the historical origins and the industry growth processes that have brought about present day conditions. Presenting the large picture is a must.
- The need for the reader to understand what is expected of: a) a mechanic in the field, b) a contractor bidding on and producing the product, c) a future engineer in a design learning mode.
- The need to understand that plumbing engineering is not an engineered science as we know it. It relies too heavily on experience, innovation, anticipation, and the proper use of established factors used to level off those conditions that show little rigidity.

The author of *Plumbing Technology* accomplishes these needs in an easy, simplified manner, all the while offering the reader a view of the complete spectrum of the plumbing industry. The author's teaching background and experience are evident as he touches on issues that, to an experienced plumbing engineer, always seem to be lacking in other texts, as well as published papers.

Justin Duncan has applied a simple, human approach to the subject without running the reader through complicated and confusing details and hypotheses and yet still maintains a technological posture. The book allows for a broad understanding of the plumbing industry as a whole and will enrich the reader pursuing participation in this industry. Time will prove *Plumbing Technology* to be a valuable learning tool for the plumbing industry.

Bernard J. McCarty, PE, CIPE
Former President of the American
Society of Plumbing Engineers

About the Author

Justin Duncan, P.E. earned his B.S. degree in mechanical engineering and his Master's degree in environmental engineering. He currently teaches courses in plumbing and fire protection at Northeastern University and is a senior mechanical engineer at Bechtel Corporation.

Other published works by Mr. Duncan include several booklets, instructional programs regarding the operation of various types of machinery, plumbing installation manuals, and articles on fire protection. Mr. Duncan has also recently written a book on fire protection, which is being published by Business News Publishing Company.

Chapter One

History of Plumbing

Plumbing systems in use today are a result of developments and improvements over the years. These developments were achieved with the contribution and accumulated experience of engineers, contractors, plumbers, committees for standards development, manufacturers, insurance companies, research laboratories, and the National Bureau of Standards (NBS).

Today's plumbing systems are strictly regulated and enforced by standards and codes. It is vital that we become familiar with the applicable state code, as well as any local regulations, before developing a new project, making any design modifications including major repairs or replacing/refurbishing an existing plumbing system.

Published codes and standards help disseminate information, leading toward the goal of uniform standards in all fifty of the United States. At the present time, each state's standards are very similar to the others, but they are not identical, due to local needs and requirements. The process of creating a national unified Building Code is now in progress, and the Plumbing Code is expected to be next.

While familiarity with the applicable codes is mandatory, just knowing the code is not enough. Plumbing technology is not limited to water, waste, and vent systems. We design, install, and maintain safe water supplies and sanitary systems to protect people's health, safety, and welfare. The entire plumbing system should perform in a manner that meets these needs.

History

Our knowledge of plumbing history is based on available historical records and published books and articles. Manufacturers' literature is another good source of plumbing history, as it recognizes past mistakes and emphasizes successes and improved techniques. Our knowledge is also based on years of hard work and imaginative thinking about how to repair or replace materials and/or systems that have proved to be unsatisfactory.

It is important to study and understand the evolutionary process of plumbing. Knowing this, history provides us with the following information:

- Knowledge and understanding of systems
- Objectivity in accepting improved systems
- Direction and guidance in design and installation
- Areas requiring attention and caution
- Practical solutions for problem solving

From primitive times, intolerable health conditions have been the primary impetus for the progressive development of plumbing and sanitary systems. Improved plumbing systems contributed to the improvement of personal hygiene, building design, and the development of new building fixtures and materials. The pressurized supply of safe drinking water, sanitary disposal of wastes, and efficient disposal of storm water all worked to improve the quality of human life.

The history of plumbing in the United States, including the development of standards, is only a few hundred years old. The origins of plumbing, however, go back thousands of years. Archeologists have discovered traces of advanced plumbing systems in many ancient cultures, including India, Egypt, Greece, Israel, Babylon (south of what is today Iran), and in the mighty Roman Empire, which extended from Rome to Jerusalem and north of Europe to France and England. Many of these areas were on the paths of well-known ancient trade routes, so knowledge of water supply and plumbing systems could have been spread by travelers along these routes.

One example of ancient, advanced plumbing systems is found in the "Hanging Gardens of Babylon," one of the Seven Wonders of the ancient world. These gardens were built on elevated terraces or rooftops and contained an extremely advanced system of internal drainage. Archeologists have also found traces of an extensive irrigation system there.

The Roman Empire is responsible for our modern word for plumbing — *artifacts plumbarius* (worker in lead). Lead was one of the early materials used in plumbing systems because of its ease of fabrication and fairly low melting temperature. Wooden pipes have been used throughout the history of plumbing, but the ancient Romans made pipes of wood, lead, clay, and stone. The remains of these systems are still being found today. (Modern iron piping fabrication originated in England around 1820.)

Through various conflicts and land seizures, the Romans acquired a large number of slaves. Though lacking their freedom, these slaves made significant contributions to technical advancement. However, the ancient Roman empire fell around 450 A.D., its civilization was destroyed, and what history calls the *dark ages* followed for about 1000 years. The dark ages brought with it a lack of hygiene, which caused disease to flourish, causing the deaths of hundreds of thousands of people. As an example of the pervasive lack of hygiene, Paris, France passed an ordinance in the year 1395 to prohibit the throwing of sewage from buildings directly onto the street. At that time, this was considered a big step forward.

Plumbing in the United States

There are traces of plumbing fixtures in this country dating far back before Columbus and the Pilgrims, but overall, progress was slow in coming to the United States.

A few important dates in the evolution of plumbing in this country are:

- 1700 — Dumping wastes onto the street is prohibited in New York City
- 1728 — First underground sewer is built in New York City. In the beginning, sewers had brick sides with the bottom made out of flat stones.
- 1776 — First water supply reservoir is built in up-state New York.

In the latter part of the 18th century and early 19th century, installation of domestic sewage vaults started. Development of standards began almost simultaneously in Boston, New York, Philadelphia, and Baltimore. Because of the existing unsanitary conditions, diseases tended to spread quickly among city populations, and health boards were established in major east coast cities to deal with this menace. In 1840, after numerous, disastrous fires, New York City acted on the need for a continuous water supply. The Croton Aqueduct was built to supply the more than 300,000 people of New York City with water.

The first fixtures developed were the kitchen sink, bathtub, and toilet. Sinks initially had one hole for the cold water supply faucet and later two holes for both cold and hot water. Later, a mixing valve was developed to mix cold and hot water. Bathtubs were originally manufactured in the shape of a shoe and were made of wood, which was either painted or lined with zinc or copper. In 1872, the cast-iron, enameled bathtub came on the market. This tub was smoother, easier to clean, more durable, and more sanitary.

Tubs today are also made of fiberglass with a non-slip bottom. This fixture is light, fairly inexpensive, and easily installed. Conveniently molded parts provide for supports, and attachments enhance its use.

The construction industry slowly began to adapt its production to meet the need for improved plumbing systems within buildings. In the beginning, the pipes within apartments were exposed; later they were boxed in. Later, traps were developed to prevent the return of foul sewer odors and gases into the habitable areas where fixtures were installed.

An important landmark in the history of American plumbing was the proposal to introduce venting into a sanitary drainage system. This proposal came about in 1874 at a conference sponsored by the plumbing trades in New York City. The main topic at this conference was the odor that occurred in buildings from siphoned-out traps. A theoretical scheme was proposed, which would equalize the atmospheric pressure in the sewer line (downstream of the trap) with the inlet pressure at the fixture outlet (normally open upstream of the trap) to prevent siphoning. The scheme was tested, it worked, and its use spread rapidly all over the country and throughout the civilized world.

In the cities of the United States, sanitary plumbing systems replaced the use of privies and privy vaults in about 1886. In the country, outdoor earth privies were in use for many more years thereafter. In the year 1890, the first wash-down type of sanitary water closet (WC) or toilet was introduced. The main reason for the rapid introduction of plumbing fixtures, in general, was their sanitary, smooth surface and reasonable cost due to mass production. Another step forward was the introduction of toilet rooms (bathrooms) in dwellings. Soon after, people demanded running hot water at their fixtures. The free market system of supply and demand brought about major advancements and a new industry developed.

After World War I, the demand for higher quality and less expensive plumbing fixtures increased in the United States. Market demands for plumbing fixtures abated somewhat with the depression of 1929, but in the years following the depression, there were many improvements made to plumbing fixtures and equipment. These improvements led to a reduction in the potential sources of contamination. Electric utilities eventually extended their territorial coverage to rural areas, and the pumping of water for domestic use as well as irrigation improved and became less expensive. Private sewage systems (septic tanks and leaching fields) were developed shortly thereafter.

From the depression to the present time, the housing industry has grown by leaps and bounds, and plumbing systems - both qualitatively and quantitatively - have grown along with it. Material shortages and new technologies have caused a variety of techniques to be implemented over the years. For example, copper became scarce in 1966, so plastic piping was used instead. Plastic piping has become a very large industry and while it does not replace copper tubing entirely, it has captured a substantial share of the market.

Conservation came into vogue in 1974 with the advent of the oil crisis. People began to realize that substantial savings in fuel and water costs could be achieved through conservation. One method of conservation was achieved through solar heating. Solar heating, limited hot water temperature, and reclaimed heat wherever possible are some of the energy conservation possibilities. Water saving fixtures were another method of conservation. Presently, more types of waterless or water-saving fixtures are being tested and installed for heavy users such as recreation facilities and race tracks. Toilets that employ natural biological decomposition to convert wastes to stable and usable products have also been developed.

Toilets

The first patent for a modern toilet belongs to an Englishman named Alexander Cumming, who invented the "S" trap in 1775. An improved version was developed and patented three years later by Joseph Bramah. Early toilets were flushed with water from a wall-mounted reservoir (tank) located approximately five feet above the toilet. This reservoir was later replaced with a low tank and then with a one piece toilet-tank.

In the United States, the chronological development of the 20th century toilet is as follows:

- 1900 — American pottery manufacturers develop glazed vitreous china
- 1905 — Wall-hung, tank-type toilet appears
- 1915 — Low tank installations are used
- 1930 — One piece toilet and tank is developed

Today, advanced toilets come equipped with jets of water and warm air, which completely eliminate the need for bathroom tissue. Naturally, these innovative products are quite expensive. The technical achievements of the U.S. stand in sharp contrast to those of many underdeveloped countries which, to this day, do not have the benefit of modern hygienic plumbing systems.

Note: There is an American Sanitary Plumbing Museum at 39 Piedmont Street, Worcester, MA 01610; (508)756-5783. The museum is open Tuesdays and Thursdays 10:00 a.m. to 2:00 p.m. and exhibits an extensive collection of old fixtures, trimmings, and much more.

Chapter Two

Codes and Standards

Engineers and technicians try to design safe and useful plumbing systems. Contractors try to install operationally sound plumbing systems. Codes and standards give these professionals the guidance they need to do the job correctly. In order to understand the difference between codes and standards, let us start with a definition. A plumbing *code* is defined as a set of rules (principles), legislation, and/or regulations adopted by a government authority to ensure a minimum sanitary requirement for habitable places. A *standard* is defined as a set of recommended principles and guidelines established by a professional organization as the basis for the design, installation, and maintenance of a certain system.

Codes

Codes provide a minimum set of rules for the design, installation, and maintenance of various building systems. Most codes establish performance objectives by providing specific requirements that must be met. It is usually left to system designers to decide how to meet those objectives, as more than one solution is often possible. Code jurisdictions are usually established within geopolitical boundaries.

Plumbing codes were developed to protect the health and welfare of the general public. Without codes and regulations, economics would govern, and public health would probably suffer. In 1848, England passed a law entitled the *Public Health Act*, which outlined the basic rules for a sanitary system. In the United States, the first tentative code or group of regulations was established in 1842. These regulations were supposed to prevent the pollution of water supplies and provide adequate drains and soil pipes to quickly discharge sanitary wastes.

In 1866, the state of New York granted authority to the Metropolitan Board of Health of New York City to control and regulate plumbing in the city. Soon after, the New York Sanitary Code (including plumbing regulations) was enacted. In 1870, Washington, DC became the first in the nation to put its plumbing regulations into a separate code. Other states then began to establish examination boards for plumbers. These boards also wrote plumbing regulations in cooperation with local health boards. Little by little, all major cities in the U.S. adopted plumbing regulations as part of their sanitary codes.

By 1913, the plumbing codes of various cities in the United States were quite comprehensive. They specified how the plumbing systems were to be installed; what fixtures were to be used and where; minimum sizes for water supply piping, drains, and vents; materials to be used; administrative procedures for securing permits; rules for inspection and testing, as well as many other details.

Wholesalers and trade associations soon pushed for the standardization of industrial products. They appealed to the United States government to initiate studies and develop a "model plumbing regulation." This effort resulted in the *Recommended Minimum Requirements for Plumbing in Dwellings and Similar Buildings*, dated July 3, 1923. These requirements were issued periodically until 1932.

There are presently four major national model codes:

- *National Plumbing Code,* published by Building Officials and Code Administration International (BOCA)
- *National Standard Plumbing Code*, published by the National Association of Plumbing-Heating-Cooling Contractors (NAPHCC)
- *Uniform Plumbing Code*, published by the International Association of Plumbing and Mechanical Officials (IAPMO)
- *Standard Plumbing Code*, published by the Southern Building Code Congress International (SBCCI)

These codes are revised and republished every two to four years. Most states either have their own published plumbing code or have legally adopted one of the codes listed above. Technicians using specific codes must keep themselves current with the modifications. At the present time, there is no one code that is officially recognized in all fifty of the United States. (Due to the ingenuity of the designer, installer, and manufacturer, new solutions come into being. Certain sections of a code may require interpretation. Two different jurisdictions might interpret the same code section differently. Written clarification is mandatory in such cases.)

Before any plumbing work (including new installations, remodeling, major repairs, or replacements) is performed, state codes and local regulations must be consulted. The installation of more plumbing fixtures than are required by code or the use of larger diameter water supply pipes in an installation are always acceptable. However, the installation becomes more expensive if additions are made to the basic code requirements.

Any state or national plumbing code contains a large amount of basic useful data; however, the codes are not intended to instruct persons on how to design a complete system or how to eliminate problems due to faulty installation or due to installation aging. One must learn these techniques in an organized way from technical books, in a specialized course, or through years of apprenticeship. The codes provide a framework of *minimum* requirements that must be considered in the design development.

Note: The study of this book will guide the user to determine what is basically needed in order to ensure the correct, economical, and proper design, installation, and maintenance of a plumbing system. The tables, charts, diagrams, etc., provided in this book are based on code requirements, but they are for illustrative purposes only. Always consult local codes before performing any work on a plumbing system.

Special Considerations

When a plumbing system is designed for an agency or department of the U.S. government, such as the General Services Administration (G.S.A.), the U.S. Army Corps of Engineers, or various military departments, specific requirements may exist that pertain to that agency or department. These requirements must be obtained ***before*** the design begins and must be closely adhered to during the design development.

In the private sector, there may be regulations governed by more than one agency in addition to the local ones. This means that true standardization in plumbing design does not exist, but there are strong similarities among them.

Massachusetts State Code

An example of a good plumbing code is the Commonwealth of Massachusetts *Fuel Gas and Plumbing Code*. It includes 22 basic principles, which are considered to be the building blocks for any plumbing system. These principles are listed here, because they constitute careful plumbing practice for any state. The principles are as follows:

1. ***All Occupied Premises Must Have Potable Water.*** All premises intended for human habitation, occupancy, or use must be provided with a supply of potable water. Such a water supply shall not be connected to unsafe or questionable water sources, nor shall it be subject to the hazards of back flow, back pressure, or back siphonage.

2. ***Adequate Water Required.*** Plumbing fixtures, devices, and appurtenances must be supplied with water in sufficient volume and at pres-

sures adequate to enable them to function properly and without undue noise under normal conditions of use.

3. ***Hot Water Required.*** Hot water must be supplied to all plumbing fixtures that normally need or require hot water for their proper use and function.

4. ***Water Conservation.*** Plumbing must be designed and adjusted to use the minimum quantity of water consistent with proper performance and cleaning.

5. ***Dangers of Explosion or Overheating.*** Devices and appliances for heating and storing water must be so designed and installed as to guard against the dangers of explosion or overheating.

6. ***Required Plumbing Fixtures.*** Each family dwelling must have at least one water closet, one lavatory, one kitchen-type sink, one bathtub or shower, and a laundry tray or connections for a washing machine, to meet the basic requirements of sanitation and personal hygiene. (In multiple dwellings housing families, one laundry tray or washing machine for every ten apartments, or fraction thereof, shall be acceptable as meeting laundry facilities in this principle.) One laundry tray or washing machine for every twenty apartments, or fraction thereof, shall be acceptable in housing for the elderly. All other structures for habitation must be equipped with sufficient sanitary facilities. Plumbing fixtures must be made of durable, smooth, non-absorbent and corrosion-resistant material and must be free from concealed fouling surfaces.

7. ***Drainage System of Adequate Size.*** The drainage system must be designed, constructed, and maintained to guard against fouling, deposit of solids, and clogging, and with adequate cleanouts so arranged that the pipes may be readily cleaned.

8. ***Durable Materials and Good Workmanship.*** The piping of the plumbing system must be of durable material, free from defective workmanship, and so designed and constructed as to give satisfactory service for its reasonable expected life.

9. ***Liquid-Sealed Traps Required.*** Each fixture directly connected to the drainage system must be equipped with a liquid-seal trap.

10. ***Trap Seals Shall be Protected.*** The drainage system must be designed to provide an adequate circulation of air in all pipes with no danger of siphonage aspiration or forcing of trap seals under conditions of ordinary use.

11. ***Exhaust Foul Air to Outside.*** Each vent terminal must extend to the outer air at the roof line and be so installed as to minimize the possibilities of clogging and the return of foul air to the building.

12. ***Test the Plumbing System.*** The plumbing system must be subjected to such tests as will effectively disclose all leaks and defects in the work or the materials.

13. ***Exclude Certain Substances from the Plumbing System.*** No substance that will clog or accentuate clogging of pipes, produce explosive mixtures, destroy the pipes or their joints, or interfere unduly with the sewage disposal process shall be allowed to enter the building drainage system.

14. ***Prevent Contamination.*** Proper protection shall be provided to prevent contamination of food, water, sterile goods, and similar materials by back flow of sewage. When necessary, the fixture, device, or appliance shall be connected indirectly with the building drainage system.

15. ***Light and Ventilation.*** No water closet or similar fixture shall be located in a room or compartment that is not properly lighted and ventilated.

16. ***Individual Sewage Disposal System.*** If water closets or other plumbing fixtures are installed in buildings where there is no sewer within a reasonable distance, a suitable provision shall be made for disposing of the sewage by some accepted method of sewage treatment and disposal.

17. ***Prevent Sewer Flooding.*** Where a plumbing drainage system is subject to back flow of sewage from the public sewer, a suitable provision shall be made to prevent its overflow in the building.

18. ***Proper Maintenance.*** Plumbing systems shall be maintained in a safe and serviceable condition from the standpoint of both mechanics and health.

19. ***Fixtures Shall Be Accessible.*** All plumbing fixtures shall be so installed with regard to spacing as to be accessible for their intended use and cleansing.

20. ***Structural Safety.*** Plumbing shall be installed with due regard to preservation of the strength of structural members and prevention of damage to walls and other surfaces through fixture usage.

21. ***Protect Ground and Surface Water.*** Sewage or other waste must not be discharged into surface or subsurface water unless it has first been subjected to some acceptable form of treatment.

22. ***Piping and Treatment of Special Wastes.*** All materials that could become detrimental to the health and welfare of the general public, which enter the sanitary system of any building, shall be carried within special piping systems and either collected and disposed of or treated prior to entering the sanitary drainage system in accordance with the requirements of the authorities having jurisdiction.

In addition to the above principles, we want to make sure that in public places the designer and the contractor will include plumbing fixtures especially designed and installed for handicapped people.

Updating Codes

One very important element in plumbing design is that standards and codes are updated, modified, and improved on a regular and ongoing basis. For example, until 1989, the code in Massachusetts allowed the clothes dryer exhaust pipe (duct) to be made of wire-reinforced canvas. The new requirement calls for this exhaust pipe to be manufactured of galvanized sheet metal with formed joints (no screws or nails inserted) to reduce or eliminate the danger of fire. An installation completed in 1991, which included the reinforced canvas dryer exhaust, was correctly rejected by the plumbing inspector since it was made of an unacceptable material. (Evidently the installer did not keep current with the latest code modifications.) Those involved in this trade must regularly obtain the latest code inserts and keep themselves informed of the latest modifications.

The BOCA Code is in the process of recommending that more fixtures be installed in public places for females versus males. Plumbing fixture use is different for the two sexes since women have different garment layers, privacy expectations, and grooming requirements. We can easily understand the rationale for more fixtures when we think of the long lines at women's toilets during intermission at theaters or other public events.

Bidding

Applicable local codes must always be consulted since they always represent the minimum legal requirements. Slightly larger water pipe sizes may be used for installations to reduce noise and possible water hammer. However, anything above and beyond the code requirements increases the installation cost. When the specifications are written to include more than the basics, the bid price will reflect it. A plumbing bid may only be successful if the installation is less expensive or more cost effective. The answer is clear cut: to supply the least expensive installation, the contractor shall provide only what is required by code for pipe sizes and materials, fittings and valves, and the installation of basic white fixtures with no frills or special trimmings. However, engineered drawings must be followed by the installer.

If the contract bid encompasses a correct installation, including the sizes and standard fixtures as required by code and the installation and the bid price does not include exaggerated overhead or profit, it is likely the bid will be successful. It is not usually necessary to include additional safety factors in a plumbing design because these have already been built into the code. This is, however, the design engineer's responsibility.

Besides inflating a bid, going beyond the code can be detrimental to the plumbing system. For example, increasing a drain pipe size beyond the code indication might be harmful to system operation due to the slowing of sewage velocity, thus decreasing the capability of the pipe to carry solids. Larger sizes for future extension of a plumbing system in a building should be considered only upon request and only in special situations. However, adding a few fixtures to a system at a later time does not usually have a detrimental effect on the system's satisfactory, operational capacity.

The considerations just discussed are usually applicable in new construction. For renovations and/or remodeling of public places, there are some basic concerns the owner must address with the plumbing specialist to determine what work must be performed. These concerns include:

- Does the number of fixtures meet the latest code requirements for both male and female occupants?
- Are fixtures provided for handicapped people?
- Should the existing toilet fixtures be replaced with modern, water-saving ones?
- Are additional toilet rooms required?
- Is new extended piping required for new or existing toilet rooms or for new equipment drains or process wastes (for industrial applications)?
- Do floor levels, drain locations, and existing sewer pipe elevations provide the necessary slopes for correct gravity drainage?
- Will any of the modifications interfere with the existing structural elements?

Only when these questions are resolved should the engineer or the contractor begin the plumbing design.

Standard design symbols should be used in preparing plumbing drawings. Standard symbols were designed to take the place of a written description of a fixture in a plumbing design. These symbols convey information and simplify a drawing. The legend, which should appear on the plumbing drawing, defines the symbols and acronyms used on that particular plumbing drawing. Technicians must become familiar with the common plumbing symbols. Some of these symbols are shown in Appendix A.

Chapter Three

The Properties of Water and Hydraulics

Next to air, water is the most indispensable substance for the survival of humans, animals, and plants. A person can live without food for a month or more, but a person can survive only a few days without water. The human body needs water to wash out and dissolve toxins, which are continually produced as waste substances. If not eliminated, these toxins will poison the body. Water is also necessary for digestion, metabolism, and many other functions in the human body.

Water in humans and other animals must be replaced due to continuous elimination through exhalation, perspiration, and urination. Lack of water may have a harmful effect upon bodily functions. As a rule of thumb, a person should consume half an ounce of water per pound of body weight every day, so a person weighing 150 lb requires 75 oz of water (nine to ten, 8-oz glasses) daily. This may seem like a lot of water, but it is the amount medically recommended for a correct balance. Water intake should be evenly distributed during the day, and **the water we drink must be safe.** The U.S. has the safest domestic water supply in the world.

Potable Water

To become potable (drinkable), water must be treated. Treatment generally includes filtration and chlorination;[1] however, there are other specific water treatment processes that reduce or neutralize excess acidity, alkalinity, or other dissolved matter.

Harmful and non-harmful organic bacteria are generally found in surface water reservoirs. If not destroyed, harmful bacteria (pathogens) may cause diseases like typhoid and cholera in humans who ingest the contaminated water. Non-harmful bacteria encourage algae and slime growth, which discolor the water and produce odors. Both non-harmful and harmful bacteria can be controlled through the use of chlorine or ozone.

Chlorine rids a water supply of contaminants through a chemical reaction, which produces hypochlorous acid. This acid has disinfecting properties. Small amounts of pathogens may be destroyed by simply boiling the water prior to consumption. Potable water must have an agreeable taste, be colorless and odorless, and not contain more than 0.4 mg of chlor per gallon.

Neither the water occurring in nature nor the water used for domestic purposes is pure. Water contains a number of naturally dissolved materials and chemicals (generally in very small amounts), as well as certain chemicals that are purposely added through treatment. One of these added chemicals may be fluoride, which could prevent or reduce the incidence of tooth decay in those who drink that water.

To prevent contamination and diseases, any new and/or refurbished potable water supply system must be disinfected after installation, in accordance with strict and detailed rules, before it is ready for consumption. The pipes, fittings, and appurtenances as they arrive from factories could possibly contain bacteria as well as dust and harmful impurities. These impurities must be neutralized and washed out before the system becomes operational. The specification located in Appendix H gives the details to be followed for this process.

Water Supply

Fresh water is a national treasure and must be managed properly rather than wasted. There are numerous countries around the world that lack surface or underground fresh water, which should force those in the U.S. to realize the importance of this natural resource.

There are two categories of water that exist in nature: fresh water and sea water. Sea water is found in oceans and open seas, which occupy approximately 75% of the earth's surface. Fresh water can be found in natural springs, rivers, and lakes (or any inland natural surface reservoir), but most fresh water is located in underground reservoirs or aquifers. Humans have traditionally lived near surface water. This location has provided a source of drinking water, as well as a supply of fish and game animals. Through the years, humans have learned to tap into the fresh water located in underground reservoirs.

Water evaporates from plants, soil, and surface bodies of water. The water accumulates in the upper part of earth's atmosphere in the form of clouds. Under certain meteorological conditions, the water returns to the earth in the form of rain or snow depending upon the geographical location and time of year. Rain and snow are the sources of surface waters as well as underground water. This movement of the water in nature is called the *hydrological cycle*, Figure 3-1.

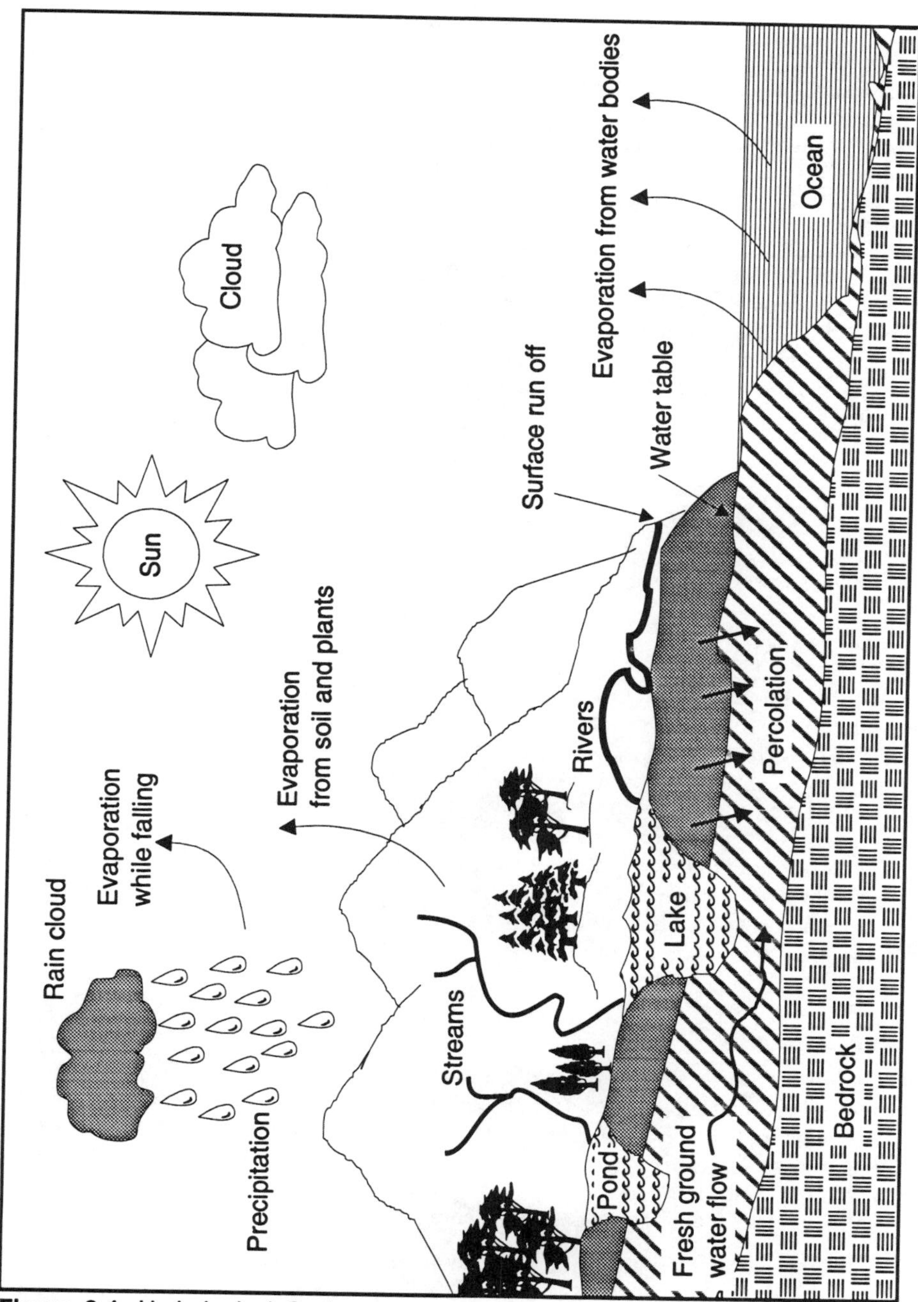

Figure 3-1. Hydrological Cycle

Precipitation from the atmosphere percolates through the ground, which acts as a filter. When the water reaches a saturated geological formation or an impermeable stratum (like clay), it forms aquifers. A geologist or hydrologist, both of whom specialize in underground geological formations, can locate subsurface water. Underground water reservoirs represent 97% of the fresh water in nature, while surface water represents only 3% of fresh water available. Underground water flows slowly toward the lowest point of discharge.

Not all the water that infiltrates the soil becomes ground water. Water may be pulled back to the surface by the capillary force of plants. This capillary force also contributes to above-ground evaporation of water. Water percolation through the upper strata of the ground combined with water's capability to dissolve materials gives the stored underground water certain qualities, which may render it more or less acceptable for human consumption or other applications.

When a source of water is needed for domestic use, there are a few elements a plumbing specialist should know:

- Amount needed
- Accessibility
- Quality
- Quantity available

Amount Needed

The amount of water needed depends upon the proposed building occupancy. In a project development process, occupancy needs must be determined first, since there is a vast difference between an industrial establishment (which might need process water), a commercial institution, or a residential user. Each user has specific requirements, which are either already known or are required by code. These requirements can be determined by calculations based on available information. In considering the demand for water, the plumbing specialist has to determine the current amount needed, as well as any increase in the future amounts, based on the owner's development plans. Practical knowledge and prior experience are beneficial when estimating and selecting this data.

Accessibility

Surface water can be observed and measured and its availability determined by studying historical data (e.g., droughts, rain, snow, temperature, etc.) for the particular location. Natural underground reservoirs are not always readily available due to depth. With today's technology, any water location can be made accessible, but if the price is excessive, then the location will be considered impractical and economically disadvantaged.

From a hydrological point of view, it is possible to illustrate a section of ground, Figure 3-2. Water found in the saturation zone is called *ground water*. Ground water can be found in natural reservoirs of various sizes and at different depths or in various layers. The formation or strata above the saturation zone from which water can be pumped for usage is called an *aquifer*.

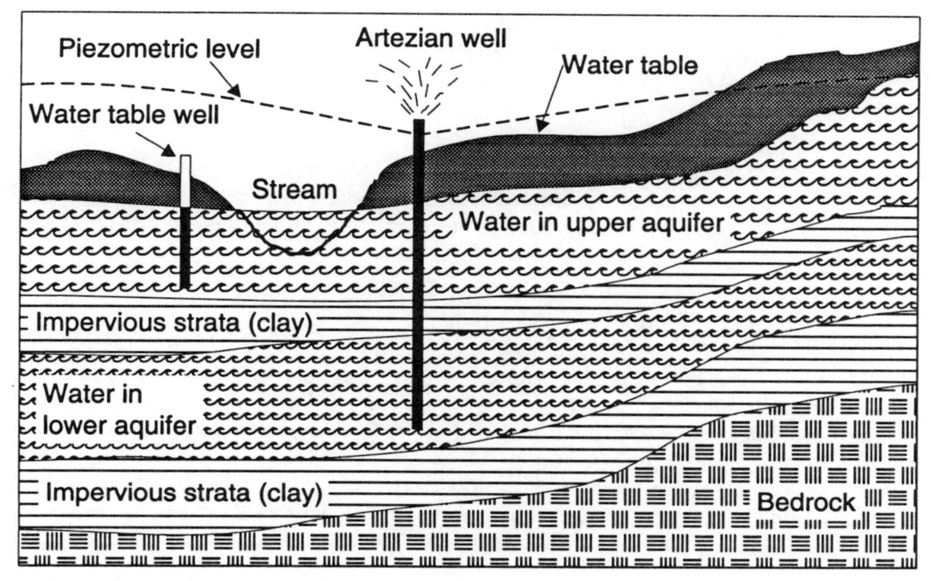

Figure 3-2. Ground Section

Water Quality/Quantity

Based on many sources of information such as satellite pictures and test wells, geologists and hydrologists can help determine the possible existence of underground water. However, the quality of water available cannot be determined without actual laboratory testing. Laboratory testing determines what substances are dissolved in water. Only after testing can the necessary treatment and cost be determined in accordance with intended purpose: industrial or potable use.

To determine the quantity of water available, pumping tests must be performed. The geologist and hydrologist usually work together to prepare a map of the underground strata, called a geological map. This map shows areas where water may exist. Test wells then confirm whether or not water is available. Some information may come from an analysis of existing water wells and their yields. However, in an area where development is planned but no wells exist, drilled test wells or test holes are required to obtain the necessary information. The samples brought to the surface during drilling provide sound geological and hydrological information for specialized persons. Together with the pumping results, the quantity of water can be determined.

The approximate cost of producing potable water from ground, surface, and sea water is given below (these are general figures for information only — locations and numerous other characteristics may influence the price):

- Ground water - $1.50/1000 gal
- Surface water - $4.00/1000 gal
- Sea water (desalinated) - $6.00/1000 gal

The water availability in certain locations may determine the development of a particular site or the preference for another. Other factors that must be considered include the proper use, protection, and conservation of the water source, as well as socio-economical conditions.

Properties of Water

In its natural cycle, water evaporates from the surface of natural bodies of water, then clouds form, and rain or snow develops. Upon reaching the ground, water percolates very slowly through the upper layers of the earth's crust, which is called permeable strata. During this slow, natural, downward movement, water comes in contact with various naturally occurring substances (minerals). This process filters the water while adding minerals, mostly in the form of salts. At some point, an equilibrium is reached.

Dissolved minerals in ground water may affect its potential usage. If the concentration of a certain mineral is excessive for the specific water usage, it may need to be removed during water treatment. Due to natural filtration, most ground water contains virtually no bacteria and no suspended matter, which means most underground water needs little treatment to become domestic (potable).

The natural chemical properties of water (H_2O) are: solubility (it dissolves a variety of substances), hardness, specific electrical conductance, hydrogen ion concentration (pH, which at the value of 7 shows that the water is neutral, neither alkaline, which is when ph is between 7 and 14, nor acid, when ph is between 7 and 0), dissolved carbon dioxide, and dissolved solids. Some of the physical properties of water include: density, viscosity, compressibility, boiling point, and freezing point. It is important to recognize these properties because in one way or another, they influence the flow of water in a plumbing system.

Density

By definition, density is the ratio of mass (weight) of a substance to the volume it occupies. Density is given in pounds per cubic foot, which can be written as lb/ft^3 or lb/cu ft. In plumbing calculations, water density is usually considered to be 62.3 lb/ft^3. This value represents the density of water at a temperature of 70°F (room temperature). Water density varies slightly with the temperature; the warmer the water, the less dense it becomes, as shown by the values below:

Water temperature (°F)	Water density (lb/ft³)
32	62.416
50	62.408
70	62.300
100	61.998
150	61.203
210	59.843

Viscosity

Viscosity, as applied to plumbing, concerns the friction of water molecules among themselves, as well as along the walls of the pipes and fittings. It is the physical property that directly influences the flow of water in pipes. The forces at work between the water molecules themselves are called *cohesion* and *adhesion*. These forces can be measured in the laboratory.

Cold water is more cohesive than warm water, thus its viscosity is greater. This greater viscosity increases the friction of the flow of cold water through pipes. Warm water flows somewhat more easily through pipes because it is not as cohesive. However, the actual difference in viscosity between domestic cold and hot water is so small that it is considered insignificant for practical purposes and is negligible in calculations.

Viscosity is measured in centistokes or centipoise. At 60°F, water has an absolute viscosity equal to 1.12 centipoise, which corresponds to a measurement of kinematic viscosity of 0.00001216 ft^2/sec.

Compressibility

As noted previously, water at ambient temperature is considered noncompressible for all practical purposes.

Boiling/Freezing Points

Water boils at 212°F (100°C) at sea level (atmospheric pressure). If the pressure varies, the boiling temperature point will also vary. The lower the pressure exerted upon the surface, the lower the boiling point. For example, atmospheric pressure is lower on top of a mountain, so water boils at a lower temperature. The changes in the boiling point as a function of pressure are as follows:

Absolute pressure (psi)	Water boiling point (°F)
1	101.8
6	170.1
14.7 (atmospheric)	212.0

Water Flow

The flow of water can be characterized as *laminar* or *turbulent.* In laminar flow, streams of water molecules flow naturally parallel to each other up to a certain velocity. Above that velocity, the flow becomes turbulent. This characteristic was demonstrated by Osborne Reynolds, who developed a simple formula to determine the *Reynolds number* (R), which classifies the flow as laminar or turbulent. If the Reynolds number (R) is less than 2000, the flow is laminar. The simplified formula is as follows:

$$R = \frac{VD}{v}$$

where:
- R = Reynolds number (no unit of measurement)
- V = water velocity (ft/sec)
- D = pipe diameter (ft)
- v = kinematic viscosity (ft^2/sec)

Example 3-1. Compute the Reynolds number based on the following information:

Water velocity = 4 ft/sec
Pipe diameter = 4" (0.33 ft)
Viscosity = 0.00001216 ft^2/sec

Solution 3-1. $R = \frac{(4)\ (0.33)}{0.00001216} = 10.9 \times 10^4$

In the above example, R is much larger than 2000 so the flow is turbulent. If the velocity in a 4" diameter pipe were less than 0.072 ft/sec, the Reynolds number would be smaller than 2000 and the flow would be laminar. This simple calculation tells us that the flow of water in pipes at "normal" velocities of 4 to 8 ft/sec is always turbulent. Turbulent flow does not adversely affect plumbing design. It is also true that the velocity of flowing water in a cross section of the pipe is not uniform.

The velocity of water is greatest at the center of a pipe. More friction exists along the walls, where water molecules rub against pipe walls, Figure 3-3. The velocity used in these problems is an average velocity, which represents 80% of the maximum velocity at the center of the pipe. To facilitate the understanding of water flow in a plumbing system, this book includes sample problems based on a pumped system and a gravity system.

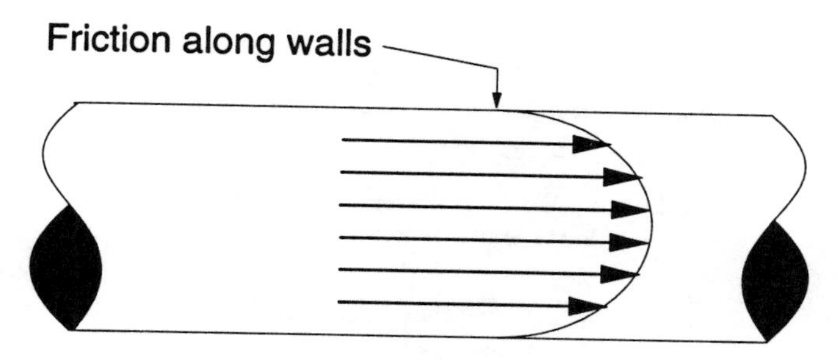

Figure 3-3. Flow of Water in Pipes

The basic formula for fluid flow is:

$$Q = AV$$

where: Q = flow (ft^3/sec)
A = cross section of the pipe (ft^2)
V = water velocity in pipe (ft/sec)

Another item often used in plumbing calculations is the *velocity head*. This is defined as the decrease in head (or the loss of pressure), which corresponds to the velocity of flow. The formula for velocity head is:

$$hv = \frac{V^2}{2g}$$

where: hv = velocity head (ft)
V^2 = velocity in pipe $(ft/sec)^2$
g = acceleration of gravity (ft/sec^2)

As a dimensional verification:

$$hv = \left(\frac{ft^2}{sec^2}\right)\left(\frac{sec^2}{ft}\right) = ft$$

When performing plumbing calculations, consideration needs to be given to factors that affect the flow of water through the pipes. These factors include friction or pressure losses that occur when water flows through pipe, pipe fittings, or equipment (e.g., water meters or heaters), Figure 3-4. Another factor is *static head,* which is the amount of potential energy due to the elevation of water above a certain reference point. Static head is measured in feet of water and can be converted into psi (pounds per square inch).

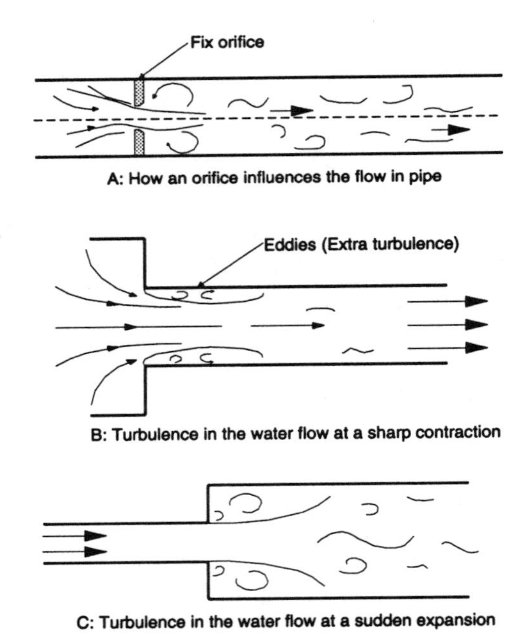

Figure 3-4. Friction in Fittings

Plumbing calculations are important to ensure pipes are sized correctly. If the pipes are not sized correctly and the flow of water stops suddenly, the dynamic force of the flow may produce water hammer, shock, or noise. If the velocity is constantly too high when water is flowing, erosion might also occur in the pipes.

HYDRAULICS

Hydraulics is part of a larger branch of physics called *fluid mechanics.* Plumbing, which deals with the flow of fluids in pipes, is a practical application of hydraulics. Hydraulic principles are based on the chemical, physical, and mechanical properties of water. These properties were discussed earlier and include density, viscosity, and type of flow (laminar or turbulent) as a function of velocity, water temperature, and pressure.

Plumbing deals with two types of fluids: gas and liquid. The main difference between these two fluids is that gas is compressible. Liquids, mainly water in plumbing applications, are essentially non-compressible at room temperature. Water flows in pipes by gravity from a higher to a lower elevation. The difference in elevation or water level in a system is called static head. If the water is required to flow in the opposite direction, it must be assisted by a mechanical device such as a pump.

A piping system or network includes pipes, fittings, and valves. As discussed earlier, water flowing through these pipes, fittings, valves, and other equipment produces friction or a loss of pressure. This pressure loss or resistance to flow occurs because the molecules of water "rub" against the walls of the pipes and fittings. The types of energy involved in the flow of water include kinetic and potential energy. Following are some examples.

Example 3-2. What is the outlet pressure for water flowing in a straight pipe having the following characteristics:

Pipe material: Schedule 40 steel
Pipe diameter: 2"
Flow: 40 gpm
Length of pipe: 50 ft
Inlet water pressure 25 ft = 10.82 psi

Solution 3-2. Remember that when water flows along the pipe, friction, loss of head, or loss of power result. Also realize that while data is provided for calculations in this book, in an actual plumbing application, this data must either be calculated (e.g., required flow) or selected (e.g., piping material, limiting water velocity in pipes, etc.).

Based on the information in Appendix B, at a flow of 40 gpm the velocity of water is 3.82 ft/sec and the head loss is 3.06 ft per 100 ft of pipe. The pipe length in this problem is only 50 ft, so the head loss is:

$$\left(50 \text{ ft}\right)\left(\frac{3.06 \text{ ft}}{100 \text{ ft}}\right) = 1.53 \text{ ft}$$

The pressure available at the pipe exit is 25 ft - 1.53 ft = 23.47 ft, or 10.16 psi, Figure 3-5.

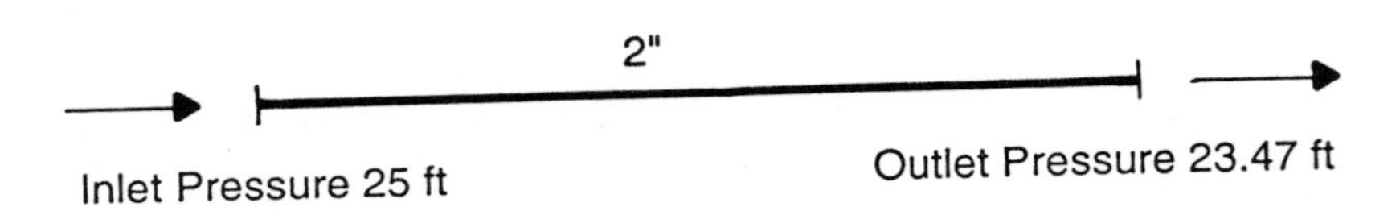

Figure 3-5. Straight Pipe

Example 3-3. Based on the data given in Example 3-2, what is the outlet pressure for water flow in pipe with fittings, Figure 3-6? (There is no difference in elevation.)

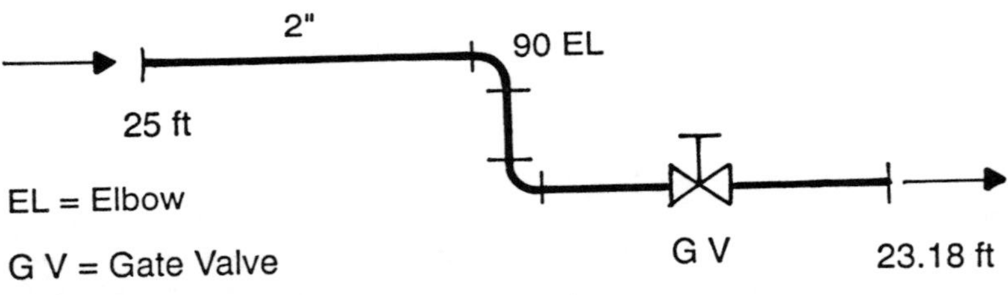

Figure 3-6. Pipe with Fittings

Solution 3-3. Remember, the total friction in the system consists of friction in the pipe plus the friction in fittings.

The velocity head, which may be found in Appendix B, is calculated as follows:

$$hv = \frac{V^2}{2g} = \frac{3.82^2}{(2)(32.17)} = 0.227 \text{ ft}$$

The friction in fittings (hf) can be found in Appendix B and is based on the velocity head calculation. The K factor is a resistance coefficient, which helps calculate the friction in fittings. It represents the coefficients for each individual fitting, the sum of which equals total friction in fittings (e.g., K1 + K2 + ..., where K represents each fitting) when multiplied with the velocity head. For the problem at hand, the K value for a 2-inch, 90°, standard elbow is 0.57. The K value for a 2-inch gate valve is 0.15. Thus, the friction loss in fittings is as follows:

$$hf = [K1\ (elbow) + K2\ (elbow) + K3\ (valve)] \bullet \frac{V^2}{2g}$$

$$= (0.57 + 0.57 + 0.15)\ (0.227) = 0.293\ ft$$

We already calculated the friction in this length of pipe (see Example 3-2), so just add it to the friction in fittings:

$$1.53\ ft + 0.293\ ft = 1.823\ ft$$

The pressure available at the pipe exit is now:

$$25\ ft - 1.823\ ft = 23.177\ ft\ (round\ to\ 23.18)\ or\ 10.03\ psi$$

In this example, the available pressure is lower because of the added friction in the fittings.

Example 3-4. Based on the data given in Examples 3-2 and 3-3, what is the outlet pressure for water flow in a pipe with a vertical portion, Figure 3-7?

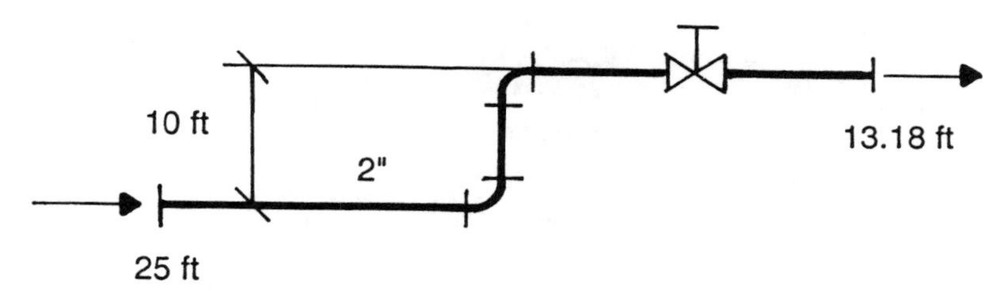

Figure 3-7. Pipe with Vertical Portion

Solution 3-4. The calculations made in Example 3-3 remain unchanged since the data is the same. However, there is now some pressure loss due to a 10-ft vertical section.

The friction in pipe and fittings remains the same at 1.823 ft. From the inlet pressure, deduct the difference in elevation:

$$25\ ft - 1.823\ ft - 10\ ft = 13.177\ ft\ (round\ to\ 13.18)\ or\ 5.70\ psi$$

Example 3-5. Storm water must be transferred from a holding tank located at a low elevation to a reservoir located at a higher elevation, Figure 3-8. Calculate the total head loss in the system in order to select the appropriate pump size to do the job. Use the following data for this calculation:

Pipe material: Schedule 40, standard weight, steel
Pipe diameter: 3"
Flow: 130 gpm
Water velocity: 5.64 ft/sec (Table 3-1) see footnote #2
Friction: 3.9 ft per 100 ft of pipe (Table 3-1)
Limiting velocity: 6 fps

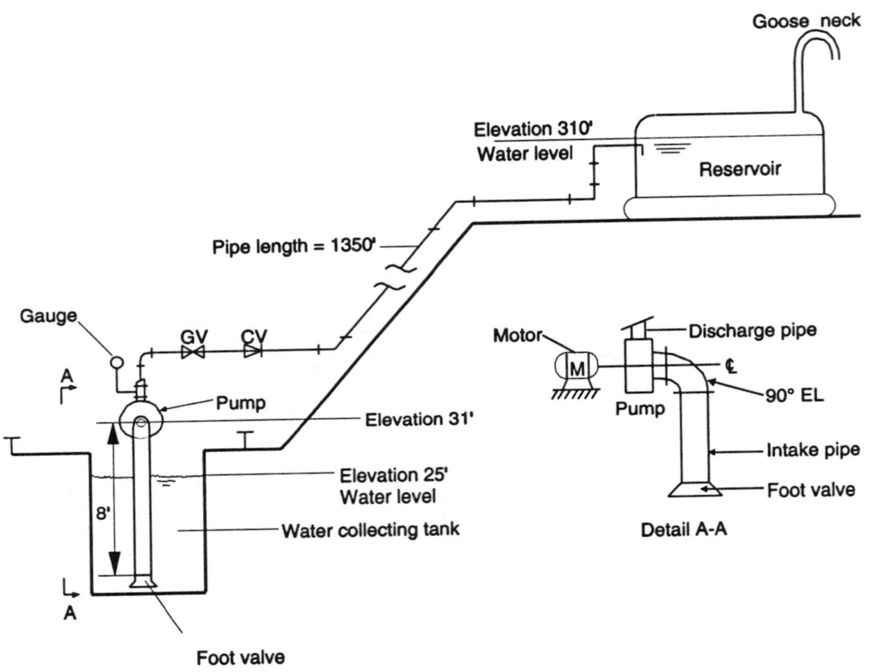

Figure 3-8. Transfer of Water from a Lower Elevation to a Higher Elevation

Solution 3-5. The system shown in Figure 3-8 can be divided in two sections: the suction side of the pump and the discharge side of the pump. The head loss in the suction side of the pump includes: friction loss in the pipe; friction loss in fittings and valves; and static head (since the flow is against gravity, it is considered a loss). Static head must be added to the friction loss, since the pump must overcome both of these factors to push the column of water up through the pipes.

Friction in 3 inch pipe
Asphalt-dipped Cast Iron and New Steel Pipe
(Based on Darcy's Formula)

Flow U.S. gal. per min.	Asphalt-dipped cast iron, 3.0" inside dia.: Velocity ft. per sec.	Velocity head-ft.	**Head loss ft. per 100 ft.**	Standard wt. steel - sch 40, 3.068" inside dia.: Velocity ft. per sec.	Velocity head-ft.	**Head loss ft. per 100 ft.**	Extra strong steel - sch 80, 2.900" inside dia.: Velocity ft. per sec.	Velocity head-ft.	**Head loss ft. per 100 ft.**	Schedule 160 steel, 2.624" inside dia.: Velocity ft. per sec.	Velocity head-ft.	**Head loss ft. per 100 ft.**
10	0.454	0.000	0.042	0.434	0.003	0.038	0.49	0.00	0.050	0.593	0.005	0.080
15	0.681	0.010	0.088	0.651	0.007	0.077	0.73	0.01	0.101	0.89	0.012	0.164
20	0.908	0.010	0.149	0.868	0.012	0.129	0.97	0.02	0.169	1.19	0.022	0.275
25	1.13	0.02	0.225	1.09	0.018	0.192	1.21	0.02	0.253	1.48	0.034	0.411
30	1.36	0.03	0.316	1.3	0.026	0.267	1.45	0.03	0.351	1.78	0.049	0.572
35	1.59	0.04	0.421	1.52	0.036	0.353	1.70	0.04	0.464	2.08	0.067	0.757
40	1.82	0.05	0.541	1.74	0.047	0.449	1.94	0.06	0.592	2.37	0.087	0.933
45	2.04	0.06	0.676	1.95	0.059	0.557	2.18	0.07	0.734	2.67	0.111	1.16
50	2.27	0.08	0.825	2.17	0.073	0.676	2.43	0.09	0.86	2.97	0.137	1.41
55	2.50	0.10	0.990	2.39	0.089	0.776	2.67	0.11	1.03	3.26	0.165	1.69
60	2.72	0.12	1.17	2.6	0.105	0.912	2.91	0.130	1.21	3.56	0.197	1.99
65	2.95	0.14	1.36	2.82	0.124	1.06	3.16	0.15	1.4	3.86	0.231	2.31
70	3.18	0.16	1.57	3.04	0.143	1.22	3.40	0.18	1.61	4.15	0.268	2.65
75	3.40	0.18	1.79	3.25	0.165	1.38	3.64	0.21	1.83	4.45	0.307	3.02
80	3.63	0.21	2.03	3.47	0.187	1.56	3.88	0.23	2.07	4.75	0.35	3.41
85	3.86	0.23	2.28	3.69	0.211	1.75	4.12	0.26	2.31	5.04	0.395	3.83
90	4.08	0.26	2.55	3.91	0.237	1.95	4.37	0.29	2.58	5.34	0.443	4.27
95	4.31	0.29	2.83	4.12	0.264	2.16	4.61	0.33	2.86	5.63	0.493	4.73
100	4.54	0.32	3.12	4.34	0.293	2.37	4.85	0.36	3.15	5.93	0.546	5.21
110	4.99	0.39	3.75	4.77	0.354	2.84	5.33	0.44	3.77	6.53	0.661	6.25
120	5.45	0.46	4.45	5.21	0.421	3.35	5.81	0.52	4.45	7.12	0.787	7.38
130	5.90	0.54	5.19	5.64	0.495	3.90	6.30	0.62	5.19	7.71	0.923	8.61
140	6.35	0.63	6.00	6.08	0.574	4.50	6.79	0.71	5.98	8.31	1.07	9.92
150	6.81	0.72	6.87	6.51	0.659	5.13	7.28	0.82	6.82	8.90	1.23	11.3
160	7.26	0.82	7.79	6.94	0.749	5.80	7.76	0.93	7.72	9.49	1.40	12.8

Table 3-1. Friction of Water (Courtesy, Cameron Hydraulic Data book)

The friction loss in the pipe can be calculated with Darcy's formula, which will be discussed later in this chapter. However, for purposes of this calculation, the corresponding values will be taken directly from Table 3-1.

On the suction side of the pump, the head loss due to friction (hp) in an 8-ft length of pipe is calculated as follows:

$$hp = \left(8 \text{ ft}\right)\left(\frac{3.9 \text{ ft}}{100 \text{ ft}}\right) = 0.312 \text{ ft}$$

To calculate the friction loss in fittings, use the same formula used in Solution 3-3; that is:

$$hf = \left(K\right)\left(\frac{V^2}{2g}\right)$$

where:
- hf = friction in fitting (ft)
- V = water velocity in pipe (5.64 ft/sec per Table 3-1)
- g = acceleration of gravity (32.174 ft/sec^2)
- K = resistance coefficient for each fitting

therefore: $$hf = \left(K\right)\left(\frac{5.64^2}{(2)(32.174)}\right) = \left(K\right)\left(0.494\right)$$

On the suction side of the pump, the fittings are a foot valve (a type of check valve) and a 90° elbow at the entrance into the pump (see detail in Figure 3-8). Based on the tables listing the values of resistance coefficients included in Appendix B, the applicable value of these K coefficients is 1.4 for the foot valve and 0.54 for the 90° elbow. Insert these values to complete the equation above:

$$hf = (1.4 + 0.54)\ (0.494) = 0.958 \text{ ft}$$

To calculate total head loss on the suction side, add the static head to the friction losses (pipes and fittings). Remember that the static head is the difference in elevation between the center line elevation of the pump and the water level[3] in the holding tank. (Since we are calculating the losses by dividing the system into the suction and the discharge sides of the pump, the pump center line becomes the reference point between the two sides in this case.)

From Figure 3-8, it is possible to calculate the difference in elevation:

$$31 \text{ ft} - 25 \text{ ft} = 6 \text{ ft}$$

The total suction head or pressure loss, which is measured in ft, thus becomes:

6 ft (static head) + 0.312 ft (friction in pipe) + 0.958 ft (friction in fittings) = 7.27 ft

Now it is necessary to calculate the friction or head loss in the discharge side of the pump. For the 1350-ft length of 3" diameter discharge pipe, the friction in the pipe is:

$$\left(1350 \text{ ft}\right)\left(\frac{3.90 \text{ ft}}{100 \text{ ft}}\right) = 52.65 \text{ ft}$$

Before calculating the friction in fittings, first tabulate the applicable K-coefficient values from Table 3-2.

Fitting	Value of K	Pipe diameter
45° EL	0.29	3"
90° Standard EL	0.54	3"
Gate valve	0.14	3"
Swing check valve	1.80	3"
Pipe exit	1.00	sharp edge

Table 3-2. Friction in Fittings Resistance Coefficient

It is then possible to calculate the friction loss in fittings:

$$hf = \left(K\right)\left(\frac{V^2}{2g}\right)$$

$$hf = (0.54 + 0.14 + 1.8 + 0.29 + 0.29 + 0.54 + 0.54 + 1.0)\left(\frac{V^2}{2g}\right)$$

$$hf = (5.14)(0.494) = 2.54 \text{ ft}$$

The static head, or difference in elevation, is 279 ft (310 ft - 31 ft). The static head, in this case, is considered a loss. (As mentioned earlier, the lower reference point in this case is the pump center line.)

Since the flow is against gravity and the static head must be overcome, the total discharge head loss (h_d) is:

h_d = 52.65 ft (loss in pipe) + 2.54 ft (loss in fittings) + 279 ft (static head) = 334.19 ft

Adding together the losses on both sides of the pump, the total pressure loss in the system (H_T) is as follows:

$$H_T = 7.27 \text{ ft (suction head)} + 334.19 \text{ ft (discharge head)}$$
$$= 341.46 \text{ ft}$$

Table Verification

In Example 3-5, we obtained the pipe friction value from a neat and orderly table. However, these tables are the result of a great deal of calculation based on the *Darcy-Weisbach formula*. Another similar one is called the *Hazen and Williams empirical formula*. The Hazen and Williams formula is referred to as empirical because it is based on laboratory and field observations. The Hazen and Williams pressure loss formula is:

$$hf = (0.002083)(L)\left(\frac{100}{C}\right)^{1.85}\left(\frac{Q^{1.85}}{d^{4.8655}}\right)$$

where: hf = friction in pipe in ft/100 ft
0.002083 = empirically determined coefficient
L = length of pipe in ft (in this case, it is 100 ft)
C = roughness coefficient based on the pipe material (Table 3-3)
Q = flow in gallons per minute (gpm)
d = pipe diameter in inches

Example 3-6. To ensure the values given in the tables for Example 3-5 are correct, calculate the friction loss using the Hazen and Williams formula and the following data:

L = 100 ft
C = 150
d = 3"
Q = 130 gpm

Solution 3-6. Plug the data into the Hazen and Williams formula as follows:

$$hf = (0.002083)(100)\left(\frac{100}{150}\right)^{1.85}\left(\frac{130^{1.85}}{3^{4.8655}}\right) = 3.8 \text{ ft}$$

The value used from the table was 3.90 ft per 100 ft, which is very close to the one just calculated (3.80 ft/100 ft). This exercise demonstrates that for all practical purposes, the table may be used with confidence.

Pipe material	Values of C		
	Range	Average value	Normally used value
Bitumastic-enamel-lined steel centrifugally applied	160-130	148	140
Asbestos-cement	160-140	150	140
Cement-lined iron or steel centrifugally applied	-	150	140
Copper, brass, or glass as well as tubing	150-120	140	130
Welded and seamless steel	150-80	140	100
Wrought iron, Cast iron	150-80	130	100
Tar-coated cast iron	145-50	130	100
Concrete	152-85	120	100
Full riveted steel (projecting rivets in girth and horizontal seams)	-	115	100
Corrugated steel	-	60	60

Value of C	150	140	130	120	110	100	90	80	70	60
For (100/C) at 1.85 power is	0.47	0.54	0.62	0.71	0.84	1	1.22	1.5	1.93	2.57

Table 3-3. Values of the Constant C Used in Hazen and Williams Formula

Alternative Solution

There is another easier and faster way to solve the hydraulic problem given in Example 3-5. It involves using *equivalent length for fittings*, Table 3-4. This easier alternative is defined as an equivalent length of straight pipe that has the same friction loss as the respective fitting or valve.

Various piping books and publications may indicate slightly different equivalent length values for the same fitting. These differences are usually small and therefore negligible. If a certain type of fitting cannot be found in an available table, an approximate value can be estimated based on a similar fitting. The equivalent value (length) can also be obtained from the fitting's manufacturer.

Example 3-7. Solve the same problem given in Example 3-5, but use the equivalent length for fittings and valves and make the calculation for the total head loss for the entire system (suction and discharge). The system data remain the same.

Nominal pipe size	Gate valve – full open	Globe valve – full open	Butterfly valve	Angle valve – full open	Swing check valve – full open	90° elbow	Long radius 90° & 45° std elbow	Close return bend	Standard tee – through flow	Standard tee – branch flow	Mitre bend 45°	Mitre bend 90°
1/2	0.41	17.6		7.78	5.18	1.55	0.83	2.59	1.0	3.1		
3/4	0.55	23.3		10.3	6.86	2.06	1.10	3.43	1.4	4.1		
1	0.70	29.7		13.1	8.74	2.62	1.40	4.37	1.8	5.3		
1 1/4	0.92	39.1		17.3	11.5	3.45	1.84	5.75	2.3	6.9		
1 1/2	1.07	45.6		20.1	13.4	4.03	2.15	6.71	2.7	8.1		
2	1.38	58.6	7.75	25.8	17.2	5.17	2.76	8.61	3.5	10.3	2.6	10.3
2 1/2	1.65	70.0	9.26	30.9	20.6	6.17	3.29	10.3	4.1	12.3	3.1	12.3
3	2.04	86.9	11.5	38.4	25.5	7.67	4.09	12.8	5.1	15.3	3.8	15.3
4	2.68	114	15.1	50.3	33.6	10.1	5.37	16.8	6.7	20.1	5.0	20.1
5	3.36	143	18.9	63.1	42.1	12.6	6.73	21.0	8.4	25.2	6.3	25.2
6	4.04	172	22.7	75.8	50.5	15.2	8.09	25.3	10.1	30.3	7.6	30.3
8	5.32	226	29.9	99.8	33.3	20.0	10.6	33.3	13.3	39.9	10.0	39.9
10	6.68	284	29.2	125	41.8	25.1	13.4	41.8	16.7	50.1	12.5	50.1
12	7.96	338	34.8	149	49.7	29.8	15.9	49.7	19.9	59.7	14.9	59.7
14	8.75	372	38.3	164	54.7	32.8	17.5	54.7	21.8	65.6	16.4	65.6
16	10.0	425	31.3	188	62.5	37.5	20.0	62.5	25.0	75.0	18.8	75.0
18	16.9	478	35.2	210	70.3	42.2	22.5	70.3	28.1	84.4	21.1	84.4
20	12.5	533	39.2	235	78.4	47.0	25.1	78.4	31.4	94.1	23.5	94.1
24	15.1	641	47.1	283	94.3	56.6	30.2	94.3	37.7	113	28.3	113
30	18.7					70	37.3	117	46.7	140	35	140
36	22.7					85	45.3	142	56.7	170	43	170
42	26.7					100	53.3	167	66.7	200	50	200
48	30.7					115	61.3	192	76.7	230	58	230

Table 3-4. Equivalent Length For Pipe Fittings

Solution 3-7. The developed length of pipe, or the actual measured length is:
1350 ft (discharge) + 8 ft (suction) = 1358 ft

For this application, and based on Table 3-4, the equivalent length for fittings and valves are listed below:

Fittings	Quantity	Equivalent length each fitting (ft)	Total equivalent length (ft)
3" Foot valve (same as a swing check valve)	1	25.50	25.50
90° Elbow	4	7.67	30.68
45° Elbow (Long Radius)	2	4.09	8.18
3" Gate valve	1	2.04	2.04
3" Swing check valve	1	25.50	25.50
Sharp pipe exit (Estimate)	1	17.50	17.50
			109.40

Now add the total equivalent length for fittings to the actual pipe length as follows:

1358 ft + 109.4 ft = 1467.4 ft = total equivalent length of pipe

Therefore, the total friction loss in pipe ***and*** fittings is:

$$\left(1467.4 \text{ ft}\right)\left(\frac{3.9 \text{ ft}}{100 \text{ ft}}\right) = 57.22 \text{ ft}$$

The total difference in elevation of the water levels (static head) is equal to 285 ft (310 ft - 25 ft); therefore, the total head (pressure) loss is 342.22 ft (57.22 ft + 285 ft).

Compare the previous value of 341.46 ft, which was obtained in Example 3-5 using the K resistance coefficients, to the value of 342.22 ft, which was obtained using the equivalent length method. Both results are close; therefore, it is easier to use the equivalent length method when solving plumbing problems. Keep in mind that these results are based on an engineering calculation in which minor approximations are acceptable.

These examples have shown that by knowing the flow, pipe diameter, system configuration, and pipe material, it is possible to calculate the head or friction loss in the system. When calculating a problem like this, consider that the piping system will age, so add in a safety factor of 10% to 15% to establish the acceptable value when selecting the pump. Thus, a value of approximately 380 ft is required for the selection of the pump head (342.22 ft + 37.78 ft). The pump head value and the pump flow (previously given as 130 gpm) are used in the selection of the pump.

Pump Selection

The objective of the examples given above was to determine the correct pump size for a system. A pump is a mechanism that is used to push a liquid with a specific force to overcome system friction loss and any existing difference in elevation. The pump produces this force with the help of a motor or driver and consumes energy in this process. The type of pump usually employed in plumbing systems is a centrifugal pump with all wetted parts made of non-ferrous material.

The pump's housing is referred to as the *casing*. The casing encloses the *impeller* and collects the liquid being pumped. The liquid enters at the center, or *eye*, of the impeller. The impeller rotates and, due to the centrifugal force created, pushes the liquid out. The velocity is the greatest at the impeller's periphery where the liquid is discharged through a spiral-shaped passage called the *volute*. This shape is designed so that there is an equal velocity of the liquid at all circumference points. The capacity (Q) of a pump is the rate of fluid flow delivered, which is generally expressed in gallons per minute (gpm).

The total head (H) developed by a pump is the discharge head (hd) minus the suction head (hs):

$$H = hd - hs$$

The suction head may be a negative or positive value depending on the pump location, water level, and the pressure of water entering the pump. The information obtained for head loss and flow of the system are the two elements used in pump selection.

To select a pump for the system described in Example 3-5, look at the pump curve in Figure 3-9. For this system, we need a pump that will deliver 130 gpm against a head of 380 ft (380 x 0.433 = 164 psi). The pump in this case will be a centrifugal pump.

On this pump curve, the flow delivery capacity in gallons per minute (gpm) is located on the horizontal axis (abscissa). The pump head (measured in ft and/ or psi) is located on the vertical axis (ordinate). The pump's efficiency percentage is shown on parallel curves on the upper part of the diagram. In our case,

DOMESTIC

CURVE CHARTS

S654B

Series B Group of 2 Ft. N.P.S.H. CENTRIFUGAL PUMP Performance Curves

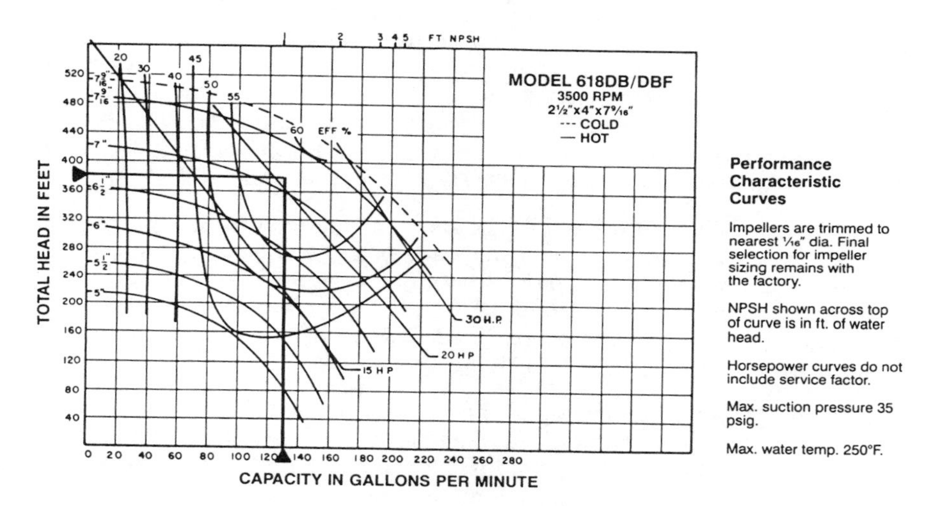

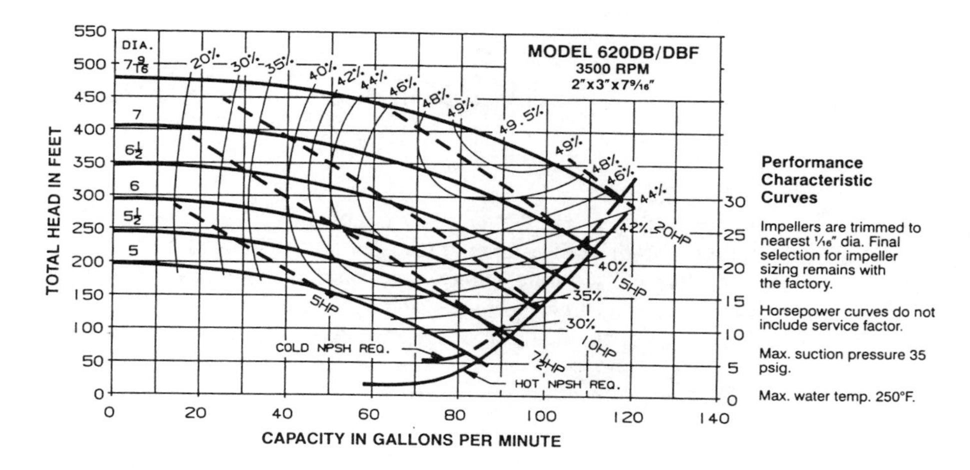

ITT Bell & Gossett
ITT Fluid Technology Corporation

Figure 3-9. Pump Curves (Courtesy, ITT Fluid Technology Corporation)

the efficiency of the selected pump is close to 60%. The electric motor horsepower (hp) is marked on the lines slanted down to the right. Since our value falls between two lines, we will choose the higher value of 30 hp. The impeller diameter along the vertical lines helps us select a 7-9/16" diameter impeller. The possibility that more capacity may be required at a later date must be considered in pump selection.

Each pump manufacturer provides a serial or model number for easy pump identification. Pump data sheets, such as the one shown in Figure 3-9, also indicate the number of revolutions per minute (rpm). In our case it is 3500 rpm. Another element in pump selection is the NPSH or Net Positive Suction Head (see Appendix B). This element is connected with the pump priming. If necessary, the pump manufacturer can offer help with the pump selection usually free of charge. For such assistance, contact the local representative.

If we calculate the flow and the head required in a plumbing system design and take into account the available city water pressure (city water pressure must be deducted from the required pump head), it is possible to determine if a pump is needed. Based on the calculations performed earlier, the pump needed for this particular system (as an example only) is a Bell & Gossett, horizontal centrifugal pump, Series B, Model 618DB/DBF.

When there is some initial, incoming pressure, but the system requires a higher operating pressure, the pump required is a *booster pump*. For example, if the pressure required by a system is 100 psi and 50 psi is received from the city, the booster pump must have a head of at least 50 psi. Add 10% of the total pressure as a safety factor, and the pump selected should be for 60 psi head.

Example 3-8. Calculate the outlet pressure in the system shown in Figure 3-10, in which water flows by gravity. The technical data is as follows:

Water flow (Q) = 50 gpm
Pipe diameter = 2"
Pipe material = Type K copper tubing
Pipe length = 380 linear feet (developed length)
Fittings = Two gate valves; one sudden contraction (from the tank into the pipe); one sudden enlargement (discharge open to the atmosphere); one 30° elbow; one 45° elbow
Difference in elevation = 150 feet (static head)

Solution 3-8. From Appendix B, the flow of 50 gpm water in a 2" diameter, Type K copper tubing (pipe) has a velocity of 5.32 ft/sec and a friction loss of 5.34 ft per 100 ft.

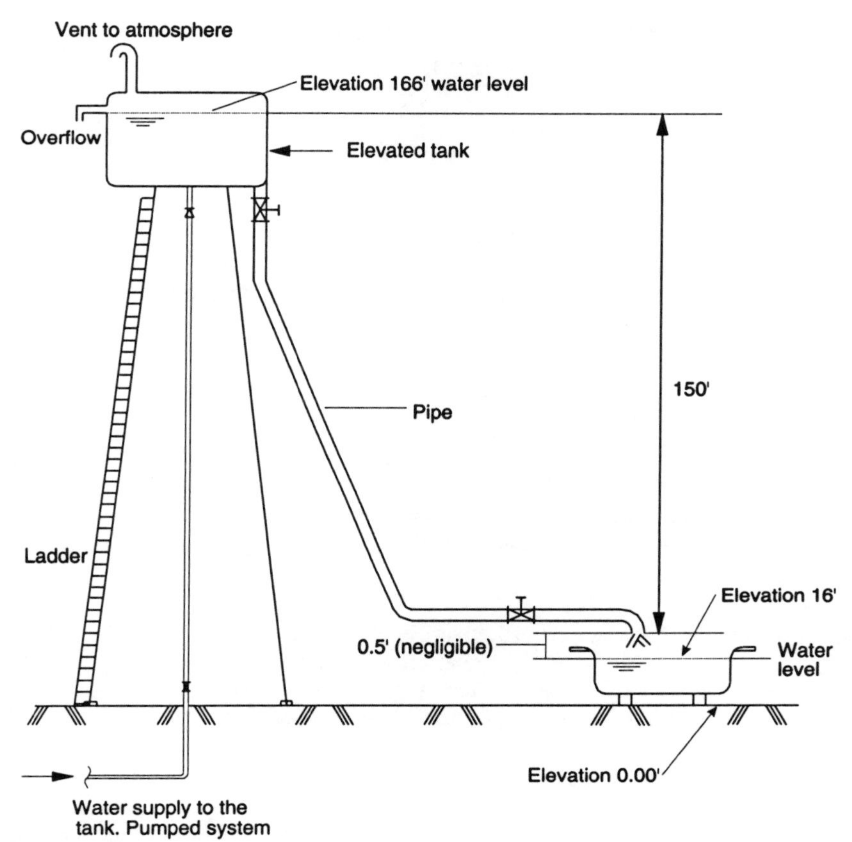

Figure 3-10. Gravity System

Use the following table to calculate the equivalent length of pipe method described earlier[4]:

Fittings	Quantity	Equivalent length each fitting (ft)	Total equivalent length (ft)
2" Gate valve	2	1.38	2.76
Sudden contraction[5]	1	10.30	10.30
Sudden enlargement[6]	1	10.30	10.30
45° Elbow	1	2.76	2.76
30° Elbow[7]	1	2.76	2.76
		Equivalent length - fittings	**28.88 (29 ft)**

The total equivalent length of the pipe and fittings is:

$$380 \text{ ft} + 29 \text{ ft} = 409 \text{ ft}$$

Friction loss in the pipe and fittings is:

$$\left(409 \text{ ft}\right)\left(\frac{5.34 \text{ ft}}{100 \text{ ft}}\right) = 21.84 \text{ ft}$$

Given the difference in the elevation of 150 feet, the outlet pressure is:

$$150 \text{ ft} - 21.84 \text{ ft} = 128.16 \text{ ft}$$

$$\frac{128.16}{2.31} = 55.48 \text{ psi}$$

The difference in elevation (static head) in this gravity, flow-type example (or in any other downhill flow) assists the flow, because the weight of the water column pushes the water down toward the discharge. This is the reason why the friction is *deducted* from the static head.

Example 3-9. Calculate the same problem given in Example 3-8, only this time with a flow of 200 gpm. All other data remain the same. Based on Appendix B, the velocity is 21.3 ft/sec, and the friction is 65.46 ft/100 ft. ***Note:*** a velocity of 21.3 ft/sec is unacceptable, but it is used here for illustrative purposes.

Solution 3-9. From Example 3-9, we determined that the equivalent length of pipe and fittings is 409 ft. Therefore, the friction loss in pipe and fittings is:

$$\left(409 \text{ ft}\right)\left(\frac{65.46 \text{ ft}}{100 \text{ ft}}\right) = 267.73 \text{ ft}$$

Given the difference in the elevation of 150 feet, the outlet pressure has a negative value:

$$150 \text{ ft} - 267.73 \text{ ft} = -117.73 \text{ ft}$$

The result of the calculations means that 200 gpm cannot flow through the system, because the pipe diameter is too small and the friction is too high for such flow. Water will flow but only at a maximum rate of 145 gpm. This value can be mathematically calculated as follows:

$$\left(\frac{x\ \text{ft}}{100\ \text{ft}}\right)\left(409\ \text{ft}\right) = 150\ \text{ft static head}$$

$$x = \frac{(150\ \text{ft})(100\ \text{ft})}{409\ \text{ft}} = 36.67\ \text{ft}$$

From Appendix B, for a 2" Type K copper tube at the friction calculated above the corresponding flow is approximately 145 gpm.

Measurements

The following is a short list of some useful units of measurement:

Acceleration of gravity = 32.2 ft/sec^2
1 ft^3 H_2O = 62.3 lb
1 gal = 0.1337 ft^3
1 gal H_2O = 8.33 lb (at 70°F)
1 ft^3 = 7.48 gal
Therefore:
1 ft^3 = (7.48) (8.33) = 62.3 lb
1 atm = 14.696 ~ 14.7 lb/in.^{2} or psi = 29.92 in. Hg (mercury)
1 atm = 33.96 ft of water
1 ft of H_2O x 0.433 = 1 psi (see footnote #8)

NOTES

[1] Standard M20, *Water Chlorination Principles and Practice*, issued by American Water Works Association (AWWA) gives all the required details for chlorination and disinfection.

[2] In most plumbing applications, a velocity of 6 ft/sec or less is advisable. Pipe sizes based on this velocity usually give years of trouble-free operation.

[3] Normally, the difference in elevation (static head) is from water level to water level or to a water discharge outlet elevation.

[4] Some manuals give slightly different equivalent lengths for copper tubing fittings than for steel, but the difference is small enough that the same table may be used.

[5 & 6] Estimated length as a standard tee through the branch.

[7] Estimated to be the same as the 45° elbow.

[8] The conversion value of 0.433 is derived from the following:

$$0.433 = \frac{14.7 \text{ psi}}{33.96 \text{ ft}}$$

Reciprocal: 1psi = 2.31 ft of H_20

Chapter Four

Project Bidding — Permits

To be able to work on a particular plumbing installation or project, the contractor must first be awarded the job. Contractors who install plumbing systems and sub-systems must consider expenses, time, materials, and profit when submitting a bid.

Projects are usually advertised, and contractors bid for these projects based on their scope, capabilities, and current workload. Contractors who produce quality work that is completed on time continue to receive work and will stay in business.

Obtaining a job can be challenging at times. Competition is fierce, and it is important to keep the following points in mind if you want to earn or keep a good reputation:

- Timely completion of the job
- Quality workmanship
- Job completion within budget

A very careful and thorough estimate is the key to producing a bid that is low enough to be competitive. In soliciting a bid, the design architect-engineer or owner prepares the design documents containing the drawings and specifications. Plumbing drawings include plans and riser diagrams or *isometric diagrams* (see Chapter 7 and Appendix F).

When required, a contractor, construction supervisor, or building manager, may prepare a job estimate. For all bids, a list of materials, or what is called a *material take off*, must be prepared, which includes all pipe sizes (measured from the engineering plans), fittings and valves (some practical experience is required to determine all or most requirements), insulation, fixtures, trims and accessories, as well as the equipment to be installed. With this list, prices can be determined for each item. The total bid will include costs for materials, labor, overhead, taxes, and a small, additional amount for unforeseen expenses.

Preparing an Estimate

When preparing an estimate, keep the following 19 points in mind (excerpted with permission from the R.S. Means Company, Inc.):

1. Use pre-printed or columnar forms for orderly sequence of dimensions and locations.
2. Use only the front side of each paper or form except for certain pre-printed summary forms.
3. Be consistent in listing dimensions: For example, length x width x height. This helps in rechecking to ensure that the total length of partitions is appropriate for the building area.
4. Use printed (rather than measured) dimensions where given.
5. Add up multiple printed dimensions for a single entry where possible.
6. Measure all other dimensions carefully.
7. Use each set of dimensions to calculate multiple related quantities.
8. Convert foot and inch measurements to decimal feet when listing. Memorize decimal equivalents to .01 parts of a foot (1/8" equals approximately .01').
9. Do not "round off" quantities until the final summary.
10. Mark drawings with different colors as items are taken off.
11. Keep similar items together, different items separate.
12. Identify location and drawing numbers to aid in future checking for completeness.
13. Measure or list everything on the drawings or mentioned in the specifications.
14. It may be necessary to list items not called for to make the job complete.
15. Be alert for: Notes on plans such as N.T.S. (not to scale); changes in scale throughout the drawings; reduced size drawings; discrepancies between the specifications and the drawings.
16. Develop a consistent pattern of performing an estimate. For example:
    a. Start the quantity take off at the lower floor and move to the next higher floor.
    b. Proceed from the main section of the building to the wings.
    c. Proceed from south to north or vice versa, clockwise or counter-clockwise.
    d. Take off floor plan quantities first, elevations next, then detail drawings.
17. Utilize design symmetry or repetition (repetitive floors, repetitive wings, symmetrical design around a center line, similar room layouts, etc.).

18. Do not convert units until the final total is obtained.
19. When figuring alternatives, it is best to total all items involved in the basic system, then total all items involved in the alternates. Therefore you work with positive numbers in all cases. When adds and deducts are used, it is often confusing whether to add or subtract a portion of an item; especially on a complicated or involved alternate.

Another point that may be added to this list is that when a shorter connection or saving possibilities are included, be sure to make a note for future reference.

Shop Drawings

After the job contract is secured based on the plumbing drawings prepared by the engineer, the contractor prepares the shop drawings. The plumbing specification usually contains the following request:

> Shop drawings shall include drawings, schedules, performance charts, instructions, brochures, diagrams, and other information to illustrate the system requirements and operation of the system. Shop drawings shall be provided for the complete plumbing system including piping layout and location of connections; schematic diagrams and connection and interconnection diagrams.

Shop drawings submitted by the contractor will inform the design engineer whether the concept was completely and correctly followed. The shop drawings serve a dual purpose:

- They allow the engineer to analyze the contractor's intentions in regard to installation details so he/she can check for any discrepancies from the contract drawings. The drawings will also be checked for clarity as well as details and compliance with the codes.
- They allow the contractor to give the architect an opportunity to develop a more practical approach for pipe routing for fixture connections. These approaches may be based on the applicable code but perhaps with a slightly different (legal) interpretation.

The contractor usually has the latitude to choose between some alternative materials listed in the specification. This alternative may produce a savings. Alternatives are usually acceptable as long as the system is in complete compliance with the governing code as well as the requirements of the local plumbing inspector. Such drawing preparation involves practical past experience. The same approach is also valid if it is done by the contractor on the engineer's drawings.

Requirement for Permit

In order to install a plumbing system in a building or structure, a plumbing permit is normally required. This requires knowledge of local conditions, applicable codes, and local regulations. In some cases, more than one organization may have jurisdiction. These organizations may include:

- Public Health and/or Safety Department
- Local Board of Health
- Public Water Authority
- Public Highway Authority
- Building and/or Water and Sewer Authority
- Gas Company
- Fire Department

These agencies have legal regulations, which must be abided by all.

Chapter Five

Plumbing Fixtures

The most visible parts of a plumbing system are the plumbing fixtures. A definition of a plumbing fixture is *a permanent piece of equipment, fixed in place serving a sanitary-hygienic purpose* (see Appendix E). Based on the current fixtures and trims available, we can safely say that fixtures also serve an artistic or decorative aspect.

The plumbing fixture is both the delivery point for the water supply and the starting point for sewage (waste system). Plumbing fixtures must contain certain qualities in order to be useful; for example, the manufacturing material must be dense, with a smooth surface, to prevent any retention of waste matter. The fixture operation should also be water conserving, easy, and quiet, in addition to serving the intended purpose and being reasonable in cost.

Plumbing specialists should have a thorough knowledge of the plumbing fixtures available and their characteristics. The selection of plumbing fixtures for a certain installation is usually proposed by the plumbing specialist in consultation with an architect and/or building owner. The plumbing contractor's responsibility is to procure all of the fixtures and materials requested by the specification and install them according to the codes, local regulations, and owner's satisfaction.

The variety of plumbing fixtures available on the market is tremendous. Showing the owner *catalog cuts* for approval and considering the practical usage as well as cost will help avoid later controversy. A list of available plumbing fixtures should include:

- Water closets
- Urinals

- Lavatories
- Sinks
- Drinking fountains
- Showers
- Bathtubs
- Bidets
- Floor drains
- Safety shower and eye wash

There are considerable variations in material, size, shape, operation, cost, and degree of water conservation for each type of fixture listed above.

Water Closet (WC)

A water closet is the same as a toilet and is designed to receive and discharge human excrement. From a sanitary point of view, there are a few elements that make a good water closet, Figure 5-1. These items are:

- Prompt discharge of the bowl content after usage
- An effective water seal
- A thorough scouring of the bowl cavity after each flush
- A trapway of suitable size (able to pass a 2-1/2" ball) — 3" or 4"

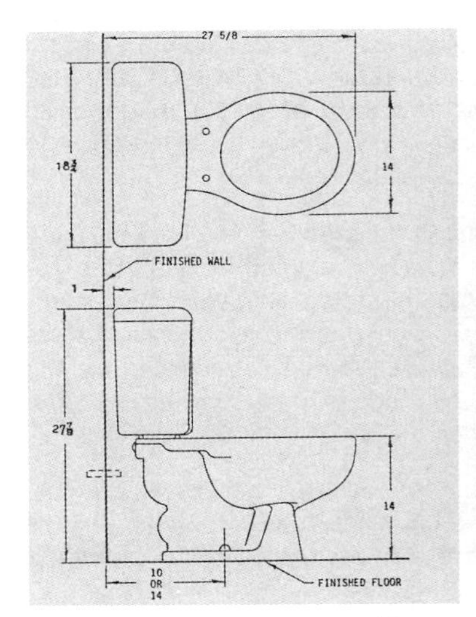

Figure 5-1. Water Closet (Courtesy, Crane Plumbing)

For private residential use, the water is supplied from a reservoir (tank), which is located at almost the same elevation as the bowl. Tanks were originally installed about 5 feet above the WC and were capable of storing 7 to 8 gallons of water for each flush. Based on water conservation requirements, today's high-efficiency water tanks hold only 1.5 to 1.6 gallons. To obtain satisfactory results with less water per flush, operational improvements have been made. This improved system includes an overflow, which discharges the extra water into the bowl. The operating pressure required for this system is about 15 psi.

Another flushing system, which is employed mainly in public facilities, includes a flush valve or flushometer instead of a reservoir. When the lever is operated, this device allows a predetermined amount of water to enter the bowl directly from the pipe and flush the bowl. Codes usually require flushometers to be of the water conserving type, and the flow rate is adjustable. When the flushing action begins, it cannot be stopped until the cycle is completed. This system should also include a vacuum breaker, which is installed on the water line to allow air to enter the pipe upon loss of water pressure. This prevents the accidental absorption of non-potable water back into the supply line. The operating pressure for this system is 20 to 25 psi.

A system using the flushometer may be floor or wall mounted. The wall-mounted fixture makes floor clean up easier and is best suited for commercial or institutional applications. Ultimately, the choice is a matter of preference, as is the color selection and bowl shape. For example, a bowl with an elongated rim with no cover on the seat offers a larger seat opening is considered a more sanitary unit. Wall-mounted plumbing fixtures must be installed on partition-type walls and be well supported to prevent possible vibration. Figure 5-2 shows fixtures installed on a partition wall with the pipes behind the partition.

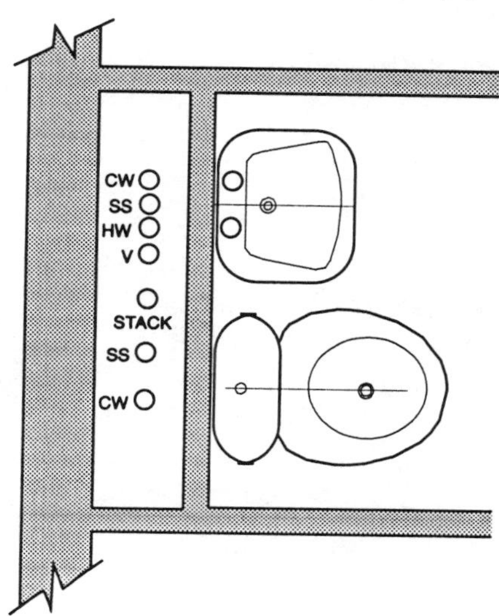

Figure 5-2. Fixtures Installed on a Partition Wall

In general, the material used for WC construction is vitreous china, but cast aluminum and stainless steel are available for special applications (hospitals, prisons, etc.). Public WCs are usually provided with flush valves rather than reserve tanks for easier maintenance. They are also less prone to vandalism.

Figure 5-3 shows four WC configurations with different flushing actions. The numbered components in Figure 5-3 are described in the following paragraphs.

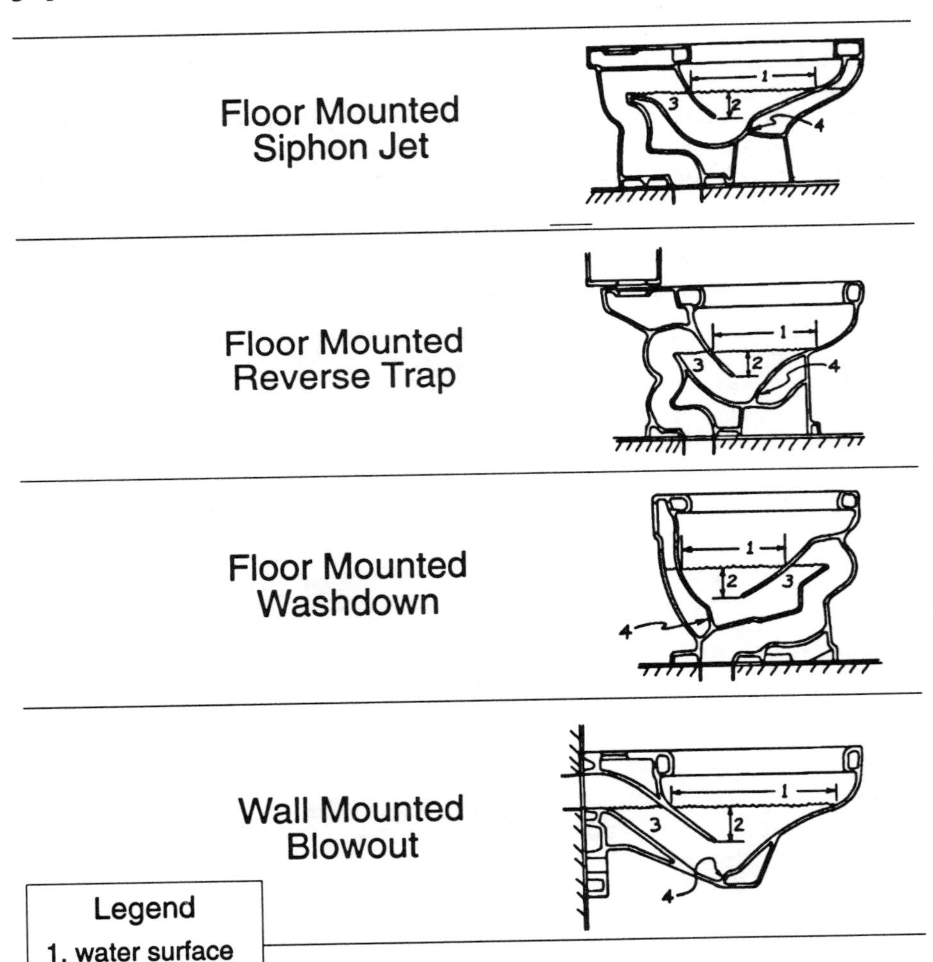

Figure 5-3. WC Installations

Siphon Jet

The flushing action of the siphon jet bowl is accomplished by a jet of water (4) being directed through the upleg (3) of the trapway. Instantaneously, the trapway fills with water, and the siphonic action begins. The powerful, quick, and relatively quiet action of the siphon jet bowl combined with its large water surface (1) and deep water seal (2) contribute to its general recognition by sanitation authorities as the preferred type of water closet bowl.

Reverse Trap

The flushing action and general appearance of the reverse trap bowl is similar to the siphon jet. However, the water surface (1), seal depth (2), and trapway size (3) are smaller, so less water is required for operation. Reverse trap bowls are more suitable for installations with flush valves or low tanks.

Washdown

The washdown bowl is simple in construction and yet highly efficient within its limitations. Proper functioning of the bowl is dependent upon siphonic action in the trapway, accelerated by the water force from the jet (2) directed over the dam. Washdown bowls are widely used where low cost is a primary factor. They will operate efficiently with a flush valve or low tank.

Blowout

The blowout bowl cannot be compared with any of the three previous types. The bowl depends entirely upon a driving jet action (4) for its efficiency rather than on siphonic action in the trapway. It is economical in water use, yet has a large water surface (1) which reduces fouling space, a deep water seal (2), and a large unrestricted trapway (3). Blowout bowls are especially suitable for use in schools, offices, and public buildings. They are operated with flush valves only.

URINALS

Urinals should have a smooth, impervious surface; an acid-resistant finish; and a minimum fouling surface. They are usually made of vitreous china, enameled cast iron, vitrified clay, or for special applications, cast aluminum and stainless steel, Figure 5-4.

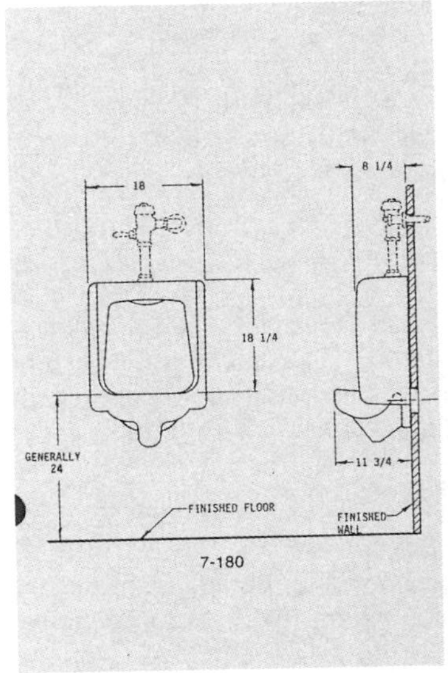

Figure 5-4. Blowout Wall Urinal (Courtesy, Crane Plumbing)

Urinals can be floor mounted, wall mounted, or free standing in a number of sizes and shapes, for both men and women. One urinal made for female use (called the *she-inal* by Urinette, Inc.) uses a flush valve (preferably with a disk handle in front for public places), water tank (for individual installation), or in the case of through urinals, a washdown pipe assembly, which provides a continuous flow of regulated water volume (practically discontinued now due to non-economical use of water).

Wall-mounted models are the most popular because of the advantages offered in cleaning and maintenance. Compared to a WC, the siphon jet and blowout urinals provide a flushing action, which removes foreign matter such as cigarette butts, gum wrappers, etc. The washout and washdown models are quieter in operation but are intended only for liquid waste removal (some washout urinals have integral strainers). From a maintenance point of view, existing or newly installed urinals with no strainers should be provided with a removable and flexible (rubber or plastic) strainer, which is inexpensive.

Lavatories

The selection of hand-washing fixtures (lavatories) is virtually unlimited. Wall-mounted units, with or without legs, plain tops and splash backs, units with pedestal bases, Pullman or drop-in models for countertops, rectangular, circular, oval, and circular segment are all standard models. Vitreous china and enameled cast iron are the most extensively used materials, but natural or man-made marble is used often also.

Lavatory units should be selected based on the intended usage. For instance, the designer should specify splash back models where the unit is to be installed against plaster or drywall construction. The selection must consider this architectural feature, Figure 5-5.

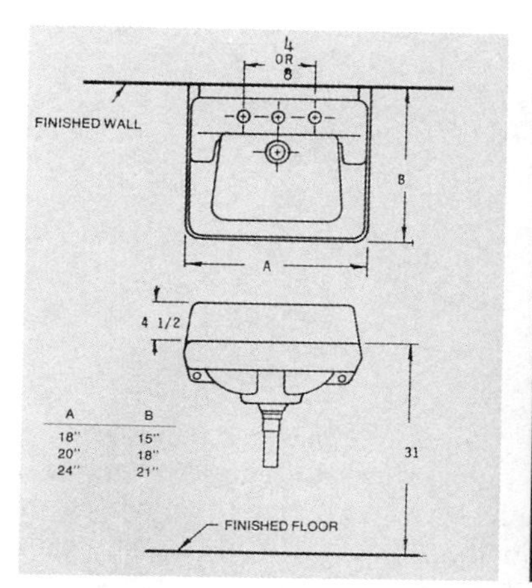

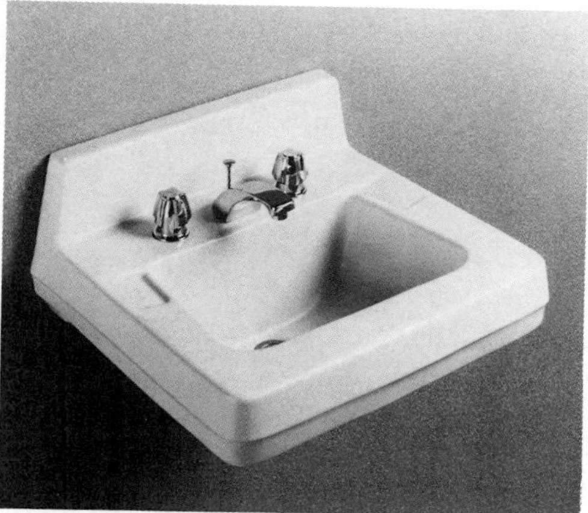

Figure 5-5. Wall Hung Lavatory (Courtesy, Crane Plumbing)

SINKS

Sinks are available in single, double, and triple bowl models, Figure 5-6. The more popular residential types have a back ledge punched to accommodate a variety of faucets available, and the sinks are conveniently installed in cabinet countertops (with usable space underneath). Flat rim types, without a ledge or punchings, are seldom used except in commercial applications where a wall mounted faucet may be preferred.

For residential use, a sink with at least one compartment measuring 15" x 18" is recommended. A 10-inch long spout is preferable to the usual 8-inch spout. The dual bowl sink is widely used, and the compartments are usually of equal size.

Laundry trays are available in both single and double compartment models, with or without splash back in combination with service sinks. Wall hung, counter-mounted models, as well as models with legs are available. Enameled cast iron is the material most commonly used for sinks as well as laundry trays, but many models in stainless steel, stone composition, or fiberglass are available on the market.

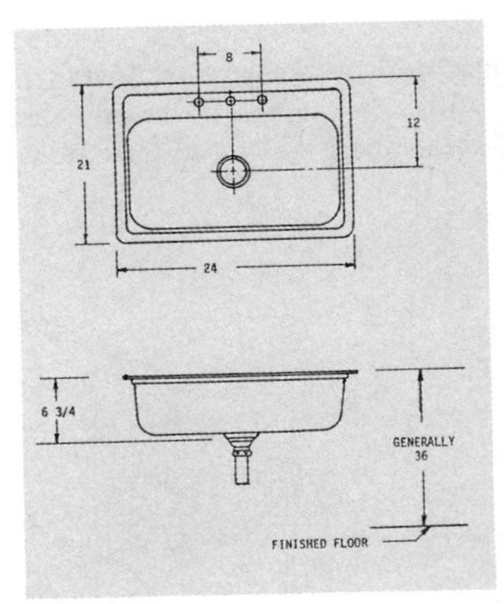

Figure 5-6. Single Kitchen Sink (Courtesy, Crane Plumbing)

Drinking Fountains

Drinking fountains (non-refrigerated) are available in a range of models, including free-standing, partially (semi) or fully recessed, pedestal type, deck type for countertops, and single or multiple bubbler types. Some fountains on the market come with an optional glass filler. Stainless steel, vitreous china, enameled cast iron, and fiberglass are just some of the materials offered by manufacturers.

Electric water coolers (refrigerated) are available in many of the same models and materials. The enclosure characteristics are especially important in ensuring that this type of fountain meets the manufacturer's requirements for housing the chiller unit.

Showers

Shower receptors are available in a variety of shapes and sizes. The most popular materials used are polyester resin, stone, reinforced fiber compositions, precast stone, and reinforced fiberglass plastics.

In the built-in shower, the drain and shower trim must be specified. A wide choice of trims are available, as are numerous types of shower heads (i.e., vandalproof, water-saving and models that can be fixed or adjusted to any angle, etc.). Mixing valves with a single handle control designated as *non-scalding* are preferable to the two-faucet type with a mixing battery.

The bottom of the shower should be of a non-slip type of surface. Fancy models with contoured walls, which allow the user to sit, are also available.

Bathtubs

Bathtubs are available in various sizes and shapes, with the most popular being a 5-foot recessed model that is built in on three sides, Figure 5-7. The interior dimension is 4-1/2 feet, which is short and sometimes uncomfortable. The newer models are larger and some have built-in whirlpool action. Due to these larger, multi-functional fixtures, the bathroom size increases, as does the bathroom importance and house/apartment value.

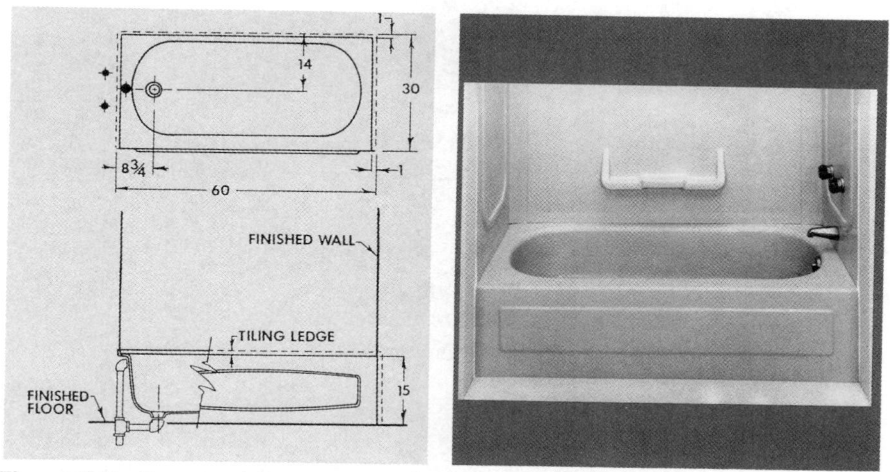

Figure 5-7. Recessed Bath (Courtesy, Crane Plumbing)

Tubs are generally made of enameled cast iron, porcelain enamel on pressed steel, or reinforced fiberglass plastic. Models with various contours are also available for the user's comfort.

It is advisable to select a slip-resistant bottom for any tub. Local codes should be checked for their requirements concerning tub trap accessibility. These requirements could govern the type of waste-overflow fitting to be installed with the tub.

Bidet

The bidet is a small bath used primarily for cleansing after water closet use. It is usually made of vitreous china and is about the same size as a water closet, Figure 5-8.

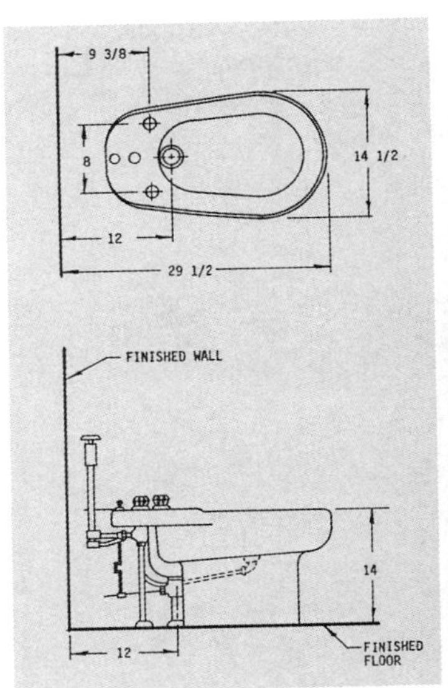

Figure 5-8. Bidet (Courtesy, Crane Plumbing)

The hot and cold water supply and drain fittings are very similar to those used with lavatories, except that water enters the bowl from a flushing rim instead of a spout. In addition to filling the bowl with tepid water, this design serves to warm the hollow china rim for comfortable sitting. A spray rinse used for external rinsing is optional on most models.

Bidets are installed in a large number of bathrooms throughout Europe. They are less popular in the United States, since most Americans use a shower for personal hygiene.

Floor Drain

Even though it deserves more attention, the floor drain is often ignored. In residential applications, floor drains are not usually installed. In multi-purpose, commercial, or industrial buildings (especially where mechanical equipment is installed), it is necessary to install floor drains. Pumps, water heaters, backflow preventers, water meters, and other mechanical equipment must have a floor drain in close proximity. To avoid contamination, the equipment is usually not directly connected into the drain (provides an air gap). Establishments housing large, commercial kitchens and large numbers of urinals and toilets need to have floor drains for cleaning, as well as fixture overflow.

Emergency Shower and Eye Bath

The emergency shower and eye bath can either be installed separately or as a unit. They must be installed in areas where caustic substances or hazardous chemicals are handled and/or stored.

These units are supplied with cold water only and have no trap or drain connected to the fixture. A floor drain may be located in the vicinity of the fixture. These fixtures are used only in case of an emergency, and in such cases, it is more important to pour water over the affected area rather than select its proper temperature.

The specification included in Appendix H indicates the requirements for these types of fixtures.

Fixture Trims

Plumbing fixtures include many accessories called fixture trims. These trims include faucets, shut-off and mixing valves, flush valves, flush tanks, vacuum breakers, and flow controls. All visible trims, pipes, valves, etc., must be chromium plated (or gold plated if requested) for a pleasant appearance.

Advanced Plumbing Fixtures

To this point, standard plumbing fixtures have been discussed. However, this book would not be complete without a mention of more advanced plumbing fixtures. In addition to the devices listed here, improved piping and fitting materials are also available.

Computer Chip Toilets

The Japanese have developed a toilet equipped with a computer chip memory, which does some of the work for the user. The user does not need to use toilet paper with this toilet, because at the press of a button, jets of temperature-controlled water are pointed upward for cleansing, followed by a hot air blast for drying. Puffs of perfumed spray are also included. This is an interesting unit but so far, prohibitively expensive and not generally used.

Infrared

Infrared beams may be used in a lavatory to automatically activate an "electronic" faucet. When the person inserts an object (such as hands) into the lavatory's operating range, it automatically opens a solenoid valve and water begins flowing. Once the hands are removed, the valve closes. This system is a good example of an electric/electronic device applied to plumbing, which reduces water consumption.

Water closets and urinals can also make use of infrared technology. A sensor controls the automatic flushing after the person leaves the fixture.

Infrared systems are in use today but are mainly found in areas of high traffic, such as airports and rest areas.

Ultra Valve

The shower ultra valve is a pressure balanced, thermostatically controlled valve, which uses a small microprocessor to monitor the water supply temperature. The valve has the capability of making continual adjustments in order to maintain a comfortable water temperature.

Electric, Waterless Toilets

An electric, waterless toilet also bears mentioning here. This is a self-contained fixture that requires no water or plumbing connections — just an electrical connection and a vent duct (pipe). This toilet incinerates waste.

According to Incinolet, one manufacturer of this toilet type, a liner made of special, moisture-impervious paper is placed in the bowl, so no waste ever touches any toilet surface. After use, the bowl liner with waste drops into the incinerator when the user steps on a foot pedal. A heater and blower come on and stay on until the incineration process is complete. The outer surface remains cool, but the heater itself attains a temperature of 1400°F to evaporate urine and dehydrate/ignite solids and toilet paper. The quality of the exhausted air is protected by a built-in emission control system. Only clean, inorganic ash remains. The ash pan has to be emptied once or twice a week depending on the frequency of usage.

The following specification data for this toilet was provided by Incinolet:

Incinolet Electric Toilet Characteristics

Height: 21 inches
Width: 15 inches
Depth: 24 inches
Venting: 140 cfm (minimum); 200 cfm (maximum)
Power: 3750 watts at 120/208 or 240 V

Incinolet may be vented with metal pipe, PVC (Schedule 80), or heavy-duty, rubber-coated, flexible hose. Each unit must be individually vented, Figure 5-9.

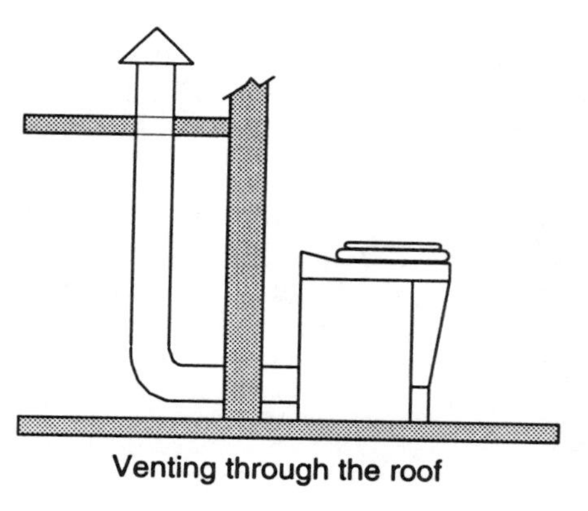

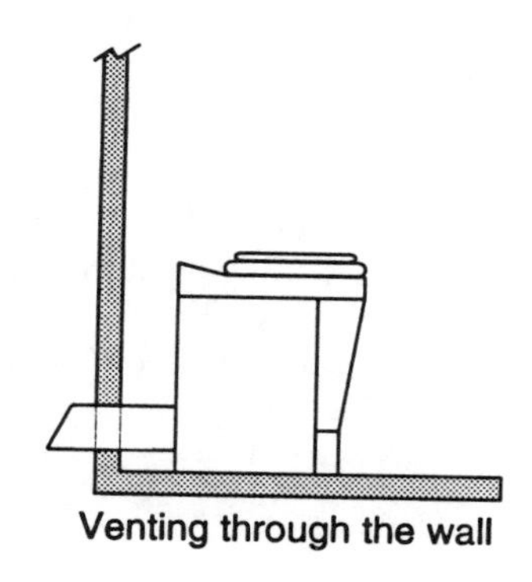

Figure 5-9. Waterless Toilet Venting

Fixtures for the Handicapped

In the early 1960s, it was recognized that regular fixtures could not be used comfortably by handicapped people. For example, the standard toilet partition space would not accommodate wheelchair movement. Consequently, modifications were made in the space allowed, as well as in the toilet height, lavatories, drinking fountains, etc., to make these fixtures accessible to the physically handicapped.

Handicapped access in toilet rooms was not implemented without regulation. State codes and standards based on standardized wheelchairs were developed and became law. In addition, a federal law entitled *Americans with Disabilities Act* (ADA) mandates handicap provisions. Today, public plumbing facilities must provide a prescribed arrangement for handicapped people. Toilet rooms for the handicapped must:

- be at least 3 feet wide.
- have a door (where doors are used) that is 32 inches wide and swings out.
- have handrails on at least one side of the toilet (installed 33 to 36 inches above and parallel to the floor). The rail must have an outside diameter of 1-1/4" to 1-1/2". There must also be a 1-1/2" clearance between rail and wall, and it must be fastened securely at the ends and the center.
- have a water closet with the seat 17 to 19 inches above the floor.

This type of water closet and its installation are also important. A wall-mounted water closet with a narrow understructure that recedes sharply is most desirable. If a floor-mounted water closet is used, it should not have a front that is wide and perpendicular to the floor at the front seat. The bowl should be shallow at the front of the seat and turned backward more than downward to allow the individual in a wheelchair to get close to the water closet using the wheelchair.

Toilet rooms should have lavatories with narrow aprons which, when mounted at standard height, are usable by individuals in wheelchairs. Hot water pipes and traps under a lavatory must be insulated so that a wheelchair-bound individual will not suffer injury.

Toilet rooms for men must have wall-mounted urinals with the opening of the basin 17 inches above the floor, or floor-mounted urinals that are level with the toilet room floor. There are additional architectural requirements that must be followed, which are detailed in the state building code. Drinking fountains and coolers with special construction features for handicapped persons' accessibility are available.

Chapter Six

Energy Conservation

When energy conservation is applied to plumbing systems, there are usually three goals in mind:

1. Water conservation
2. Reduction of hot water consumption and/or its temperature
3. Reduction in fuel consumption (reduction in heat losses)

In spite of the fact that three-quarters of the planet's surface is covered with water, fresh water is somewhat expensive, because it requires treatment and conveyance. As a result, there are potential savings to be made by using less water or wasting less water.

People became aware of the need for water conservation mainly during drought periods when water became scarce and consumption was limited through local regulations. However, the world-wide economy can no longer afford to waste natural resources, including water.

Water conservation and energy conservation are ultimately interconnected. A reduction in water consumption means a reduction of energy usage. Before water gets to consumers, it must first be treated, then conveyed through the supply system network. Both processes require energy consumption. Less water means less pumping is required, which means less electricity (power) is used. If the need for less energy consumption means reduced water consumption, the project designer must take this into account. This reduction will translate into smaller pipe sizes and less insulation of hot water pipes, resulting in reduced construction costs.

Lowered domestic hot water demand or reduced temperature of hot water means less fuel consumption, which means a direct reduction of the building's operating expenses. It may also mean reduced equipment size, which will lower the initial cost. Planning for potable water demand and consumption as well as reduction in hot water consumption also means a direct reduction in the initial construction cost as well as the building operating expenses.

Another consequence of reduced water consumption is a reduction in sewage flow. In most municipalities, residents pay a proportional rate for the amount of sewage discharged. Reducing the sewage amount lowers the operating cost. The sewage must be treated before it is discharged into waterways, so a reduced flow will result in less power used. Keep in mind that a reduction in water flow does not have to be done at the expense of safety. Water carries wastes, and if water flow is reduced, modifications in equipment design are needed.

Water Heaters

To reduce energy consumption, the efficiency of the water heating equipment must be maximized. This includes running the water heater only at the times when needed.

The water heater tries to keep the temperature of the hot water at the set point (normally 140°F). If a water heater is continuously in operation, it will call for heat (burner or hot water boiler for indirect heaters) any time the water temperature drops a few degrees. In commercial and industrial buildings, turning the water heater off when there is no activity (nights 11pm - 5am) means fuel economy. Such activity may be easily controlled by a time setting clock.

Another element that should be considered is the reduction in heat loss. Consuming fuel to heat domestic water can be reduced if the heat losses to the environment are reduced to a minimum. To minimize these losses, hot water pipes must be insulated. Water heaters are furnished with internal insulation; however, for more fuel economy, a layer of fiberglass insulation on the outside of the water heater is recommended. The sample specification in Appendix H covers the materials recommended for insulating hot water pipes.

Bare hot water pipes are an important source of heat loss. Pipe insulation represents an added expenditure during the construction phase, and a specialized subcontractor usually performs the installation per specification requirements. The subcontractor must make sure the insulation:

- has low thermal conductivity.
- is noncombustible.
- is adequately strong.
- will not deteriorate with time.
- has a neat appearance.

The insulation efficiency depends on the material used and its heat transmission coefficient (R). The higher the insulation R factor, the lower the heat loss.

FIXTURES

Toilets with a 1.6-gallon tank capacity help conserve water, as do atomizing shower heads, which use less water than regular heads. In addition to reducing the amount of water consumed, atomizing shower heads also reduce the energy required to heat the water for showers. The same situation exists with the 0.5 gpm lavatory. A self-closing faucet is included for commercial and industrial (public) applications. All of these modifications are fine as long as consumers are satisfied with the results.

A building with clean plumbing fixtures, which are in good working condition, is a desirable place in which to live. Water and sewer are already expensive, and water and waste treatment plants are costly to build, operate, and maintain. Ultimately, water consumers pay for these costs. In no case, however, should a cut in energy and water consumption have a negative impact on general hygiene.

Here are a few hints on how to help conserve energy, reduce water consumption, and reduce sewage:

- Take showers instead of baths. Showers average 25 gallons per use as opposed to baths, which require 36 gallons.
- Do not throw wastes (tissues, cigarette butts, etc.) into the water closet for flushing. Household wastes should go into the trash.
- If you need cold water for drinking, keep a bottle of plain water in the refrigerator and use as needed.
- Run a full load in the washing machine and/or dishwasher — a washing machine uses approximately 50 gallons of water per load.
- In work places with five-day/eight-hour schedules, shut down the hot water circulating pump during the night and over the weekend.
- Adjust drinking water coolers to supply 55°F water instead of the usual 50°F water.
- If a gravity water supply tank is installed, use it during the day to save pumping at peak electrical demand.

With a little care, every household, industry, or commercial establishment can contribute to water savings with benefits received on the individual level, as well as on a national level.

Chapter Seven

Water Supply and Demand

For domestic use, it is important that the water source is safe, reliable, and of potable quality. The water supply must not be subjected to contamination from illegal connections, backflow, or siphonage (see Appendix C for requirements of water supply protection). The water supply system must be installed in strict accordance with the governing codes and the local authority having jurisdiction.

Water Meters

A water meter can be found in every domestic water supply system entering a building. In a new development that includes several different buildings, it is mandatory for each building to have its own water meter. The reason behind this is simple; if water consumption suddenly increases in one building, it is possible to pinpoint the origin.

The water meter may be installed outside the building (in close proximity) or inside the building (near to the pipe entering the building). Outside the building, the meter must be installed in a watertight, frostproof pit. This pit must be accessible to allow for meter reading. If it is inside the building, the meter must be located along a wall for protection against damage. It must also be installed approximately three feet above the floor for easy reading and inspection. Whether outside or inside the building, the meter is provided with shut-off valves on both sides and a valved T outlet downstream for local flow checking.

Figure 7-1 shows various meter sizes, their flow rates, and their pressure loss. The exact pressure loss for a particular installation must be obtained from the meter manufacturer.

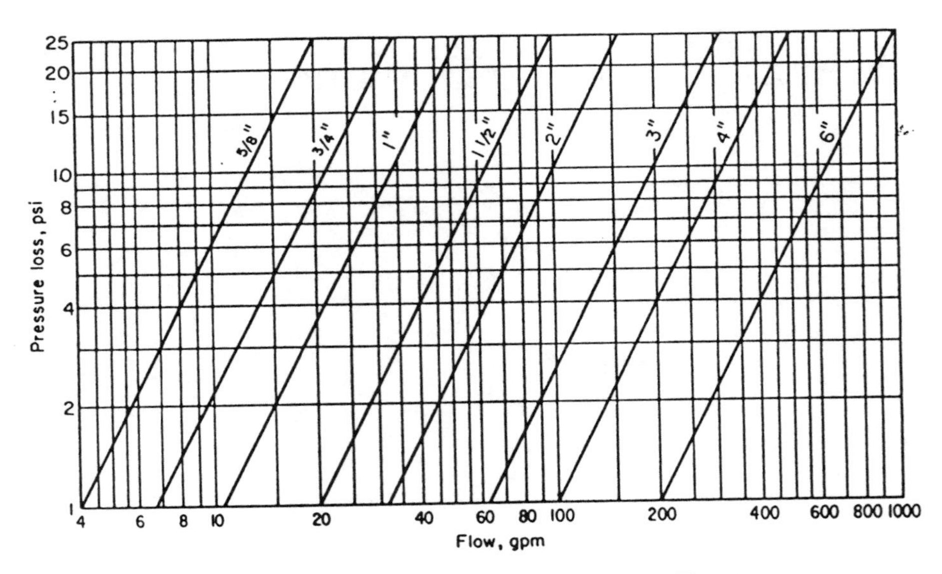

Figure 7-1. Pressure Loss Through Meters During Water Flow

There are several different water meter types commercially available. These include the following:

- Disk meter — sizes are normally 5/8", 3/4", 1-1/2", and 2". They are manufactured to meet requirements of AWWA Standard C700. They have a maximum working pressure of 150 psi, and they measure flow in one direction. This meter is often used in residential and small commercial installations.
- Compound meter — sizes are normally 2", 3", 4", and 6". They are manufactured to meet requirements of AWWA Standard C700. They have a maximum working pressure of 150 psi, and they measure flow in one direction. This type of meter is used when most of the time the flow is low, but occasional high flow rates are anticipated.
- Turbine meter — sizes are 2", 3", 4", 6", and 10". This type of meter has the characteristics of a compound meter, but it is more suitable when a variety of flow rates are expected.
- Propeller meter — sizes are 2" through 72". Propeller meters are used only for high rate flow.

WATER STORAGE AND PRESSURE

Under no-flow conditions, the maximum theoretical pressure at a plumbing fixture is 80 psi, which is a very high pressure for normal fixture operating conditions. If this pressure exists, it requires some kind of pressure regulator (i.e., pressure reduction valve, restriction orifice, etc.) to reduce the pressure to levels of about 30 to 40 psi for a group of fixtures. The minimum pressure for a system with reserve (storage), tank-type water closets, or urinals is approximately 10-15 psi for each fixture, Table 7-1. When flushometers are used to operate the WCs and urinals, 20 to 25 psi is required while the rest of the fixtures should have a normal working pressure of 8 to 12 psi.

Fixture	Pressure (psi)
Lavatory faucet	8
Lavatory faucet, self-closing	12
Sink faucet, 3/8 in.	10
Sink faucet, 1/2 in.	5
Laundry tub cock, 1/4 in.	5
Bathtub faucet	5
Shower	12
Water closet, ball cock type	15
Water closet, flush valve type	10-20
Urinal, flush valve	15
Dishwasher, domestic	15-25
Garden hose (50 ft and sill cock)	30

Table 7-1. Minimum Operating Pressures for Plumbing Fixtures

If city water is the supply source, it can be assumed that the supply pressure to the system will be almost constant. However, this information must be "officially" obtained for a reliable installation. There may be great variations in the water demand, depending on the time of day (for example, in high rise buildings with a large number of consumers). The system might require water storage or a reserve tank, located on the building's roof, to supplement the quantity and pressure during the peak consumption hours, which might just coincide with low pressure in the city line.

Storage tank capacities can be determined based on some practical rules. In general, the rough determination of storage capacity is based on 8 to 12 gallons per person per day, plus the constant consumers such as air conditioning, lawn watering, etc. For multi-apartment buildings, an estimate should consider two persons per bedroom or four persons per apartment (whichever is largest). Once the number of persons is established for a high rise residential building, it is possible to determine the consumption of water per person as a function of the number of apartments, Figure 7-2.

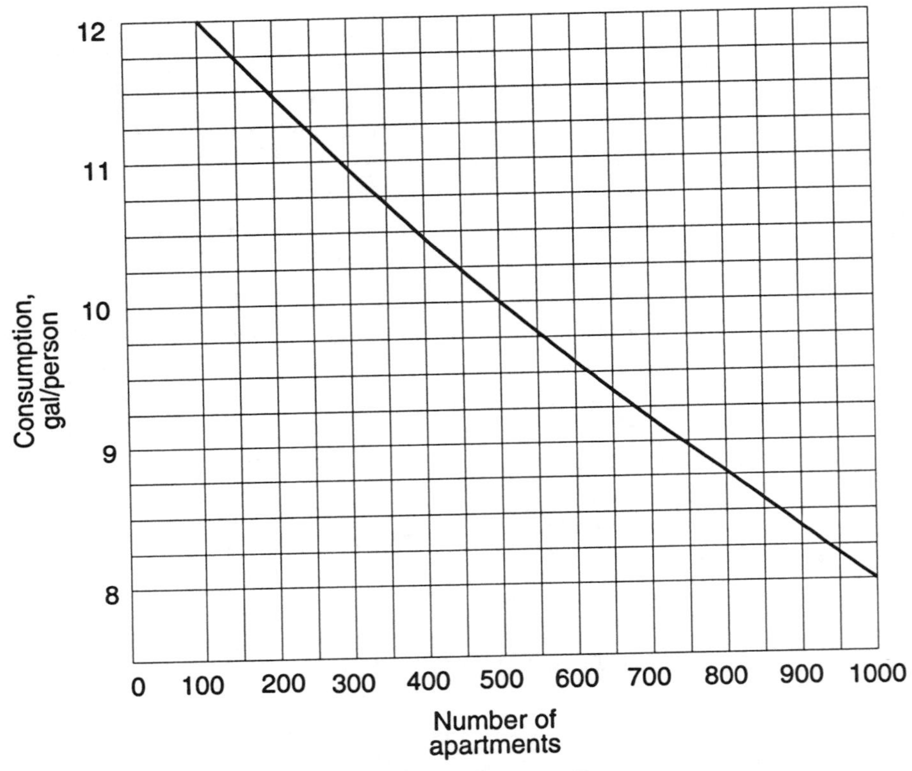

Figure 7-2. Estimated Domestic Water Consumption

Another possible way to calculate the water requirements or storage capacity is to base the estimate on a value of 0.024 gallons/habitable sq ft. For office buildings, the determination is based on 100 to 200 sq ft/person times the number of occupants, depending on the building occupancy and luxury standards.

When water storage is necessary, consider these requirements in the construction of water storage tanks:

- Maximum capacity for a single tank may not exceed 30,000 gallons
- Tank must be leak proof
- Tank must be vermin proof
- Tank must resist corrosion

- Tank must have noncombustible legs or supports
- Atmospheric connection shall be protected against intrusions by a screen

In cases where a tank is used for the dual purpose of storing potable water and water for fire suppression, the connections must be made as shown in Figure 7-3. This arrangement ensures that a certain amount of water is always available for fire suppression. The water supply entering into the storage tank must be located at a minimum of 4" above the maximum water level in the tank. (This air gap is required to prevent contamination.)

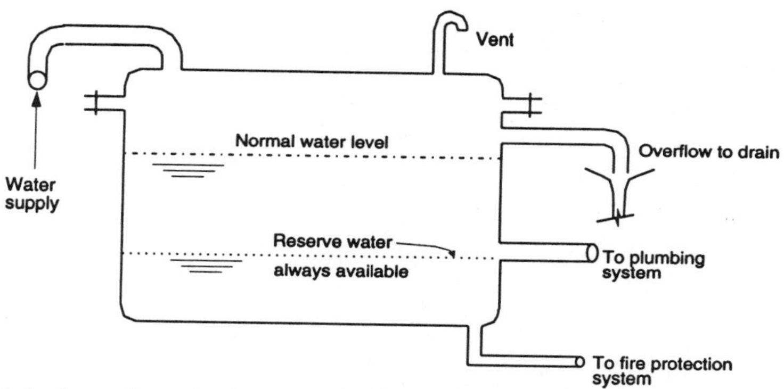

For general construction requirements, refer to NFPA (National Fire Protection Assocation) Standard No. 22 (Vol. I)

Figure 7-3. Water Storage Tank

PIPING

Each plumbing fixture in a given system must have enough water flow at the appropriate pressure. It is necessary to make sure energy and water conservation are built into the system in accordance with applicable state, local, or national codes.

The pipe sizes shown in Table 7-2 are minimum sizes. One size larger may be used for certain fixtures provided that the design results in a good working system. However, the cost of installation increases if pipes are larger, and this must be taken into consideration.

Fixture or device	Nominal pipe size (in.)	
	cold	hot
Lavatory	3/8	3/8
Bathtub	1/2	1/2
Shower, single head	1/2	1/2
Kitchen sink, residential	1/2	1/2
Kitchen sink, commercial	3/4	3/4
Combination sink and tray	1/2	1/2
Laundry tray	1/2	1/2
Sinks (service, slop)	1/2	1/2
Sinks, flushing rim	3/4	3/4
Water closet, tank type	3/8	-
Water closet, flush valve type	1	-
Urinal, flush tank	1/2	-
Urinal, direct flush valve	3/4	-
Drinking fountain	3/8	-
Dishwasher, domestic	-	1/2
Hose bibbs	1/2	-
Wall hydrant	1/2	-

Table 7-2. Minimum Sizes of Fixture Water Supply Pipes *(Please note that this table and any other table showing sizes, consumption, etc., must be checked against the governing state plumbing code and local governing regulations.)*

Each fixture type has a specific demand at its water outlet. This demand considers the maximum water flow in gpm with the valve fully open, Table 7-3.

The water supply used for potable water is often the same as that used in fire suppression systems. However, the water in a fire suppression system is usually stagnant. This means polluted water (rusty/black water or water containing antifreeze) could potentially backflow into the potable water system. **This must be prevented.** To prevent potable water contamination, backflow preventers in the building's main water supply are now frequently required by local regulations and state codes (see Appendix C). The backflow preventer is an assembly of two check valves in series (one downstream from the other) with shut-off valves at each end and a series of orifices and test points to periodically determine its proper operation. This device requires regular testing and scheduled maintenance. In such a common water supply, the backflow preventer is installed in the fire protection line.

Once there is an adequate supply of potable water at the required pressure, which corresponds to the amount needed during peak hours, it is necessary to

Outlet	gpm
Lavatory faucet	3.0
Lavatory faucet, self-closing	2.5
Sink faucet, 3/8 in. or 1/2 in.	4.5
Sink faucet, 3/4 in.	6.0
Bathtub faucet, 1/2 in.	5.0
Shower head, 1/2 in.	5.0
Laundry tub cock, 1/4 in.	5.0
Water closet, ball cock flush tank	3.0
WC flush valve, 1 in. and at 25 psi flow pressure	35.0
WC flush valve, 1 in. and at 15 psi flow pressure	27.0
WC flush valve, 3/4 in. and at 15 psi flow pressure	15.0
Drinking fountain jet	0.75
Dishwashing machine, domestic	4.0
Laundry machine, 8 lb or 16 lb	4.0
Hose bibb or sill cock, 1/2 in.	5.0

Table 7-3. Water Demand

ensure that the distribution system or piping network is designed and installed correctly. Piping must be sized correctly to achieve the necessary requirements, which include proper quantity, pressure, reduced noise levels, and a correct velocity to prevent water hammer.[1] At the same time, economical sizes must be selected to prevent overdesign and unwarranted costs. The most economical pipe sizes are the ones recommended by code (minimum requirements).

Pipe Selection

Practical experience and knowledge are required in pipe selection and installation. The pipes in a plumbing system must be:

- the correct size.
- well supported.
- able to accommodate the expansion and contraction due to the hot water circulation.
- able to accommodate minor vibration.
- tight (no leakage).
- disinfected.

The underground piping system must also be protected from corrosion by using cathodic protection or plastic pipes, which will not corrode when installed in "aggressive" soil.

When installing the water supply piping system there are some practical rules to consider, which include the following:

- Water lines must be installed parallel to the walls in an organized and geometrically regular system.
- Pipes have to be pitched to allow for drainage, if necessary.
- Supply pipes must not have traps, sags, or be vertically bowed.
- Water pipes may not be installed in stairwells, hoistery, elevator shafts, or in front of windows, doors, or any other wall opening.
- Vertical pipes must be located in chases behind the plumbing fixtures. Enough depth of the chase to accommodate pipes and fittings is necessary. Table 7-4 lists dimensions for clearances depending on the fixture type.
- Horizontal pipes located above ceilings must be checked (first on the drawings) to ensure they do not interfere with support beams, ducts, or electrical fixtures.
- In specialty rooms (e.g., telephone rooms, electrical rooms, control rooms, etc.), plumbing system pipe installation is not permitted.
- Pipes may not be installed in areas where they may be subject to freezing or corrosion unless they are protected.
- Underground pipes must be installed a minimum of 3 feet away from the building foundation (except for those that penetrate the foundation).
- Pipes penetrating through the building foundation must be installed within steel sleeves. The space between the pipe and sleeve must be filled with a waterproof, elastic caulking material to prevent building settlement from shearing the pipe.
- Underground pipes must be installed below the frost level and laid in firm, prepared beds.

In summary, pipes must be reasonably protected from injury or an aggressive environment.

Pipe Materials

Pipes may be fabricated of ferrous (iron base metals such as various forms of steel) or non-ferrous (copper, brass, aluminum, plastic, ceramic, glass) materials. Pure metals are not usually used, because alloys are more resistant to the environment. Naturally, each type of material has unique characteristics, which make it more desirable for a particular application. If the proper material is selected for the piping system, few problems are anticipated. Incorrect piping material could present many problems that may render the system vulnerable to hazardous situations.

In recent years, the use of lead pipe and lead containing solder in domestic water systems has been discontinued. Studies showed that lead dissolves in the water, and it has been determined that the lead is poisonous to humans, especially children.

Fixture	Combination and support	Space required (in.)*
Water closet:		
Floor type	Single and back-to-back	4
Floor type	Battery and back-to-back	12
Wall hung	Single and back-to-back residential carrier Smith 500	8
Wall hung	Single commercial carrier Smith 400	10
Wall hung	Single back-to-back commercial carrier Smith 400	13
Wall hung	Battery carrier Smith 200	16
Wall hung	Back-to-back carrier Smith 200	19
Lavatory	Single and back-to-back	6
	Battery and battery back-to-back	11
Urinal	Single	6
	Battery	12
Sinks	Single and back-to-back	6
	Battery	11
Flushing rim clinical sink	Single	10
Shower	Single	4
	Back-to-back	6
Slop sink		8
Drinking fountains (water coolers):		
Projecting	Stud support	4
Projecting	Chair carrier	6
Simulated semi-recessed with cooler	Stud support	4
Simulated semi-recessed with cooler	Chair carrier	6
Semi-recessed	With or without cooler	6
Recessed	With or without cooler	12
Wal-Pak		12
Unit water cooler	Chair carrier	6

*Space required means space between the back sides of the two gyp boards. No allowance is included for stack requirements.

Table 7-4. Minimum Wall Space Requirements for Plumbing Fixtures

There are a number of factors contributing to piping material selection for a plumbing installation. Some of them are applicable to overhead pipes, some to underground pipes, and some to both. These factors include the following:

- Initial material cost
- Installation cost
- Expected service life
- Weight
- Ease of making joints
- Chemical resistance
- Susceptibility to corrosion
- Thermal expansion
- Friction loss coefficient
- Pressure rating
- Rigidity
- Resistance to crushing
- Combustibility (except metal piping)
- Chemical characteristics of water supply
- Ease of replacement (availability in the area)
- Water pressure available (influenced by pipe roughness coefficient)

Piping materials commonly used for various piping system applications include the following:

Application	Pipe material
Domestic cold water	Copper, plastic
Domestic hot water	Copper, galvanized steel
Underground domestic water	Copper (Type K), plastic
Compressed air	Black or galvanized steel, copper
Vacuum system	Copper, black steel
Fuel oil	Black steel, copper
Distilled water	Stainless steel, plastic, glass
Drains, wastes, and vents	Plastic, glass (special applications), copper, glazed clay, cast iron

More details regarding piping material alternatives and their specific applications can be found in the sample specification in Appendix H.

Copper Piping

Although plastic pipes are used extensively in plumbing systems, due to desirable physical and chemical properties, copper pipes are usually used for water supply. There are three categories of copper pipes or tubing:

1. Copper Type K — heaviest pipe wall (thickest) for outdoor or underground installation
2. Copper Type L — medium pipe wall for indoor installation
3. Copper Type M — light wall for indoor use where permitted by the local regulations

Copper tubing is widely used for a number of reasons. It is lightweight, has good mechanical strength, and provides ease in making connections. Copper also has a pleasant physical appearance. However, copper tubes may corrode under certain conditions. The corrosion may never occur, or it may show up only 4 to 6 years after the installation is completed. The contractor should be aware of the elements that may have an adverse effect on copper tubing. These elements include:

- ***Water Quality*** — A laboratory test will yield information about water behaving aggressively towards copper. It is also important to determine the water aggressiveness function of temperature. If the water is neutral (pH=7), the pipes are safe. Aggressive water might require special treatment, or it will quickly corrode the pipes. In the long run, any type of water will corrode a pipe made of any metal.
- ***Soil*** — The soil surrounding a copper pipe installation might, under certain conditions, induce pipe corrosion and ultimately pipe failure. Soils with very low electrical resistance seem to produce more frequent corrosion. The soil resistance can be measured, and if found inadequate, pipe protection is recommended. The cost of protection is minimal compared with early failure.
- ***Installation*** — The installation quality might be negatively influenced if it is executed poorly. Backfill conditions may also be a cause, along with poor installation or poor workmanship. To ensure quality, the backfill operation must be closely supervised by an experienced person. Pipe connections (soldering) may produce corrosion when the flux contains corrosive chemicals and the pipes are not properly flushed after installation.

Plastic vs Metal Pipes

Plastic pipes were introduced in the construction field around 1950 for drain, waste, and vent, as well as potable water supply. The plastic pipe has certain advantages over the steel pipe:

- It is light weight (about one-sixth that of steel)
- It has a smooth interior, which reduces the resistance to water flow

- It is easy to install
- It is resistant to corrosion, scaling, and pitting
- It does not promote the growth of algae, fungi, or bacteria
- It is not subject to galvanic or electrolytic action, thus eliminating the need for cathodic protection in buried systems
- It is inexpensive

The disadvantages of plastic include the following:

- Plastic experiences greater expansion and contraction with temperature changes (perhaps as much as four times compared to steel).
- Plastic is weaker than metal pipe and requires protection from external physical damage, water hammer, and pressure fluctuations.
- Plastic loses its strength rapidly as temperatures rise and must be protected from heat; for example, the pressure rating for PVC piping must be derated 30% when its temperature rises from 70° to 140°F.
- Water flow noise in gravity waste systems.

The designer or contractor must check with local authorities to find out what limitations, if any, exist for plastic pipe installation.

Pipe Installations

Pipes may be connected to pipes, fittings, or valves in any of the following ways, depending on the material listed:

- Flanged (cast steel or cast iron)
- Welded (forged steel)
- Screwed (black or galvanized steel)
- Brazed (copper, brass)
- Hub and spigot (cast iron) joint with packed oakum and molden lead or gaskets
- Hubless (cast iron) joint with clamps
- Glued-Special Adhesive (plastic pipes)
- Special joints between pipes of dissimilar materials

Pipes are fabricated in standardized lengths, which must be assembled to become an installation. Steel pipe has threaded ends, so the pieces and fittings are screwed together; copper tubes are soldered together; plastic pipes are either screwed or glued together.

When hot water circulates through pipes, the pipes first expand and then contract as the water cools. Expansion and contraction must be calculated and the piping arrangement made so that movement is limited to a maximum of 1-1/2". Plumbing pipes are not installed in a straight, continuous line in ordinary buildings, so expansion loops are not required, because every bend helps to absorb small pipe movements. However, if the length is large, expansion calculations are required to determine the need for expansion loops. These are dependent on pipe length, flüid temperature, and the piping material. Calculations for expansion are based on formulae and specialized tables, which detail shape, size, and other installation elements. These tables are not included in this book, but they can be found in specialized piping manuals.[2]

Plumbing pipes that convey cold and hot water must be insulated. Cold pipes are insulated to avoid condensation, and hot water pipes are insulated to reduce heat loss. Both are also insulated to protect against freezing in cold climates. Piping insulation is called thermal insulation. There is a large selection of materials to be used for insulation, and the plumbing specialist must make an appropriate choice. Insulation material and its thickness must be compared to cost. Table 7-5 lists the common types of insulation that may be used in plumbing as well as other systems, and Table 7-6 includes recommended insulation thickness. The sample plumbing specification in Appendix H gives practical details on the insulation types and their installation.

Type of insulation	Applicable ASTM standard	Forms of insulation	Service temperature range (°F)	Thermal conductivity,* Btu in./(hft² °F) mean temperature	Common application
Calcium silicate	C-533, type I	Pipe, block	Above ambient to 1200	0.41	High-pressure steam, hot water, condensate; load bearing
Cellular glass	C-552	Pipe, block	-450 to 800	0.38	Dual temperature, cold water, brine; load bearing
Cellular elastomer, flexible	C-534	Pipe, sheet	-40 to 200	0.27	Dual temperature to 200°F, cold water, runouts, non-load bearing
Cellular polystyrene	C-578, type I	Board (pipe)	-65 to 165	0.25	Dual temperature to 165°F, cold water, hot water to 165°F; limited load bearing
Cellular polyurethane	C-591, type I	Board (pipe)	-40 to 225	0.16	Dual temperature to 200°F, cold water, hot water to 225°F; limited load bearing
Mineral fiber; fiberglass	C-547, class 1	Pipe	-20 to 450	0.23	Dual temperature to 450°F, steam, condensate, hot and cold water; non-load bearing
Mineral fiber; rock or slag	C-547, class 3	Pipe	Above ambient to 1200	0.23	Steam, condensate, hot and cold water; non-load bearing

*The ASTM standards listed give maximum thermal conductivity values, which are approximately 10% higher than the nominal values given by most insulation manufacturers.

Table 7-5. Various Types of Thermal Insulation

System**	Water temp (°F)	Insulation thickness,* for nominal pipe size of		
		Up to 2½ in.	3 to 6 in.	8 in. and above
Domestic hot water, general purpose	95	½	½	1
Domestic hot water, general purpose	110	½	½	1
Domestic hot water, general purpose	120	½	1	1
Domestic hot water, utility systems	140	1	1	1
Sanitizing systems	180	1½	1½	2
Chilled water***	45 to 55	½	½	1

For an energy conservation objective of no more than 12 Btu/(hft² of outer insulation surface) where energy conservation codes may exempt need for insulation.

*Table based on use of insulation having a conductivity in the range of .22 to .25 Btu in./(hft² °F) at 75°F mean temperature.

**Thickness selected to have heat transfer rate not exceeding 12 Btu/(hft² of outer surface) for energy conservation. Thicknesses for other insulations having thermal conductivities outside the range in the note above must be determined to meet this limit.

***Condensation control required. If relative humidity exceeds 75%, additional thickness may be required to prevent condensation on the outer surface of the insulation.

Table 7-6. Minimum Piping Insulation Thickness

Be aware of the following rules when making any pipe connections:

1. There should be no connection between potable and non-potable water systems.
2. There should be no connection between the potable water system and the drainage or vent system. (For cross connections questions and answers see Appendix C.)
3. Certain equipment may not be connected directly to potable water system under any circumstances. This includes:
   a. operating tables (hospitals)
   b. dissection tables (laboratories, etc.)
   c. embalming tables
   d. mortuary tables
   e. cooling water jackets (generators, motors)
   f. any system containing health hazardous materials

All water connections at any plumbing fixture must be made above the flood (overflow) rim level, Figure 7-4. There should always be an air gap equal to 2 or 3 times the supply pipe diameter between the maximum water level in the fixture and the water supply pipe. For example, a 1" supply pipe must have an air gap of 2" to 3". This installation rule became law in 1933 when a disease, which originated from the potable water supply that was contaminated by a below-the-rim connection, raced through Chicago, Illinois.

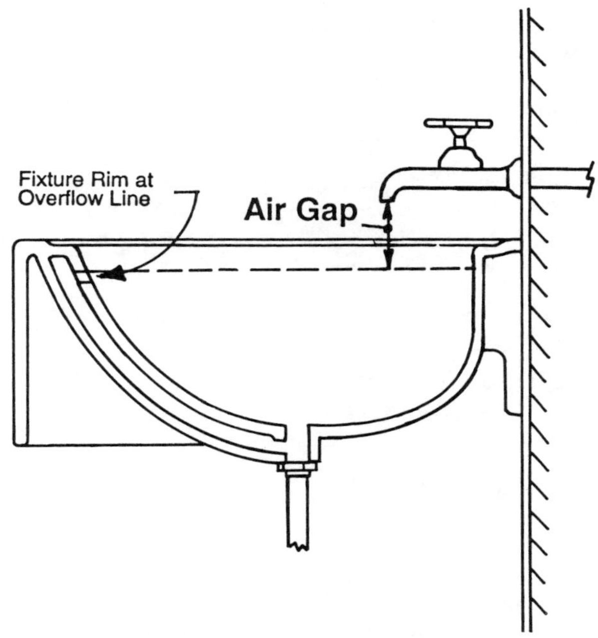

Figure 7-4. Water Supply to Fixture

If a certain fixture must have a connection below the rim, then that fixture must have an approved local vacuum breaker and/or backflow preventer (one for each fixture) installed. The vacuum breaker prevents backsiphonage by allowing air to enter the system to break the vacuum. Vacuum breakers should be properly maintained to make sure they operate as required.

Piping Supports [3]

Pipes are commonly supported by hanger rods, which either connect to the structure above or are clamped to building support beams. Standard hardware (clamps) is available for making connections between support rods, beams, and piping, as well as the building structure.

In addition to supporting the pipe, pipe supports must also allow for expansion movements without introducing restraining forces, which result in unplanned stresses either in the pipe or in the support structure. Pipes must be supported around equipment connections in such a manner as to avoid transmitting forces from piping to the equipment. Such forces could cause deformation and misalignment within the machinery, reduced reliability, and premature failures. The equipment should not be expected to support the pipe weight.

Supports for insulated pipe cannot bear directly on the insulation. Typically, heavy-gauge, sheet-steel, half sleeves (called shields) are placed between the hanger and the insulation to spread the load. Rigid insulation with the same thickness as the primary insulation is then placed between the shield and the pipe.

Pipe support hardware should be carefully protected against corrosion. Pipe support failure may cause more damage than failure of the pipe itself. Protective coating or plating of support hardware is recommended wherever atmospheric (or surrounding) corrosion could occur.

Particular care and expertise in designing pipe support systems is required in the following situations:

- Where higher pressures are used
- When water hammer can occur
- In high-rise buildings
- When pipe sizes exceed 12" in diameter
- When large anchor loads or external forces bear on the pipes

Pipe anchors must be provided at regular intervals to control and contain piping movements. A pipe anchor should be constructed so that the entire pipe circumference is clamped. A half-pipe clamp can cause line misalignment, which introduces unbalanced stresses that may cause the anchor to distort or fail.

Valves

One of the more important fittings in a plumbing system is the valve. In general, valves control (start, stop, regulate, or prevent reversal) the flow of water in pipes or at fixtures. Valve components include:

- the body
- a disk, which influences the flow
- a stem, which controls the disk
- a seat for the disk
- a bonnet to hold the stem
- end connections

There are numerous types of valves, and some are better suited for a specific task than others. There is not one type of valve that can be used for all functions. The selection depends on the pressure drop requirements, water temperature, replacement ease, and the control accuracy required.

Valves are connected to pipes with threaded, flanged, welded, grooved, or soldered ends. Solder is usually used in plumbing systems for connections to copper tubing.

Most valves are rated by their capability to withstand pressure (measured in psi). Valve ratings will also correspond to a maximum operating temperature. Table 7-7 lists the valve types and their primary functions.

Valve	Subdivisions	Function
Gate valve	Solid wedge disks Split wedge disks Double disk parallel sit Rising stem Nonrising stem Quick-opening type	Stops or starts fluid flow. The "gate" is raised to open and lowered to close the valve. Used either fully open or closed. Used for high temperatures and/or corrosive materials.
Globe valve	Standard angle Angle Plug	Throttles (partially open) or controls the flow quantity. (Frequent operation of flow adjusting.)
Ball valve	Full port size Reduced port size Standard port size (one size smaller than the pipe)	Same as globe valve.
Butterfly valve		Same as globe valve.
Check valve	Swing type Lift type Foot valve	Prevents backward flow in pipes. (Constructed so that reversing the flow closes the valve.)
Safety or relief valve (adjustable)	Pressure Temperature Pressure and temperature	Relieves the pressure and/or temperature by allowing steam or water to escape.

Table 7-7. Valve Types

There are numerous subdivisions for each type of valve. These depend on the space available, where the valve operates, and/or the operational specifics such as the basic rising or nonrising stem, screwed or flanged end, renewable or nonrenewable seat, and full or reduced port (ball valves).

ESTIMATED WATER DEMAND

Estimated water demand is a very important and highly empirical (as opposed to mathematically exact) factor in sizing a system and determining the consumption (demand) for a particular plumbing installation. Water demand cannot be determined exactly because of the large variety of fixture types and usage variables (e.g., duration/frequency of use, intermittent operation, etc.).

What does need to be established is an average rate of operation to satisfy peak demand (maximum consumption for a certain period of time). The total demand depends on the number and type of fixtures, as well as the simultaneity of use (determination based on years of practice or tests using specialty curves). If the method used to determine estimated demand is good, it will produce estimates that:

- exceed the average water demand during the period of heaviest demand.
- satisfy the peak demand.
- are adaptable to different fixtures and occupancy patterns.

The preferred method is to assign each type of fixture a certain number or weight, which can then be added together. When such numbers are placed on curves (built to include a simultaneity coefficient), they can be translated into rates of water consumption in gpm. This method is called *Hunter's curves*.

Hunter's curves have been in use since 1923. This method has been improved over the years, and it usually yields favorable results. Recent innovations, including reduced flow toilets, lavatories, and showers, have resulted in somewhat inflated consumption estimates, but no better system has been developed. Only many years of practice with good results would justify using other criteria.

This method of assigning a weight to each fixture type is also called *load value of fixtures* and is expressed in *water supply fixture units* (WSFU). Table 7-8 gives the WSFU for various fixtures. Fixtures that have both cold and hot water supplies should have separate values that equal three-fourths of the total value assigned to the fixture. For example, a private laundry machine has a total value of 2 WSFU; therefore, the separate demands on the hot and cold water supply are 1.5 WSFU (1.5 WSFU x 2 WSFU = 2 WSFU). *Note: Mathematically this equation is equal to 3, but the total maximum fixture unit used in this case is equal to 2.*

Fixture group	Occupancy	Type of supply and control	(Fixture units)		
			Hot	Cold	Total
Water closet	Public	Flush valve	-	10	10
Water closet	Public	Flush tank	-	5	5
Pedestal urinal	Public	Flush valve	-	10	10
Wall urinal	Public	Flush valve 1"	-	5	5
Wall urinal	Public	Flush tank .75"	-	3	3
Lavatory	Public	Faucet	1.5	1.5	2
Drinking fountain	Public	.375" valve	-	0.25	0.25
Bathtub	Public	Faucet	3	3	4
Shower head	Public	Mixing valve	3	3	4
Service sink	Office, etc.	Faucet	2	2	4
Kitchen sink	Hotel or restaurant	Faucet	3	3	4
Water closet	Private	Flush valve	-	6	6
Water closet	Private	Flush tank	-	3	3
Lavatory	Private	Faucet	0.75	0.75	1
Bathtub	Private	Faucet	1.5	1.5	2
Shower head	Private	Mixing valve	1.5	1.5	2
Separate shower	Private	Mixing valve	1.5	1.5	2
Bathroom group*	Private	Flush valve WC	2.25	6	8
Bathroom group*	Private	Flush tank WC	2.25	4.5	6
Kitchen sink	Private	Faucet	1.5	1.5	2
Dishwasher	Private	Automatic	1	-	1
Laundry machine	Private	Automatic	1.5	1.5	2
Laundry machine	Commercial (8 lb)	Automatic	2.25	2.25	3
Laundry machine	Public (16 lb)	Automatic	3	3	4

* Bathroom groups includes a bathtub, a lavatory, and a toilet.

Table 7-8. Weight of Fixtures in Fixture Units

Figures 7-5, 7-6, and 7-7 are curves that can be used to convert WSFU into gpm (flow). Figure 7-5 is for small WSFU numbers, and Figure 7-6 is for large WSFU numbers. Figures 7-5 and 7-6 include two sets of curves for installations with flush tanks and flush valves. Figure 7-7 illustrates a mixed system conversion of fixture units to gpm. Table 7-9 includes data taken from the curves illustrated in Figures 7-5, 7-6, and 7-7.

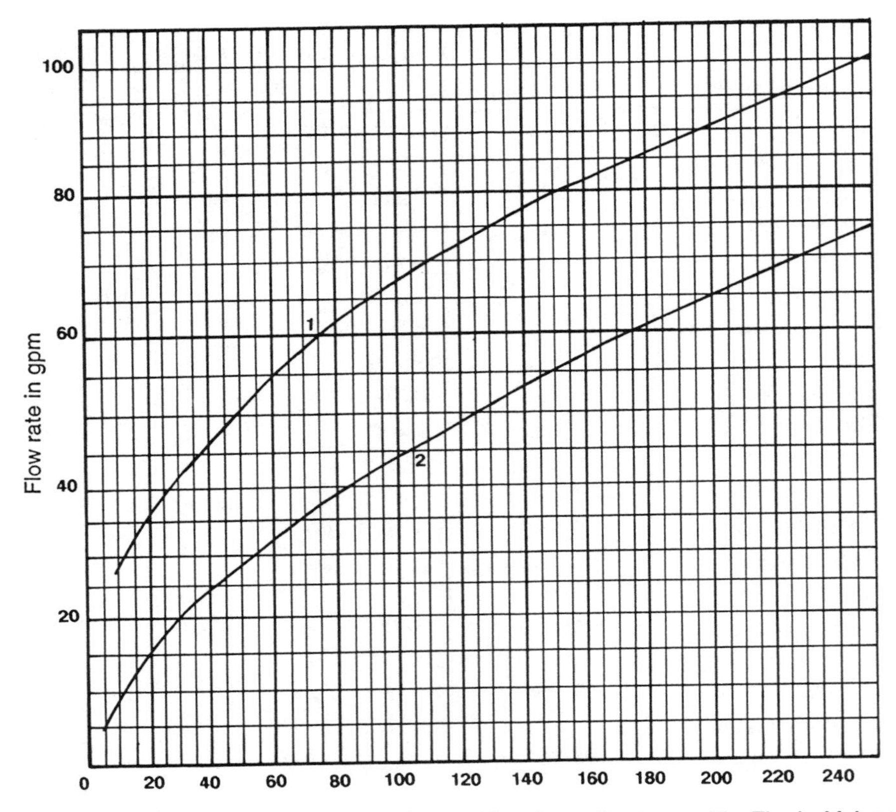

Figure 7-5. Estimated Demand (*Curve No. 1 — System with Flush Valves, Curve No. 2 — System with Flush Tanks)*

It is important to note that there may be plumbing fixtures or equipment that are used continuously, such as lawn sprinkler systems, irrigation pumps, or water-cooled equipment. These are not included in the values read from curves. Their respective consumption in gpm must be added separately at the end of the water consumption calculation. (The curves do not include an allowance for continuous users.)

To perform a calculation, it is necessary to begin with the basic data of the application. Some data is based on site information, some on the existing building or architectural drawings for new construction, and some are the selections and decisions of the plumbing specialist. Other data includes:

- size, location, and depth of all adjacent site utilities including water, sanitary sewer, storm sewer, and gas line as applicable.
- information origin, including the source, place, and date of where information was obtained.

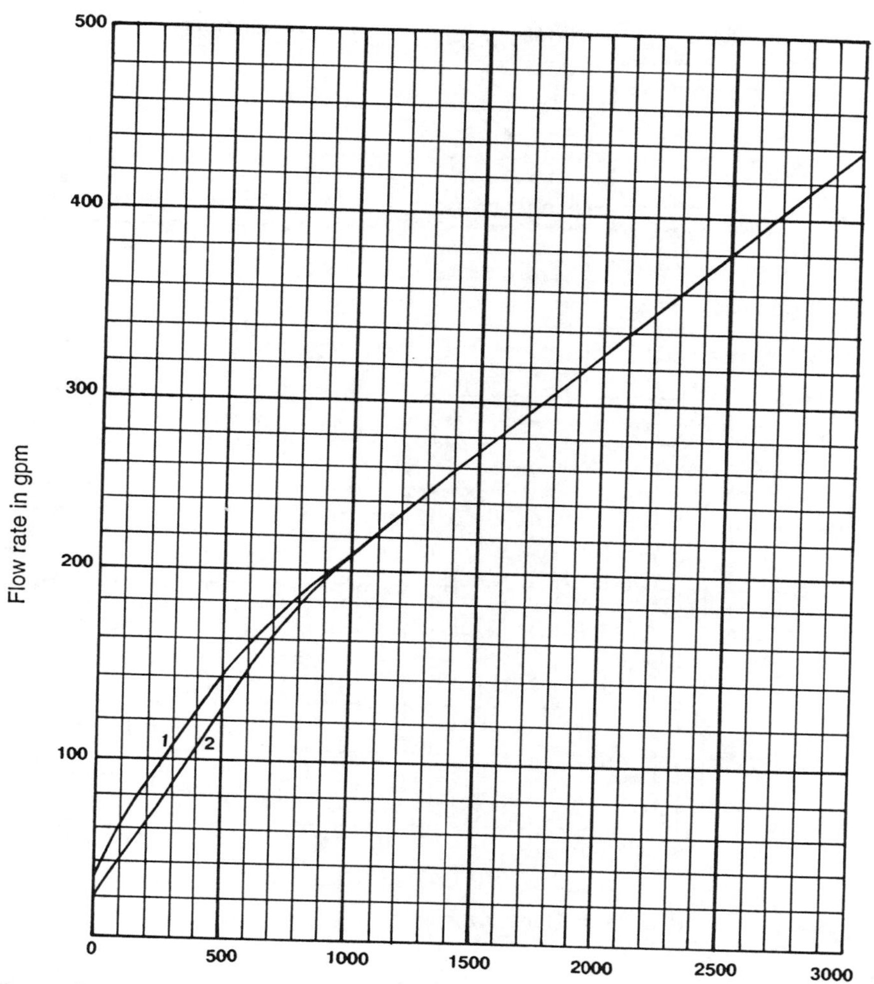

Figure 7-6. Estimated Demand II (*Curve No. 1 — System with Flush Valves, Curve No. 2 — System with Flush Tanks)*

Before gathering any data, check all applicable requirements derived from the governing plumbing code, state regulations, and special local regulations. If the project is an addition or a remodeling of an existing installation, check all existing equipment and systems for capacity and code compliance.

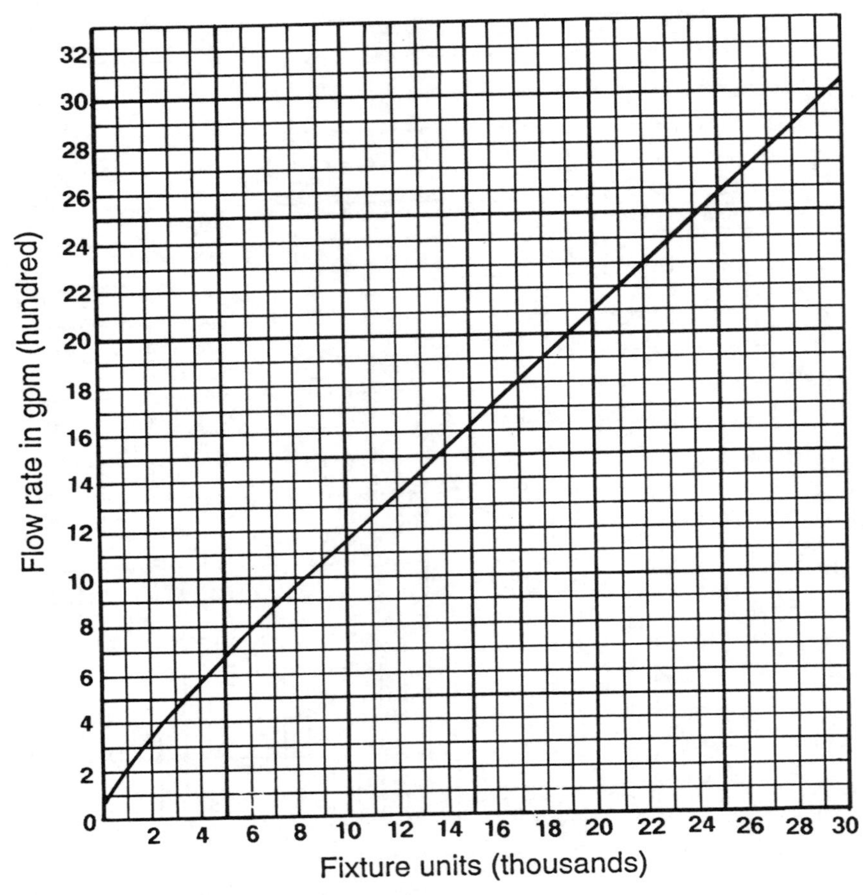

Figure 7-7. Mixed System Conversion

Example 7-1. Design a plumbing system for a residential building with a basement and two floors above. (For these purposes, a specific city and/or state will not be named.)

The data for this problem is as follows:

- Origin of water supply — city street main
- Water supply pipe size — 10" diameter
- City water pressure — constant 50 psi certified based on tests (reports available at the Fire Department or City Hall). This information needs to include the pipe elevation where the reading was performed.
- Pipe entrance into the house — level with street water line elevation
- Height of the top fixture above the street main — 30 feet (determined from building elevation drawings)

	Fixture units			Fixture units			Fixture units	
Flow gpm (L/s)	Flush tank	Flush valve	Flow gpm (L/s)	Flush tank	Flush valve	Flow gpm (L/s)	Flush tank	Flush valve
1	0	-	46	111	39	145	611	521
2	1	-	47	115	42	150	638	559
3	3	-	48	119	44	155	665	596
4	4	-	49	123	46	160	692	631
5	6	-	50	127	48	165	719	666
6	7	-	51	130	50	170	748	700
7	8	-	52	135	52	175	778	739
8	10	-	53	141	54	180	809	775
9	12	-	54	146	57	185	840	811
10	13	-	55	151	60	190	874	850
11	15	-	56	155	63	200	945	931
12	16	-	57	160	66	210	1018	1009
13	18	-	58	165	69	220	1091	1091
14	20	-	59	170	73	230	1173	1173
15	21	-	60	175	76	240	1254	1254
16	23	-	62	185	82	250	1335	1335
17	24	-	64	195	88	260	1418	1418
18	26	-	66	205	95	270	1500	1500
19	28	-	68	215	102	280	1583	1583
20	30	-	70	225	108	290	1668	1668
21	32	-	72	236	116	300	1755	1755
22	34	5	74	245	124	310	1845	1845
23	36	6	76	254	132	320	1926	1926
24	39	7	78	264	140	330	2018	2018
25	42	8	80	275	148	340	2110	2110
26	44	9	82	284	158	350	2204	2204
27	46	10	84	294	168	360	2298	2298
28	49	11	86	305	176	370	2388	2388
29	51	12	88	315	186	380	2480	2480
30	54	13	90	326	195	390	2575	2575
31	56	14	92	337	205	400	2670	2670
32	58	15	94	348	214	410	2765	2765
33	60	16	96	359	223	420	2862	2862
34	63	18	98	370	234	430	2960	2960
35	66	20	100	380	245	440	3060	3060
36	69	21	105	406	270	450	3150	3150
37	74	23	110	431	295	500	3620	3620
38	78	25	115	455	329	550	4070	4070
39	83	26	120	479	365	600	4480	4480
40	86	28	125	506	396	700	5380	5380
41	90	30	130	533	430	800	6280	6280
42	95	31	135	559	460	900	7280	7280
43	99	33	140	585	490	1000	8300	8300
44	103	35						
45	107	37						

Table 7-9. Conversion Table

- WC fixtures — tank type operation
- Required pressure at the last fixture (apply judgment) — 8 psi (lavatory)
- Restrictions and special requirements about the water meter, underground piping material — none
- Name and address of local water utility and plumbing inspector — (record as required)
- Storm and sewer systems — systems are separate
- Size and location on the street — (see utility plan)
- Elevation — H feet
- Existence of gas main under the street; name of utility that supplies it — (record as required)
- Gas pressure — 0.5 psi
- Water quality — quality within normal limits: pH = 7.0; dissolved carbon dioxide (ppm); solids (ppm); pipe corrosion develops above 150°F
- Piping selected for installation — copper tubing Type K underground; copper tubing Type L inside
- Water meter selected — disk type. Pressure drop to be obtained after line sizing from a particular manufacturer, which will be specified.
- Hot water heater selected — natural gas model with storage capacity of 60 gallons; produces 140°F water [4]
- Maximum water velocity — 6 feet per second (fps)

To complete the design, reference the list of symbols available in Appendix A.

Solution 7-1. The water riser diagram shows the elevation of all fixtures and their pipe connections, so begin the project by developing this diagram, Figure 7-8. After the diagram is produced, the next step is to prepare tables showing all the fixtures, Table 7-10, and fixture units, Table 7-11, and to determine the total water demand. Based on this information, the applicable code(s), and job costs, plumbing plans are then developed using architectural shells and actual fixture locations.

As noted earlier, each plumbing fixture has its own designation or mark number. In this manner, it is possible to identify each particular fixture, including those that are identical. If a number and letter designation is used for each fixture, confusion during installation and/or maintenance repair is eliminated (mainly in large installations).

The next step is to determine the required flow based on the calculation of the fixture units. Table 7-11 shows that the total WSFU for this installation is 21. Table 7-9 illustrates that for installations with toilets having flush tanks (small residential), the corresponding flow is 15 gpm. Per Figure 7-8, it is necessary to add the demand for two hose bibbs — one in operation at a time, at 5 gpm. This brings the total required flow to 20 gpm.

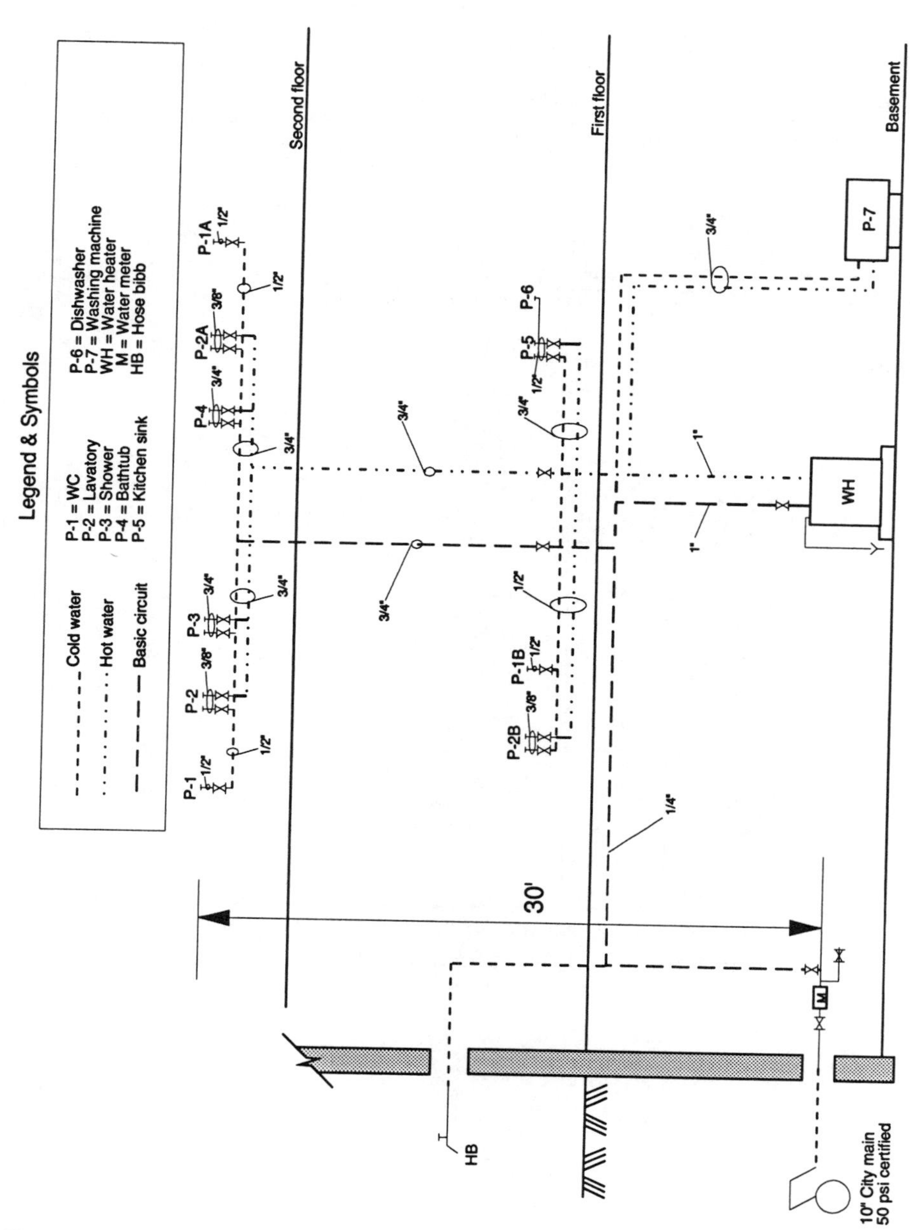

Figure 7-8. Riser Diagram

Area	Fixture designation	Type	WSFU Cold	WSFU Hot	Total	Connecting pipe diameter Min	Connecting pipe diameter Selected*
2nd floor	P-1	WC	3.0	-	3.0	3/8	1/2
	P-2	Lavatory	0.75	0.75	1.0	3/8	3/8
	P-3	Shower	1.5	1.5	2.0	1/2	3/4
	P-4	Bathtub	1.5	1.5	2.0	1/2	3/4
1st floor	P-5	Kitchen sink	1.5	1.5	2.0	1/2	3/4
	P-6	Dishwasher	-	1.0	1.0	1/2	1/2
Basement	P-7	Washing mach.	1.5	1.5	2.0	1/2	3/4

*For certain fixtures, the pipes selected are one size larger than minimum to reduce noise.

Table 7-10. Fixtures

Fixture designation*	Type**	No. installed	WSFU x No.	Total WSFU
P-1, P-1A P-1B	WC	3	3 x 3	9
P-2, P-2A P-2B	LAV	3	1 x 3	3
P-3	SH	1	2 x 1	2
P-4	BT	1	2 x 1	2
P-5	SINK	1	2 x 1	2
P-6	DW	1	1 x 1	1
P-7	WM	1	2 x 1	2
				21

*Note: Each fixture has its own designation or mark number.

**For abbreviations see symbols list in Appendix A.

Table 7-11. Total Fixture Units

If the requirement were that each fixture must run simultaneously, then the total gpm would be 45.5, as listed in Table 7-12. By using the Hunter's curves, it is possible to determine that water consumption is only 44% of the maximum total shown earlier (20 ÷ 45.5 = 0.44 or 44%).

Fixture	No. installed	gpm per fixture	Total gpm
WC	3	3.0	9.0
LAV	3	3.0	9.0
SH	1	5.0	5.0
BT	1	5.0	5.0
SINK	1	4.5	4.5
DW	1	4.0	4.0
WM	1	4.0	4.0
			40.5
		Add one hose bibb	5.0
			45.5 gpm

Table 7-12. Fixtures Running Simultaneously

To determine the system pressure loss, it is necessary to know the longest pipe run, which corresponds to what is called the *basic design circuit.* This is the distance (length of pipe) in feet between the (point of entrance) water meter and the most remote fixture in the system. The measured length is actually called *developed length.* The *equivalent length,* as discussed in Chapter 3, is the developed length plus an equivalent length for the fittings and valves.

Every pipe connected to fixtures shown is based on the selected size table. As mentioned earlier, some pipes may be one size larger than the minimum size. For purposes of this example, the main pipe size selected is based on a flow of 20 gpm, which includes a maximum velocity of 6 fps. This leads to a selection of 1-1/4" diameter pipe (Type L copper tubing), in which the water has a velocity of 5.10 fps and the head loss is 8.46 ft/100 ft (see Appendix B).

Basically, the sizing of a water pipe is influenced by four factors:

1. Flow rate
2. Water velocity in the pipe (maximum limit selected)
3. System pressure available
4. Piping material

Another item used in calculations is flow pressure, which is the pressure that exists at any point in the system when water is flowing. To help determine the flow at any faucet by measuring the flow pressure at the outlet, use the following formula:

$$g = 20d^2p^{1/2}$$

where: g = flow at outlet in gpm
d = pipe diameter in inches
p = flow pressure in psi

For example: pipe diameter (d) = 3/8"
flow pressure (p) = 9 psi

therefore: $g = (20)\ (3/8)^2\ (9)^{1/2}$
$g = (20)\ (.1406)\ (3) = 8.44$ gpm

If a certain fixture is closer to the water supply pipe entering the building and the supply pressure is too high, it must be reduced. To reduce the pressure at a fixture, a flow control orifice, Figure 7-9, or a throttling valve may be used. The orifice can be fixed or adjustable. If a valve is used to reduce the pressure, the valve can be a globe valve or a pressure-reducing valve (factory set).

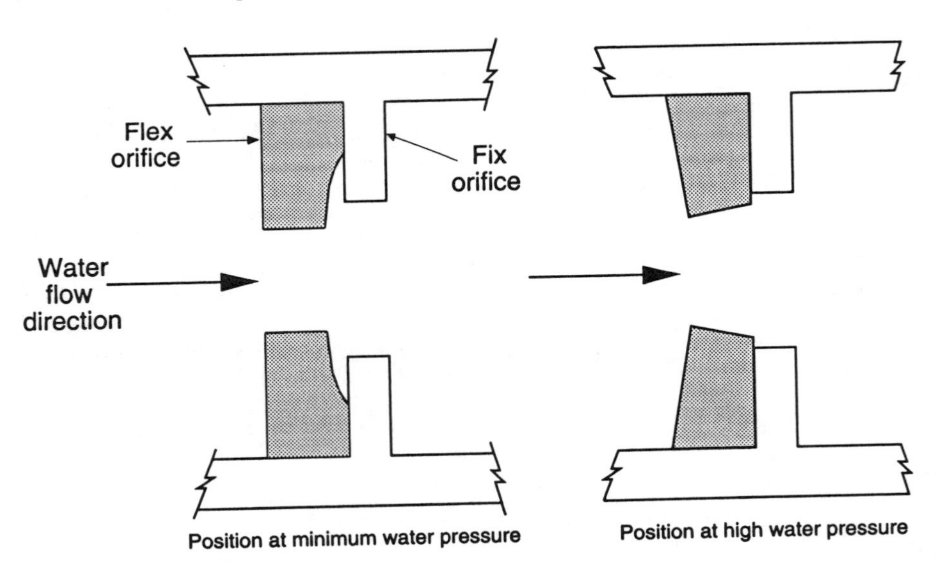

Figure 7-9. Flexible Orifice

Example 7-2. Determine the water supply pipe size and type for an office building equipped with flush valve WCs.

Solution 7-2. The number of WSFU for this office building equipped with WCs with flush valves is as follows:

No. of fixtures	WSFU	Hot water WSFU	Cold water WSFU	Total WSFU
50 WC	10	0	500	500
10 UR	5	0	50	50
40 LAV	1.5	60	60	80
Total		60	610	630

Hot water demand for 60 WSFU = 33 gpm (required for water heater sizing)

Cold water demand for 610 WSFU = 145 gpm

Cold and hot water demand for 630 WSFU = 160 gpm

There are two continuous users of cold water in this building: 1) a water-cooled condenser for the air conditioning unit, and 2) two hose bibbs. The water-cooled condenser demand is 40 gpm, and the demand for two hose bibbs (at 5 gpm each) is 10 gpm. Therefore, the total constant consumption is 50 gpm. This brings the total water required to 210 gpm (160 gpm + 50 gpm).

Select from the pipe table* for a maximum velocity of 6 fps, Schedule 40 galvanized steel pipe. The selection is a 4" diameter pipe in which the water has a velocity of 5.29 fps and a pressure loss of 2.47 ft/100 feet, which is equal to 1.07 psi/100 feet (2.47 ÷ 2.31 = 1.07 psi/100 ft).

Example 7-3. Determine the main pipe size for a 14-story office building. In this building, the difference in elevation from the pump location to the highest plumbing fixture is 160 feet. Pumps are located at the same level as the (water pipe) street connection and the street main elevation. Other given data for this problem are as follows:

- Water supply pressure for toilets and urinals — 20 psi (system with flush valves)
- Total WSFU (cold and hot) — 4200 WSFU
- Basic circuit length — 300 ft
- Piping material — Type K copper (local authority's request)
- City water pressure — 50 psi

*

Flow in U.S. gallons	4" Diameter standard steel Sch 40	
	Velocity (ft/sec)	Head loss (ft/100 ft)
200	5.04	2.25
210	5.29	2.47
220	5.54	2.7

Solution 7-3. The fittings for this problem are estimated at 30% of developed length.

The following calculations are required for this problem:

- Static head — 160 feet ÷ 2.31 = 69.3 psi
- Selected uniform pressure drop — 2.31 ft/100 ft or 1.0 psi/100 feet
- Equivalent pipe length for 300 ft (pipe) and fittings (1.3 x 300) = 390 ft
- Pipe and fittings pressure drop — 390 ft x 1.0 psi/100 ft = 3.90 psi
- Pressure required at the last fixture — 20 psi

Therefore, the total pressure required from the source is:
69.3 + 3.90 + 20 = 93.20 psi (The water meter was not considered.)

To determine the flow, look at Table 7-9. There are gpm values for 4070 and 4480 WSFU, but this system needs the gpm value for 4200 WSFU. It is possible to calculate this value using the information found in Table 7-9:

4480 WSFU corresponds to 600 gpm

4070 WSFU corresponds to 550 gpm

The difference between the two is 410 WSFU and 50 gpm.

4200 - 4070 = 130 WSFU

$$\frac{(130\ \text{WSFU})(50\ \text{gpm})}{410\ \text{WSFU}} = 15.8\ \text{gpm}$$

Now the flow is: 550 + 15.8 = 565.8 gpm. For the total WSFU of 4200, the flow corresponds to 565.80 gpm.

From the street, the water enters the building at a pressure of 50 psi. Since this is not enough to satisfy the application requirements, which were calculated to be 93.20 psi, a pump needs to be included in the system.

Pump selection is based on a requirement of 565.80 gpm (rounded to 570 gpm), and the head (pressure) calculated as:

93.2 psi - 50 psi (street pressure) = 43.2 psi

Even though the pressure from the street main is deducted from the total head (because it helps the pump discharge pressure), the total discharge head is still 93.2 psi. It is always necessary to add a 10% safety factor when selecting a pump, because pressure loss in the pipes increases as the installation ages, and the 30% equivalent length for fittings might not be conservative enough. Therefore, the pump selected will be based on 570 gpm and a total system pressure of 102.5 psi (93.2 psi + 9.3 psi) with a final pump head selection of 53 psi (102.5 psi - 50.0 psi = 52.5, round to 53 psi).

The main pipe size (copper) will be 6" diameter*, which will have a head loss of approximately 1 psi/100 ft and a water velocity of about 7.1 fps (close to the established criteria).

Example 7-4. Determine the flow needed for a public occupancy such as a small, combined office/residential building, in which toilets with flush valves are used, Table 7-13.

Fixture	No. of fixtures	WSFU x No.	WSFU total
LAV	13	2 x 13	26
WC	10	10 x 10	100
BT	1	4 x 1	4
SH	2	4 x 2	8
Kitchen sink	6	4 x 6	24
DW (same as private)	1	1 x 1	1
WM (8 lb)	1	3 x 1	3
			166

Table 7-13. Fixtures in Mixed Occupancy Building

*

Flow in U.S. gallons	6" Diameter type K copper tubing	
	Velocity (ft/sec)	Head loss (ft/100 ft)
550	6.82	2.29
570	7.1	2.36
600	7.45	2.68

Solution 7-4. Per Table 7-13, the total number of fixture units in this example is 166. Figure 7-5 shows that the approximate gpm value for 166 fixture units is between 80 and 85 gpm. Table 7-9 shows values of 82 gpm (for 158 WSFU) and 84 gpm (for 168 WSFU). Use the same method used in Solution 7-3 to calculate the gpm that corresponds to 166 WSFU:

158 WSFU corresponds to 82 gpm

168 WSFU corresponds to 84 gpm

The difference between the two is 10 WSFU and 2 gpm.

166 - 158 = 8 WSFU

$$\frac{(8 \text{ WSFU})(2 \text{ gpm})}{10 \text{ WSFU}} = 1.6 \text{ gpm}$$

Now the flow is: 82 gpm + 1.6 gpm = 83.6 gpm, which corresponds to 166 WSFU.

Figure 7-10 illustrates two water riser diagrams and is included for the reader's convenience.

SUMMARY

In any of the WSFU tables shown in this chapter, it is interesting to note that cold water WSFU plus hot water WSFU does not add up to the total WSFU. The two values give a total that is less than the sum of the two, because both cold and hot faucets cannot deliver maximum of the individual flow.

Another important element to consider is that when determining the total water demand for an installation with two separate risers, first add the two subtotal numbers of water supply fixture units. After adding these together, read the total gpm from the curves or the table for the corresponding total flow in gpm to determine the pipe size. For example, if the north riser in a building has 520 WSFU and the south riser has 800 WSFU on a system equipped with flush valves, the total water demand is 520 WSFU + 800 WSFU = 1320 WSFU, which corresponds to 248 gpm. If the numbers were considered separately, the incorrect total of 329 gpm would result:

520 WSFU corresponds to 145 gpm

800 WSFU corresponds to 184 gpm

145 + 184 = 329 gpm

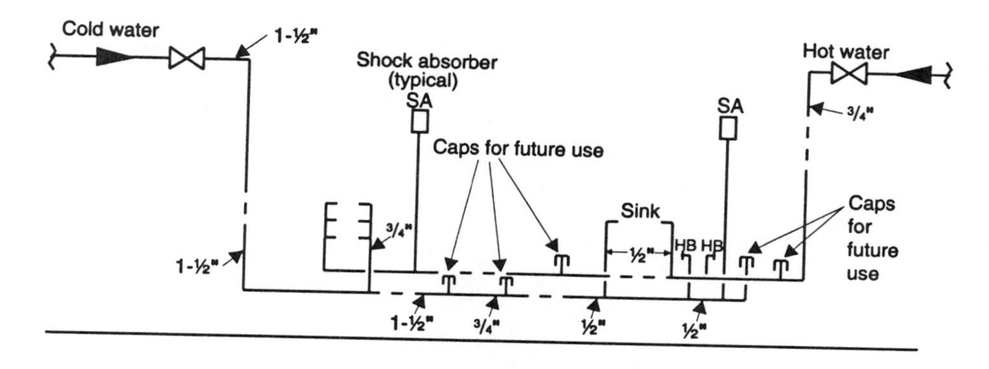

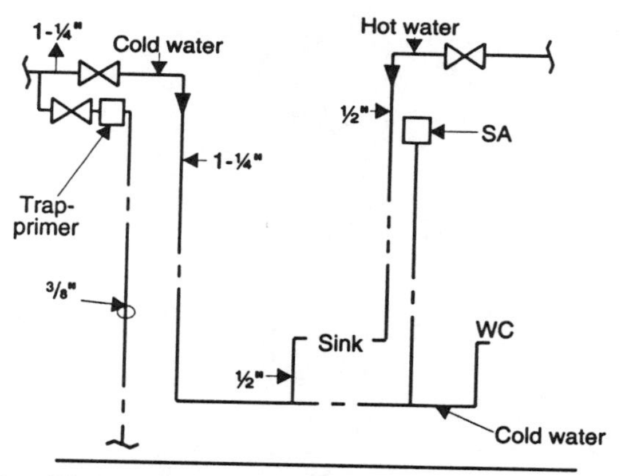

Figure 7-10. Water Riser Digrams

NOTES

[1] Water hammer is a condition in which the water velocity in the pipes is high and a valve suddenly closes. The noise that results is similar to a hammer banging in the pipes.

[2] See the *Piping Handbook* by Sabin Crocker for these tables. McGraw Hill 1967.

[3] The American Society of Plumbing Engineers (ASPE) Data Book shows a number of piping hanger and support configurations and tables with application recommendations.

[4] The next chapter will show how to size a hot water heater; in this case, the water heater was selected arbitrarily.

Chapter Eight

Hot Water Supply

A hot water system requires slightly more attention than a cold water system. This is due to the fact that when water is heated, it behaves in a slightly different manner than cold water. Any service performed on a hot water system should be done carefully, since temperatures in this type of system might cause burns on the human body if improperly handled. System safety must also be built into the equipment to protect the system and building occupants from excess pressure and temperature.

Due to the temperature increase, pipes that convey hot water to a fixture expand lengthwise. The expansion amount depends on the piping material. When the pipe cools, it returns to its original length. Table 8-1 indicates expansions for various piping materials.

Piping material	Total expansion of 100 ft of pipe length for a temperature change of 100°F	
	in.	ft
Steel	0.780	0.065
Wrought iron	0.816	0.068
Cast iron	0.714	0.059
Copper	1.140	0.095
Red brass	1.248	0.104
Plastic pipe:		
ABS, type I	6.720	0.560
PVC, type I	3.360	0.280
PVC, type II	6.600	0.550

Table 8-1. Expansion of Piping with Temperature Change

The system must be installed in such a way as to allow for the hot water pipes to move slightly. Any bend or corner will absorb part of the movement. If there is a long stretch of straight pipe, then expansion-formed joints must be included. Figure 8-1 illustrates several different expansion loops.

Figure 8-1. Various Expansion Loops

The hot water temperature must be suitable for the nature of the service it provides. For example, domestic hot water temperature should be between 110° and 140°F. Table 8-2 gives more examples of recommended hot water temperatures. Once set, the proper water temperature is maintained automatically with the help of a thermostat located in the upper half of the water heater tank. This thermostat controls the heat source output to maintain and/or recover the stored water temperature.

Usage	Temperature (° F)
Lavatory, hand washing	105
Lavatory, face washing	115
Shower and bathtub	110
Dishwasher, washing	140
Dishwasher, sanitizing	180
Laundry, commercial and institutional	180
Dishwasher & laundry, private	140
Surgical scrubbing	110
Occupants usage in commercial buildings (per ASHRAE Standard 90)	110

Table 8-2. Recommended Hot Water Temperatures

Water Heaters

When water is heated, its volume increases by 1.68%. This means the volume in a 60-gallon tank will increase by about one gallon when the water is heated 100°F (from 40°F incoming water to 140°F heated water)[1]. This expansion knowledge is very important, because precautions must be taken to ensure there is enough room in the system to accommodate the expanded volume of water. For this reason, water heaters are equipped with pressure and temperature relief valves or a combination pressure-temperature relief valve (two functions in one valve).

In general, domestic water heater tanks are designed for a pressure of 125 psi. The relief valve in the water heater is usually set at 25 psi over the maximum service pressure (operating pressure) but not over 125 psi. The P-T (pressure-temperature) relief valve should have levers for manual testing and should close automatically after relief.

The discharge pipe from the P-T valve must not be directly connected to the sewer line. Instead, the discharge should be directed in a funnel, which is then connected to the sewer system. This creates an air gap between the hot water and sewer line. This air gap provides a visual indication of potential problems, which may include frequent discharge of hot water into the drain. If this occurs, it means the temperature actuator (thermostat) is not operating properly.

An air gap is required between the sewer and the potable water stored in the hot water tank. It is not recommended to discharge (wastes) into the sanitary sewer water at temperatures warmer than 140°F, because sewer pipe connections might come apart.

When hot water heaters are installed, the name plate must be visible. The following is an example of information found on a name plate:

Test pressure: 300 psi
Working pressure: 125 psi
Fuel: Natural gas
Gas pressure in inches wc (water gauge or inches of water column)
Inlet minimum gas pressure: 5 in. wc
Maximum gas pressure supply: 14 in. wc
Input: 60,000 Btu/hr
Tank capacity: 60 gallons
Recovery capacity: 46 gal/hr

Selection

When selecting a hot water heater, there are certain guidelines that should be kept in mind:

- Availability of a certain fuel type, electricity, or another heat source in the installation area
- Equipment cost
- Desired water temperature
- Chemical characteristics of water (hardness)

Chemical characteristics of water include solids and other salts that are naturally dissolved in the water. At temperatures above 140°F, these solids precipitate out of the water and deposit at the bottom of the tank and on tank walls. This process may reduce heater efficiency, resulting in more fuel being consumed for the same amount of hot water produced.

It is advisable to select water heaters fabricated of corrosion-resistant materials. Otherwise, they should be lined with a corrosion-resistant material such as glass. Most of the hot water tanks that are glass lined for corrosion resistance are insulated internally. (For energy conservation, insulation is also recommended for the water heater tank exterior and hot water distribution pipes.) Hot water storage tanks are constructed from one of the following materials:

- Copper
- Monnel
- Galvanized steel
- Special alloys

A water heater has a certain heating capacity. By definition, this is the number of gallons of water that can be raised 100°F in one hour. By knowing the heating capacity of the water heater, it is possible to calculate the number of gallons of water that can be heated per hour:

Weight of 1 gallon of water = 8.33 lb
Water temperature raised 100°F (from 40° to 140°F)

$$(8.33 \text{ lb})\ (100°\text{F})\ (1 \text{ hour}) = 830 \text{ Btu/hr}$$

Assume the heater is 83% efficient:

$$830 \div 0.83 = 1000 \text{ Btu/hr}$$

The Btu rating input is the number of Btu required to heat one gallon of water 100°F per hour. In the example just shown, this would be 1000 Btu/hr/gal. If a water heater had a rating of 50,000 Btu/hr, the heating capacity would be 50 gallons of water (50,000 Btu/hr ÷ 1000 Btu/hr). In other words, this heater can produce 50 gal of hot water heated 100°F in one hour.

Hot Water Recovery

In a water heater with a storage tank, only 60% to 80% of its capacity is usable hot water. The rest of the water is cold and must be reheated. For example, a water heater with a 1000-gal capacity has a usable amount of about 700 gallons. If the required amount of hot water is 1000 gallons, the tank capacity must be at least 1429 gallons (1000 gallons ÷ 0.7, or 70%).

Hot water recovery is the amount of water that should be reheated to 140°F within a certain period of time in a water heater. If a water heater has 46 gallons per hour that must be heated 100°F and the water enters into the tank at 40°F and is delivered at 140°F, then this rise in temperature must occur in one hour.

The required recovery capacity is expressed as the difference between the peak hourly demand and the amount of hot water available in the tank (measured in gph). In a 1000-gallon tank with a peak demand of 450 gallons for a 2-hour period, the available water heater capacity should be:

(1000) (0.7) = 700 gallons

700 gallons ÷ 2 hours = 350 gph hot water available

450 gph - 350 gph = 100 gph deficiency or recovery capacity required/hour

The heating element must produce 100 gallons of reheated water within one hour.

Direct and Indirect Heating

There are two different methods that can be used when heating domestic water: direct heating and indirect heating.

In the direct heating method, water is in contact with the heated surface. This surface must be of a relatively high temperature to produce the desired results. The heat is then transferred from the fuel combustion (or electric resistance) directly to the water. Those direct heating methods that burn fuel require flues and chimneys. Another version of direct heating is an instantaneous, tankless, water heater. This type of water heater is used in small installations. Tankless heaters must not be installed where water contains more than 10 grains/gallon of dissolved solids (based on lab tests).

Figure 8-2 shows the following piping connections in a direct water heater (connections may vary depending on manufacturer):

- Cold water supply to tank
- Hot water supply from tank to system
- Drain to allow tank to be emptied
- Connection that allows tank to be flushed (to prevent mineral and salt deposits)
- Connection to pressure-temperature (P-T) relief valve

Within a hot water tank, the cold water supply and the hot water outlet need to be remote from each other. The cold water inlet must discharge close to the tank bottom. If both connections are at the top of the tank, then the cold water supply pipe must (unseen) extend downwards close to the bottom and terminate approximately 8" above the bottom of the tank, as shown in Figure 8-2.

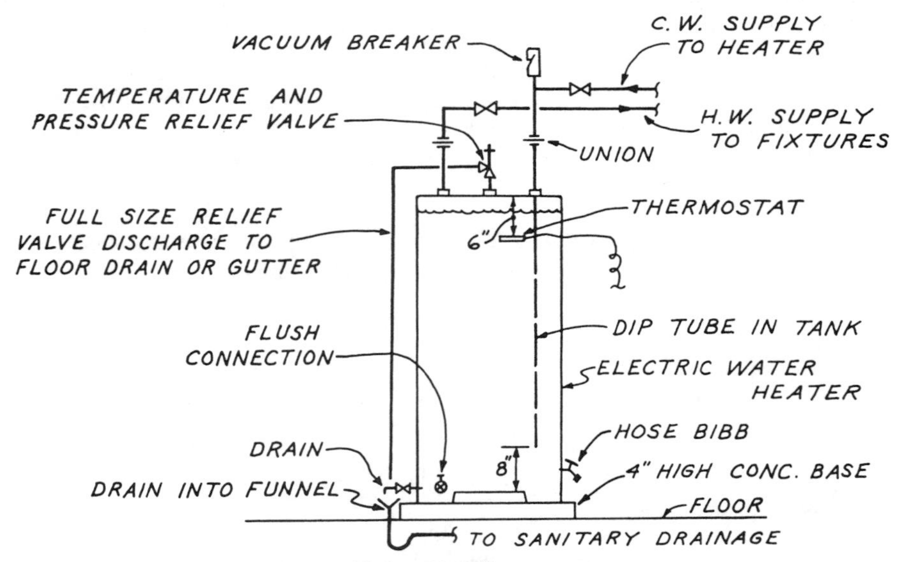

Figure 8-2. Direct Water Heater Piping Detail

A tankless system is provided with a cold and hot water tempering valve. The capacity of a tankless system depends on the heat transfer surface of the coils in which water circulates (enters cold and leaves hot).

In the indirect heating method, a heating coil consisting of copper tubes conveys hot water or steam from a boiler (or another remote heat source) to the hot water tank, Figure 8-3. This type of water heating includes a tank in which the water is heated and stored. Indirect heating methods may use either fuel or electricity, and they differ from the direct heating method in that combustion takes place in the boiler.

The pressure drop or friction loss through a water heater with a storage tank is less than the one in a tankless system. Water heater manufacturers provide this information upon request.

For energy conservation purposes, a domestic water heating installation should include a tankless system (consisting of a coil), with the water heating source originating at the heating boiler for winter operation. For summer operation, a separate direct water heater should be included. The two systems must be valved in such a manner so they can be easily interchangeable. In such an installation, the boiler is sized to include the required capacity for the building's comfort heating requirements plus the capacity to heat the domestic hot water. During the warm season, the boiler, which usually has a large heating capacity, does not need to be fired if a separate domestic water heater is installed. This will result in fuel consumption economy.

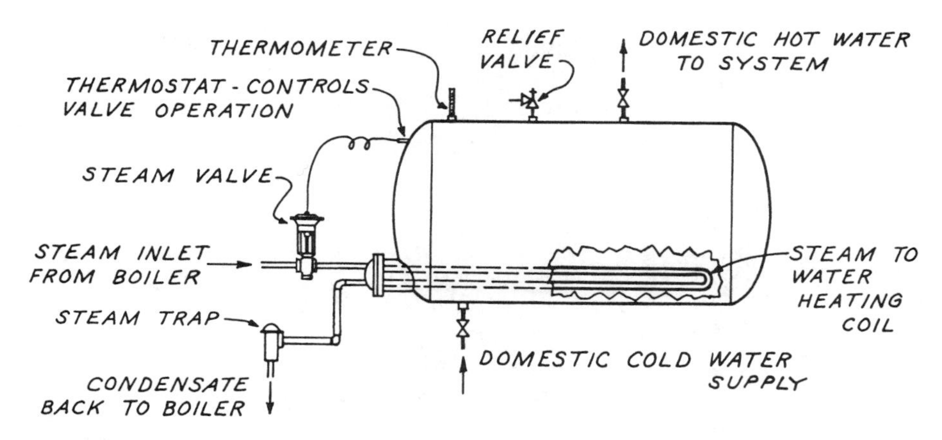

Figure 8-3. Indirect Water Heater with Storage Tank

As mentioned previously, the pressure drop or head loss through a water heater is either indicated by the manufacturer upon request or listed on the name plate. A good rule of thumb is that the pressure drop through a water heater is equal to the pressure drop in six, 90° elbows (the size of the entering cold water pipe).

Pumps

Positive control of hot water circulation is obtained by including a hot water circulating pump with all wetted parts of non-ferrous materials (an in-line pump for small installations, Figure 8-4). This arrangement also reduces water waste. When the hot water is turned on, there is no waiting period, so water is not wasted until desired hot water is present. To preclude air pocket formation, the pump must be placed in the lower circulating piping (return pipe to the tank).

A hot water circulating pump is required when the distance between the hot water source and the most remote fixture served is more than 80 to 100 ft; however, a pressure drop calculation is required before such a decision is made.

To determine the flow in gpm for the circulating pump, divide the total heat loss in the pipes and tank (measured in Btu/hr) by 10,000 (this number is derived by multiplying 8.3 lb/gal of water by 60 min/hr by 20°F, which is the usual temperature drop in pipes)[2]:

$$Q = \frac{\text{Heat loss (measured in Btu/hr)}}{(8.3)(60)(20)} = \text{flow in gpm}$$

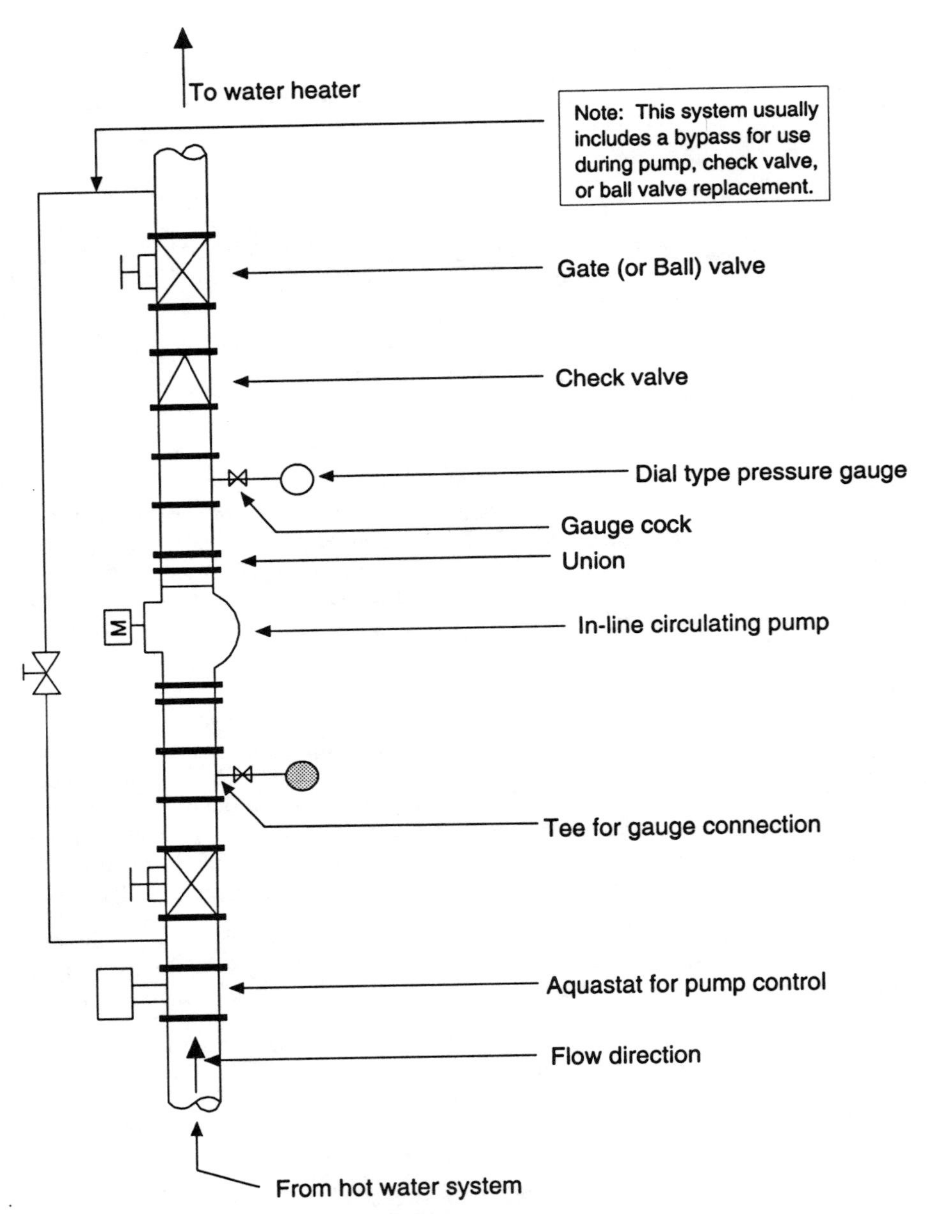

Figure 8-4. Installation of Return Pipe and Circulating Pump

The criteria for sizing hot water pipes (pressure drop, water velocity, etc.) are the same as those used for cold water pipes in earlier chapters. The WSFU for hot water was given in Table 7-8.

Distribution Systems

The hot water in a building may be supplied using different distribution systems. These systems include upfeed, Figure 8-5, downfeed, Figure 8-6, or a combination of the two, Figure 8-7. Different installations may have different types of distribution for the hot water, however, all give acceptable results.

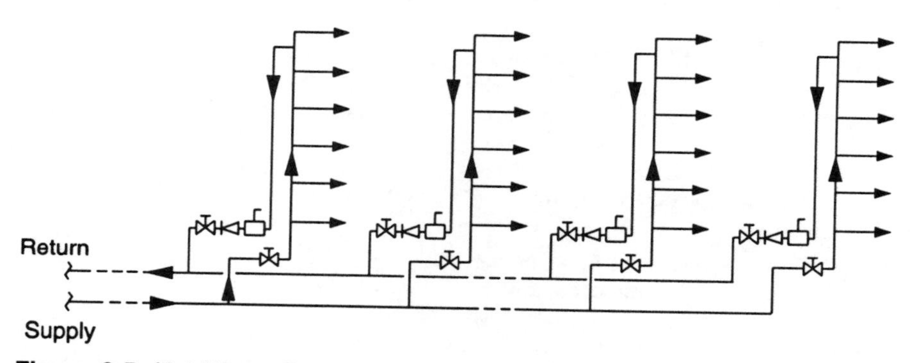

Figure 8-5. Hot Water Supply System — Upfeed

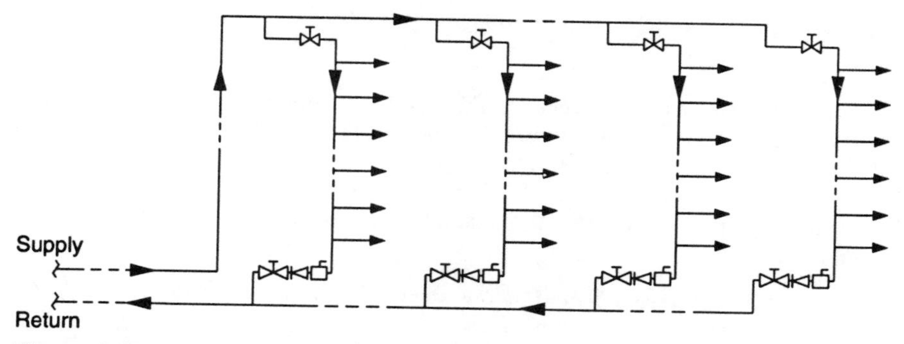

Figure 8-6. Hot Water Supply System — Downfeed

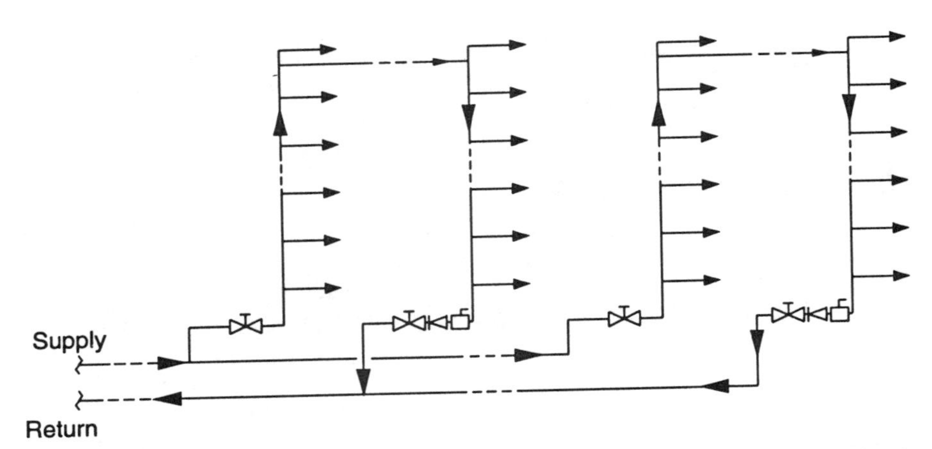

Figure 8-7. Hot Water Supply System — Combination Upfeed and Downfeed

The demand for hot water varies, and there are a few factors that influence this demand:

- Building (classification) occupancy
- Number of building occupants
- Number and type of fixtures installed
- Equipment installed that continuously uses hot water
- Time of day when peak demand occurs
- Length of peak demand (depending on the occupancy)
- Season of the year when the hot water consumption is greatest
- Existence of water conservation devices
- If there is a charge for hot water consumption

Peak demand can be defined as the maximum hot water demand rate (at 140°F) measured in gpm or gph (gallons per hour). The goal of a system installation is to fulfill these factors to satisfy the consumers' hot water needs. While there are no precise rules for calculating this demand, there are guidelines, which include some proven tables, curves, and coefficients, that help give favorable results.

Tables 8-3, 8-4, and 8-5 give recommended hot water consumption requirements for various types of occupancies.

Sizing Calculations

When performing hot water sizing calculations for a large installation, first determine the developed length of all hot water pipes (supply <u>and</u> return). Then, calculate the piping heat losses in Btu. If only an approximate result is necessary (e.g., in a system with pipes up to 2" diameter), then multiply the total developed

Type of building	Maximum hour	Maximum day	Average day
Dormitories:			
Men's dormitories	3.8 gal/student	22.0 gal/student	13.1 gal/student
Women's dormitories	5.0 gal/student	26.5 gal/student	12.3 gal/student
Motels: no. of units			
20 or less	6.0 gal/unit	35.0 gal/unit	22.0 gal/unit
60	5.0 gal/unit	25.0 gal/unit	14.0 gal/unit
100 or more	4.0 gal/unit	15.0 gal/unit	10.0 gal/unit
Nursing homes	4.5 gal/bed	30.0 gal/bed	18.4 gal/bed
Office buildings	0.4 gal/person	2.0 gal/person	1.0 gal/person
Food service establishments:			
Type A - Full meal restaurants and cafeterias	1.5 gal/max meals/hr	11.0 gal/max meals/hr	2.4 gal/avg meals/day*
Type B - Drive-ins, grilles, luncheonettes, sandwich and snack shops	0.7 gal/max meals/hr	6.0 gal/max meals/hr	0.7 gal/avg meals/day*
*Apartment houses:***			
no. of apartments			
20 or less	12.0 gal/apt.	80.0 gal/apt.	42.0 gal/apt.
50	10.0 gal/apt.	73.0 gal/apt.	40.0 gal/apt.
75	8.5 gal/apt.	66.0 gal/apt.	38.0 gal/apt.
100	7.0 gal/apt.	60.0 gal/apt.	37.0 gal/apt.
130 or more	5.0 gal/apt.	50.0 gal/apt.	35.0 gal/apt.
Elementary schools	0.6 gal/student	1.5 gal/student	0.6 gal/student*
Junior and senior high schools	1.0 gal/student	3.6 gal/student	1.8 gal/student*

*Per day of operation

**Judgement must be applied. These numbers are for information only.

Reprinted with permission from ASHRAE Handbook, 1987.

Table 8-3. Hot Water Demands for Various Building Types

	No. of rooms	No. of bathrooms 1	2	3	4	5
Apartments and private homes	1	60	-	-	-	-
	2	70	-	-	-	-
	3	80	-	-	-	-
	4	90	120	-	-	-
	5	100	140	-	-	-
	6	120	160	200	-	-
	7	140	180	220	-	-
	8	160	200	240	250	-
	9	180	220	260	275	-
	10	200	240	280	300	-
	11	-	260	300	340	-
	12	-	280	325	380	450
	13	-	300	350	420	500
	14	-	-	375	460	550
	15	-	-	400	500	600
	16	-	-	-	540	650
	17	-	-	-	580	700
	18	-	-	-	620	750
	19	-	-	-	-	800
	20	-	-	-	-	850
Hotels	Room with basin				10	
	Room with bath - transient				50	
	Room with bath - resident				60	
	2 rooms with bath				80	
	3 rooms with bath				100	
	Public shower				200	
	Public basins				150	
	Slop sink				30	
Office buildings	White-collar worker (per person)*				2.3	
	Other workers (per person)				4.0	
	Cleaning per 10,000 sq ft				30.0	
Hospitals	Per bed				80-100	

*The value for white-collar workers is for office occupancy only, not including allowance for employees' lunch rooms, dining rooms, etc.

Table 8-4. Maximum Daily Requirements for Hot Water in Gallons

Table 8-5. Hot Water Demand Per Fixture

	Apt. house	Club	Gym	Hospital	Hotel	Industrial plant	Office building	Private res.	School	Y.M.C.A.
Basins, private lavatory	2	2	2	2	2	2	2	2	2	2
Basins, public lavatory	4	6	8	6	8	12	6	-	15	8
Bathtubs	20	20	30	20	20	-	-	20	-	30
Dishwashers*	15	50-150		50-150	50-200	20-100	-	15	20-100	20-100
Foot basins	3	3	12	3	3	12	-	3	3	12
Kitchen sink	10	20	-	20	30	20	20	10	20	20
Laundry, stationary tubs	20	28	-	28	28	-	-	20	-	28
Pantry sink	5	10	-	10	10	-	10	5	10	10
Showers	30	150	225	75	75	225	30	30	225	225
Service sink	20	20		20	30	20	20	15	20	20
Hydro-therapeutic showers	-	-	-	400	-	-	-	-	-	-
Hubbard baths	-	-	-	600	-	-	-	-	-	-
Leg baths	-	-	-	100	-	-	-	-	-	-
Arm baths	-	-	-	35	-	-	-	-	-	-
Sitz baths	-	-	-	30	-	-	-	-	-	-
Continous-flow baths	-	-	-	165	-	-	-	-	-	-
Circular wash sinks	-	-	-	20	20	30	20	-	30	-
Semi-circular wash sinks	-	-	-	10	10	15	10	-	15	-
Demand factor	**0.30**	**0.30**	**0.40**	**0.25**	**0.25**	**0.40**	**0.30**	**0.30**	**0.40**	**0.40**
Storage capacity factor**	**1.25**	**0.90**	**1.00**	**0.60**	**0.80**	**1.00**	**2.00**	**0.70**	**1.00**	**1.00**

Gallons of water per hour per fixture, calculated at a final temperature of 140°F

*Dishwasher requirements should be taken from manufacturers' data for the model to be used, if this is known.

**Ratio of storage tank capacity to maximum demand per hour. Storage capacity may be reduced where an unlimited supply of steam is available from a central street steam system or large boiler plant.

length by 30 Btu/linear ft to obtain heat losses (for insulated pipes) or 60 Btu/linear ft (for non-insulated pipes). This gives the number of Btu/hr to be produced by the water heater. Then, apply the formula shown previously:

$$Q = \frac{\text{Heat loss}}{10,000} = \text{Hot water flow in gpm}$$

Table 8-6 indicates the calculated heat loss for insulated and non-insulated pipes (3/4" to 4" diameter) with water flowing at 140°F. In contrast to Table 8-6, the approximate heat loss formula indicated above includes sizable safety factors to cover the unknown.

Heat loss Btu/hr linear ft at 70° F temperature difference

Pipe size (in.)	Non-insulated pipe	Insulated pipe (fiberglass)	
		½"	1"
3/4	30	17.7	8.2
1	38	20.3	8.6
1-1/4	45	23.4	11.3
1-1/2	53	25.4	11.6
2	66	29.6	12.9
2-1/2	80	33.8	14.8
3	94	39.5	17.5
4	120	48.4	21.0

Table 8-6. Heat Loss

There are some practical reduction coefficients that should be included in case hot water consumption values are given in gpm. They are based on non-simultaneous use of hot water, and they differ based on occupancies. As previously calculated for the cold water supply, it is possible to determine the WSFU (from tables or curves) and the hot water flow in gpm. This hot water quantity will be corrected by the following recommended (reduction) coefficients:

Occupancy	Coefficient[3]
Hospitals/Hotels	0.25 - 0.50
Residential	0.3
Offices	0.3

Example 8-1. The total WSFU for a hotel translates into a requirement of 120 gpm, and the hot water temperature is 140°F. Use the reduction coefficient to determine the required heating capacity.

Solution 8-1. Apply the reduction coefficient:

Heat input = (120 gpm) (0.3) (60 min/hr) (8.33 lb/gal) (100°F[4])
= 1,799,280 Btu/hr heating capacity required

Sizing Water Heaters

Figures 8-8 through 8-15 are hot water curves that illustrate recovery and storage capacities for several different applications.

Figure 8-9, which illustrates recovery and storage capacities for dormitories, has a curve for men and a curve for women. The curve for the women's dormitory requires a slightly larger recovery capacity in gallons per hour per student, because studies show there is a slight difference in the actual hot water consumption for males and females.

Figure 8-10 also has two curves, Type A and Type B. The Type A curve includes consumption of hot water for restaurants and cafeterias with full meals. Type B includes snack food establishments such as grilles, luncheonettes, and sandwich shops.

Example 8-2. Calculate the required water heater size for a 250-student, women's dormitory for the following three situations:

1. Storage with minimum recovery rate (1.1 gph per Figure 8-9)
2. Storage with recovery rate of 3.0 gph per student
3. a) Include an additional hot water requirement for a cafeteria serving meals to all students for a minimum recovery rate (using data calculated from Situation No. 1).

   b) Same as a), except the recovery rate must be 0.75 gph (using data calculated from Situation No. 2).

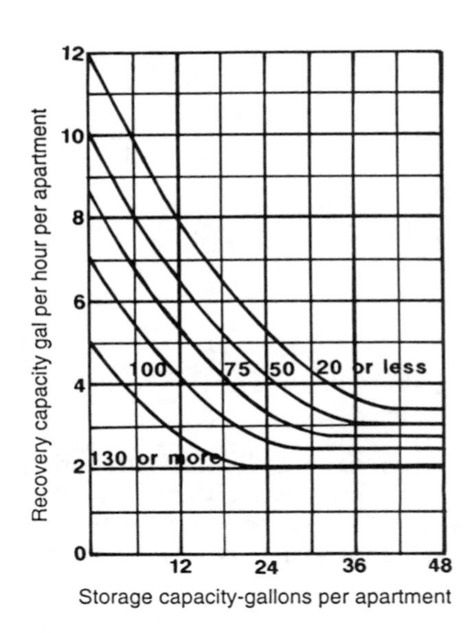

Figure 8-8. Recovery and Storage Capacities for Apartments

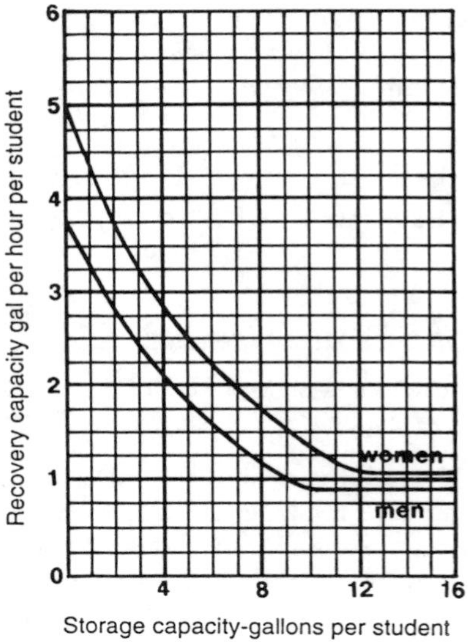

Recovery capacity gal per hour per student

women

men

Storage capacity-gallons per student

Figure 8-9. Recovery and Storage Capacities for College Dormitories

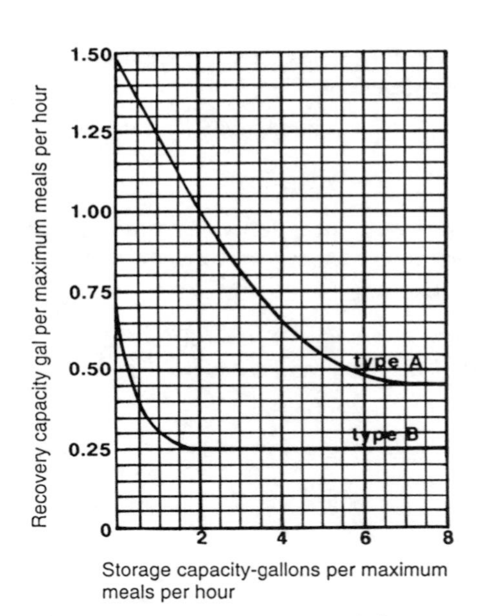

Figure 8-10. Recovery and Storage Capacities for Food Service

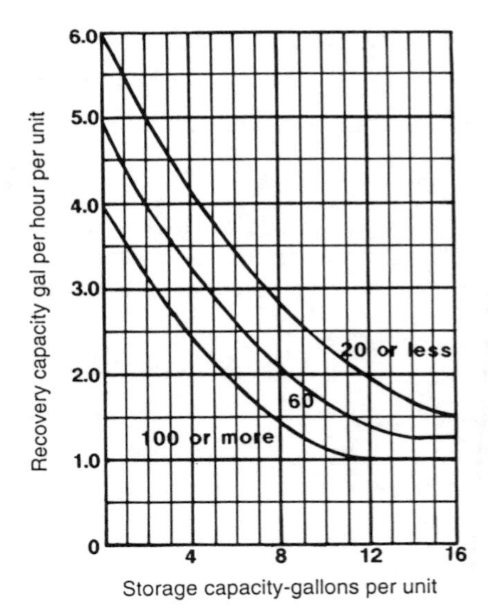

Figure 8-11. Recovery and Storage Capacities for Motels

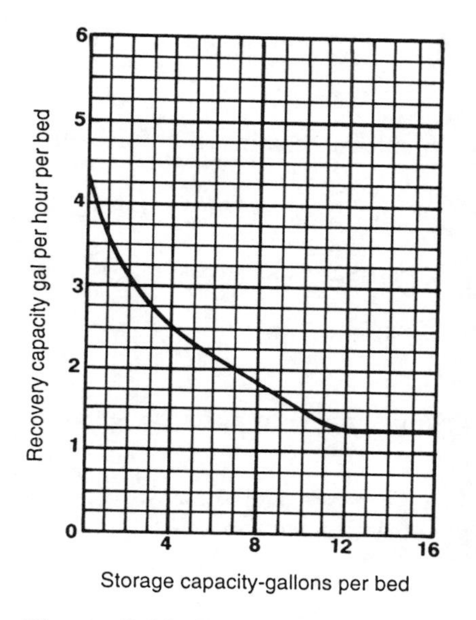

Figure 8-12. Recovery and Storage Capacities for Nursing Homes

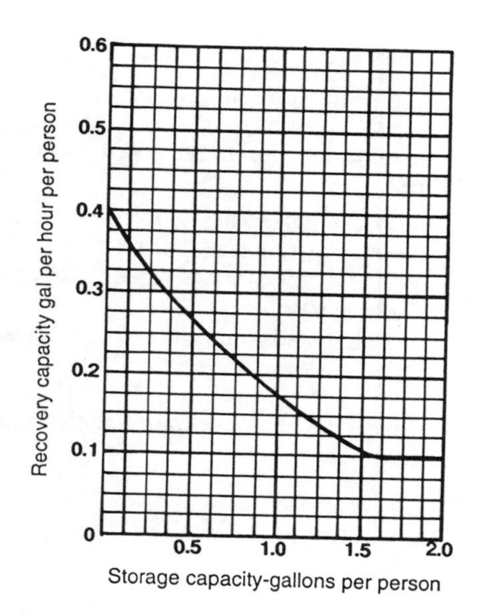

Figure 8-13. Recovery and Storage Capacities for Office Buildings

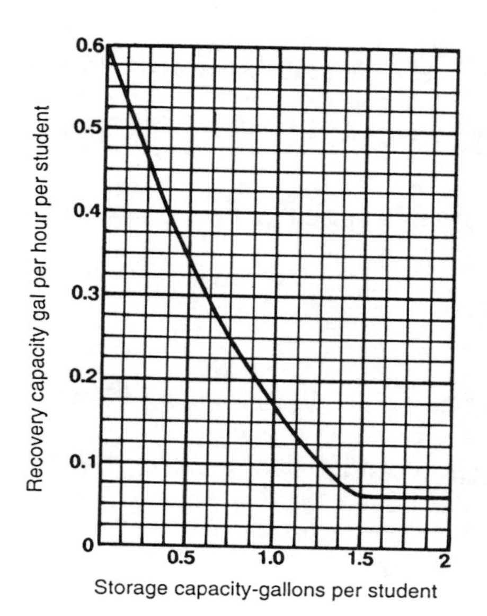

Figure 8-14. Recovery and Storage Capacities for Elementary Schools

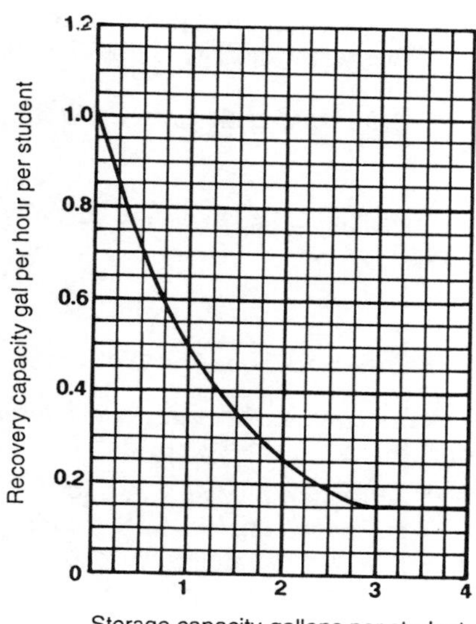

Figure 8-15. Recovery and Storage Capacities for High Schools

Solution 8-2.

Solution to Situation No. 1

Based on the data given, it is possible to determine the following information:

- Water heater has a recovery rate of 275 gph (250 students x 1.1 gph)
- Storage required per student is 12 gal (per Figure 8-9)
- Hot water storage for 250 students is 3000 gal (250 students x 12 gal/student)
- Calculated tank capacity is 4286 gal (3000 ÷ 0.7)

Since the calculated tank capacity is 4286 gal, round to a storage tank size of 4300 gal.[5]

There are other points on the curve in Figure 8-9, and these values may yield a selection of smaller capacity tanks with larger recovery rates.

Solution to Situation No. 2

Based on the data given, it is possible to determine the following information:

- Water heater has a recovery rate of 750 gph (250 students x 3 gph)
- Storage required per student is 3.75 gal (per Figure 8-9)
- Hot water storage for 250 students is 938 gal (250 students x 3.75 gal/student)
- Calculated tank capacity is 1339 gal (938 ÷ 0.7)

Since the calculated tank capacity is 1339 gal, round to a storage tank size of 1340 gal.

Solution to Situation No. 3(a)

This situation requires a cafeteria. Based on the data given, the following information can be obtained:

- Minimum recovery rate for the cafeteria is 0.45 gal (Table 8-10, Type A)
- Recovery rate for 250 students is 112.5 gph (250 students x 0.45 gal)
- Hot water storage is 1688 gal (250 students x 6.75 gph)
- Storage tank capacity is 2450 gal (1688 gal ÷ 0.7)

Based on calculations made in Situation No. 1, the dormitory and cafeteria would require a total storage tank capacity of 6750 gal:

275 gph + 112.5 gph = 387.5 gph recovery
4300 gal + 2450 gal = 6750 gal hot water storage tank capacity

Solution to Situation No. 3(b)

In this situation, the selection is for a tank with a smaller capacity but with a larger recovery rate (more fuel burned in the unit of time). In this case, the tank is only about one-third the size of the water heater selected for Situation No. 1, but the recovery rate is 750 gph (taken from Situation No. 2). Based on the data given, the following information can be obtained:

- Cafeteria recovery rate is 187.5 gph (250 students x 0.75 gph)
- Hot water storage is 875 gal (250 students x 3.5 gph)
- Storage tank capacity is 1250 gal (875 gal ÷ 0.7)

Based on calculations made in Situation No. 2, the dormitory and cafeteria would require a total recovery rate of 950 gph and a storage tank capacity of 2600 gal:

750 gph + 187.5 gph = 937.5 gph (round to 950 gph)
1340 gal + 1250 gal = 2590 gal hot water storage tank capacity (round to 2600 gal)

Example 8-3. Determine the water heater size necessary for a 1000-student high school for the following two situations:

1. Storage system with a minimum recovery rate
2. Storage system with a 3000-gallon maximum tank capacity

Solution 8-3.

Solution to Situation No. 1

Based on the data given, it is possible to determine the following information:

- Minimum recovery rate is 0.15 gph/student (per Figure 8-15)
- Recovery rate for 1000 students is 150 gph (1000 students x 0.15 gph)
- Storage required per student is 2.75 gal (per Figure 8-15)
- Hot water storage for 1000 students is 2750 gal (1000 students x 2.75 gal/student)
- Storage tank capacity is 3928 gal (2750 gal ÷ 0.7), round to 3950 gal

Note that in this case, 3950 gal is the minimum storage capacity. Larger tanks may be required if there are other buildings (e.g., gymnasium) used on a regular basis.

Solution to Situation No. 2

This situation challenges the designer to come up with a storage system with a maximum tank capacity of 3000 gal. Based on the data given, it is possible to determine the following information:

- Net storage capacity is 2100 gal (3000 gal x 0.7)
- Storage required per student is 2.1 gal/student (2100 ÷ 1000)
- Recovery rate per student is 0.25 gph (per Figure 8-15)
- Recovery rate for total number of students is 250 gph (1000 students x 0.25 gph)

Therefore, for a 1000-student high school with a maximum tank capacity of 3000 gal, the recovery rate must be 250 gph.

Example 8-4. Determine the water heater size necessary for an apartment building (in the design stage) with the hot water demands shown in Table 8-7.

No. & type of fixture		gal/hr Fixture	gal/hr Total
50	Lavatories	2	100
10	Bathtubs	20	200
40	Shower heads	30	1200
50	Sinks	10	500
50	Dishwashers	15	750
10	Washing machines	20	200
Calculated (maximum) demand			**2950 gal**

Table 8-7. Hot Water Demand for an Apartment Building

Solution 8-4. Earlier in this chapter, it was shown that a simultaneity coefficient must be considered in the hot water demand calculation when hot water consumption is in gpm or gph/fixture. For an apartment building the coefficient is 0.3, the calculated demand for this building is 2950 gal (per Table 8-7), so the probable peak consumption is 885 gal/hr (2950 gal x 0.3).

For storage capacity, a safety factor of 25% must be added, then the result is divided by 0.7:

$$\frac{(885)(1.25)}{0.7} = 1580 \text{ gal}$$

Example 8-5. Determine the water heater size necessary for a small department store that includes the following fixtures connected to hot water:

Location	Fixtures	No.	gph (each)	Demand factor	Probable maximum demand (gpd)
Public toilet	Lav	9	6	0.3	16.2
Tailor shop	Sink	1	10	1.0	10
Auto center	Lav (round)	1	6	0.4	2.4
	Wash fountain	1	20	0.4	8
Candy store	Sink	1	10	1.0	10

Add the probable maximum demands together for a total of 46.6 gph. Consider the peak demand to be two hours:

(46.6 gph) (2 hours) = 93.2 gal (round to 94 gal)

One of the types of water heaters that may be selected is a natural gas model manufactured by Bradford-White, Table 8-8.

The tentative selection is a 40-gal tank, so the following information can be obtained:

- Hot water capacity in a 40-gal tank is 28 gal (40 gal x 0.7)
- Hot water deficit is 66 gal (94 gal required - 28 gal available)
- Recovery rate of 33 gph (66 gal ÷ 2 hours)

Based on Table 8-8, the tank selection may be a residential type (Model 40T5LN), which has a recovery rate of 34.8 gph. Remember that **water heater selection must include a safety capacity.** The next size hot water tank size is Model 50T-50-3N, which has a recovery rate of 45 gph. This will be the actual selection, because it is necessary to include additional safety factors to make sure there is enough hot water even during periods of heavy consumption.

Example 8-6. Determine the water heater size necessary for a beauty salon that includes the following fixtures connected to hot water:

Fixtures	No. installed	Hot water demand in gph (each)	Total (gph)
Shampoo sink	3	20	60
Washing machine	1	20	20
Sink	1	10	10
			Total 90

Solution 8-6. It is assumed that two of the three shampoo sinks and the washing machine will work simultaneously during the peak demand, so the demand coefficient is a value equal to approximately 1.0.[6] Therefore, the shampoo sinks have a total hot water demand of 40 gph (2 x 20 x 1.0), and the washing machine has a hot water demand of 20 gph (1 x 20 x 1.0). Add these two numbers together for a total hourly demand for hot water of 60 gph.

For this type of occupancy, there is a peak demand of 6 hours:

(60 gph) (6 hours) = 360 gallons

The selection is based once again on Table 8-8, but the tentative selection is for a 75-gal, gas-fired water heater. Based on this data, the following information may be obtained:

- Hot water capacity for a 75-gal tank is 52.5 gal (75 gal x 0.7)
- Hot water deficit is approximately 308 gal (360 gal - 52.5 gal)
- Recovery rate is approximately 51 gph (308 gal ÷ 6)

The selection in this case is the gas-fired, Bradford-White water heater, Model 75T-75-3N, which has a recovery rate of 64 gph.

For further reference, Table 8-9 lists the expected cold, hot, and total water consumption for various occupancies. This table lists (oversized) consumption and is to be used only for rough, preliminary information.

Table 8-8. Gas Water Heater Technical Data (Courtesy, Bradford-White)

MODEL NUMBER	GAL. CAP.	INPUT. BTU	GPH REC. 90°F. RISE	MODEL NUMBER	GAL. CAP.	INPPUT. BTU	GPH REC. 90°F. RISE
ATMOSPHERIC COMBUSTION							
M-I-MH-0T5LN	30	32,000	32.7	M-I-MH30T5LX	30	32,000	32.7
M-I-MH40T5LN	40	34,000	34.8	M-I-MH40T5LX	40	34,000	34.8

DIMENSIONS INCHES

FL. TO FLUE CONN.	JACK. DIAM.	VENT. SIZE	FL. TO T & P CONN.	FL. TO GAS CONN.	FL. TO CW INLET	APPROX. SHIPPING WT. LBS.
59-1/4"	16"	3"	51-1/4"	13-1/2"	23"	106
59-3/4"	18"	3"	51-1/4"	13-1/2"	23"	126

MODEL NUMBER	NAT. BTU INPUT	REC. 100°F. RISE	STG. CAP. US GAL.
M-I-50T-50-3N	50,000	45	50
M-I-75T-75-3N	75,000	64	75
M-I-100T-82-3N	82,000	76	100

	DIMENSIONS IN INCHES														WEIGHT	
MODEL NUMBER	HT.	DIA.	SIDE HW CONN.	TOP HW CONN.	CW INLET	GAS CONN.	TOP VENT HT.	SIDE VENT HT.	VENT DIA.	SIDE VENT C/L	SIDE VENT DR.DIV. RAD.	WTR. CONN.	GAS CONN. DIA.	REL. VALVE OPN.	STD.	ASME
M-I-50T-50-3N	57-1/4	20	--	58	58	14	60-1/2	--	4	--	--	3/4	1/2	3/4	148	N/A
M-I-75T-75-3N	65-1/2	24-1/2	--	66-3/4	66-3/4	21	68-3/4	--	4	--	--	3/4	1/2	3/4	265	N/A
M-I-100T-82-3N	71-3/4	28	--	74-1/4	74-1/4	20-1/2	75	--	4	--	--	1-1/4	1/2	3/4	392	N/A

Type of occupancy	Cold water demand (gpd)	Hot water demand (gpd)	Total water demand (gpd)
Residence (four person occupancy)	125	125	250
Each apartment (two person occupancy)	75	75	150
Office building (8 hr/day occupancy per person)	17	3	20
Industrial plant (per 8 hour shift)	20	15	35
Schools with cafeteria but no showers (per student)	16	4	20
Schools with cafeteria and showers (per student)	15	10	25
Cafeterias and restaurants (per meal)	6	4	10
Hospitals (per bed)	175	125	300
Nursing homes (per bed)	45	30	75
Assembly halls	2.5	0.5	3
Hotels (per unit)	25	50	75
Laundry (commercial, per pound of wash)	1.5	2	4

gpd = gallons per day.

Table 8-9. Hot and Cold Water Demand

Fuel for Water Heaters

Natural gas is an odorless, colorless gas that naturally accumulates in the upper part of oil and gas wells. From the collection points, this gas is conveyed to consumers and used as fuel for hot water heaters, boilers, and other appliances. Raw natural gas is a mixture of methane, other higher hydrocarbons, and noncombustible gases. The chemical formula for methane (which is a natural combustible gas) is CH_4 :

```
     H
     |
  H—C—H
     |
     H
```

Fuel-gas is normally supplied to consumers through a network of pipes by a utility company. Generally, there is a master distribution network system which includes major lines over land with underground street mains and branches in cities.

Another combustible fuel is LPG (liquefied petroleum gas), which consists primarily of propane and butane. LPG is a byproduct of oil refinery operation. It is used when there are no gas utility supply lines in an area where a development is planned. Specialized companies may provide (lease) an LPG or liquefied petroleum bottled (storage tank) installation. To determine the required amount of fuel for an installation, it is necessary to know the customer's demands. Table 8-10 shows some of these demands.

Gas demands are typically expressed in Btuh (Btu per hour) input, which can be easily converted in cu ft of gas/hr or cfh, as shown in Table 8-11. For information purposes, Table 8-12 is a comparative heat content equivalent table.

Gas System Design

Tables 8-13 and 8-14 illustrate the consumption diversity factors, or simultaneous coefficients required when designing a fuel system.

General gas systems must be designed and installed in accordance with NFPA (National Fire Protection Association) Standard No. 54, as well as state plumbing codes, fuel gas codes, and the rules and regulations of the local authority having jurisdiction. To install a gas system, a permit is required from the authority having jurisdiction.

NFPA Standard No. 54 is a comprehensive standard, which must be consulted before any fuel gas design is developed. It gives directives about system design, piping materials, installation, venting procedures, how to put the equipment in operation, as well as numerous other related recommendations and guidelines.

There are some general considerations that must be observed when designing a fuel-gas supply system:

- Pressure drop in pipes, under maximum flow conditions, shall be such as to ensure slightly more pressure at the equipment than the minimum required.
- Maximum simultaneous gas demand
- Pipe developed length and the number and type of fittings
- Specific gravity of gas
- Foreseeable future demand
- Pressure of gas supplied versus the working pressure of equipment and/or fixtures that consume gas
- Appliance and/or equipment input rating, which shall be indicated on the name plate
- Maximum design pressure for gas piping operation located within a building shall not exceed 5 psi.
- Piping material shall be Schedule 40 steel (minimum).
- Every piece of gas burning equipment shall be provided with a flue or vent. (This does not include residential cooking ranges.)
- Gas supply pressure in residential applications must not exceed 0.5 psi (equal to 8 oz or to 14" water column).
- Each apartment or customer shall have a meter to register gas consumption (by appliances or equipment installed).

Appliance	Input-Btuh*
Commercial kitchen equipment:	
Small broiler	30,000
Large broiler	60,000
Combination broiler and roaster	66,000
Coffee maker, 3 burner	18,000
Coffee maker, 4 burner	24,000
Deep fat fryer, 45 pounds (20.4 kg) of fat	50,000
Deep fat fryer, 75 pounds (34.1 kg) of fat	75,000
Doughnut fryer, 200 pounds (90.8 kg) of fat	72,000
2 deck baking and roasting oven	100,000
3 deck baking oven	96,000
Revolving oven, 4 or 5 trays	210,000
Range with hot top and oven	90,000
Range with hot top	45,000
Range with fry top and oven	100,000
Range with fry top	50,000
Coffee urn, single, 5 gal (18.9 L)	28,000
Coffee urn, twin, 10 gal (37.9 L)	56,000
Coffee urn, twin, 15 gal (56.8 L)	84,000
Residential equipment:	
Clothes dryer	35,000
Range	65,000
Stove top burners	40,000
Oven	25,000
30 gal (113.6 L) water heater	30,000
40 to 50 gal (151.4 to 189.3 L) water heater	50,000
Log lighter	25,000
Barbeque	50,000
Miscellaneous equipment:	
Commercial log lighter	50,000
Bunsen burner	30,000
Gas engine, per horsepower (745.7 W)	10,000
Steam boiler, per horsepower (745.7 W)	50,000

*The values given here should only be used when the manufacturer's data are not available.

Table 8-10. Approximate Gas Demand for Common Appliances

1 cu ft = Approx. 1000 Btu
1 c. cu ft = 100 cu ft = 1 Therm
c. cu ft = hundred cu ft
1 Therm = 100,000 Btu = 100 cu ft = 0.1 m. cu ft
10 Therms = 1 m. cu ft
1 m. cu ft = 1000 cu ft = 10 c. cu ft = 10 Therms
m. cu ft = million cu ft
1 Quad = 10^9 m. cu ft = 10^{10} Therms = 10^{15} Btu

Table 8-11. Energy Conversion Factors for Natural Gas

Comparative thermal values	1.0 million Btu	24.0 million Btu	0.0916 million Btu	0.125 million Btu	0.139 million Btu	0.15 million Btu	0.003412 million Btu
Natural gas 1000 Btu/cu ft	1000 cu ft	24,000 cu ft	91.600 cu ft	125.000 cu ft	139.000 cu ft	150.000 cu ft	3.412 cu ft
Coal 12,000 Btu/lb	83.333 lb	2000 lb	7.633 lb	10.417 lb	11.583 lb	12.500 lb	0.2843 lb
Propane 91,600 Btu/gal	10.917 gal	262.009 gal	1 gal	1.365 gal	1.517 gal	1.638 gal	0.0373 gal
Gasoline 125,000 Btu/gal	8.000 gal	192.000 gal	0.733 gal	1 gal	1.112 gal	1.200 gal	0.0273 gal
Fuel oil #2 139,000 Btu/gal	7.194 gal	172.662 gal	0.659 gal	0.899 gal	1 gal	1.079 gal	0.0245 gal
Fuel oil #6 150,000 Btu/gal	6.666 gal	160.000 gal	0.611 gal	0.833 gal	0.927 gal	1 gal	0.0227 gal
Electricity 3412 Btu/kWh	293.083 kWh	7033.998 kWh	26.846 kWh	36.635 kWh	40.739 kWh	43.962 kWh	1 kWh

Figure 8-12. Comparative Heat Content Equivalent

Figure 8-16 is an illustrative example showing the sizing of a gas piping system based on the standard engineering method. This method can be used successfully where the maximum demand is over 250 cfh and the maximum developed length between the meter and the most remote consumer is over 250 ft. This type of system would be suitable in schools, apartment buildings, convalescent homes, etc.

No. of units	Multiplier
1	1.00
2	0.70
3	0.60
4	0.55
5	0.50
6	0.45
7	0.42
8	0.40
9	0.37
10	0.35

Table 8-13. Diversity Factors for Residential Cooking Appliances in Multi-Family Dwelling Units

No. of outlets	Multiplier
1 to 8	1.00
9 to 16	0.90
17 to 29	0.80
30 to 79	0.60
80 to 162	0.50

Table 8-14. Diversity Factors in Laboratory Outlets

As shown in Figure 8-16, the distance from Consumer 1 to the gas meter is 600 ft (150 ft + 200 ft + 100 ft + 150 ft = 600 ft). In this case, assume the gas pressure is 6 ounces, and the specific gravity of the gas is 0.65. Table 8-15 shows different pipe sizes for pipe carrying natural gas. This table is based on a water column pressure drop of 0.5 in. (For this example, use Table 8-15. For any gas system design, the applicable state gas code must be used, as well as the appropriate applicable table.)

Table 8-16 shows the pipe size needed for each section, based on information taken from Figure 8-16 and Table 8-15. The main pipe for each section is considered as the whole length of pipe.

Gas service in buildings is generally delivered in the low pressure range of 7" water column (wc). The maximum pressure drop allowable in piping systems at this pressure is generally 0.5" wc, but this is subject to regulation by local building, plumbing, and gas appliance codes. Consult these codes when designing gas piping systems. (See the National Fire Protection Association Standard No. 54.)

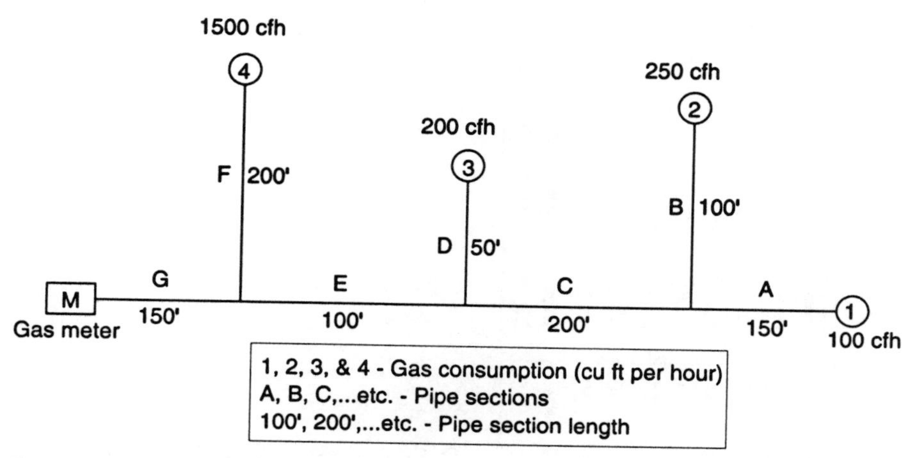

Figure 8-16. Gas Piping System

Where large quantities of gas are required or long lengths of pipes are used (as in industrial buildings), low pressure limitations result in large pipe sizes. Local codes may allow, and local gas companies may deliver, gas at higher pressures (2, 5, or 10 psig) or gas pressure busters are installed. However, under these conditions, an allowable pressure drop of 10% of the initial pressure is used, and pipe sizes can be reduced significantly. Appliance gas pressure regulators must be specified to accommodate higher inlet pressures. Specialty tables provide information on pipe sizing for various inlet pressures and pressure drops at higher pressures.

When gas consuming equipment requires higher pressure than that delivered by the gas company, booster gas pressure must be installed.

Pipe Size (in.)	Length in feet 10	20	30	40	50	60	70	80	90	100	125	150	200
1/2	170	118	95	80	71	64	60	55	52	49	44	40	34
3/4	360	245	198	169	150	135	123	115	108	102	92	83	71
1	670	430	370	318	282	255	235	220	205	192	172	158	132
1-1/4	1320	930	740	640	565	510	470	440	410	390	345	315	270
1-1/2	1990	1370	1100	950	830	760	700	650	610	570	510	460	400
2	3880	2680	2150	1840	1610	1480	1350	1250	1180	1100	1000	910	780
2-1/2	6200	4120	3420	2950	2600	2360	2180	2000	1900	1800	1600	1450	1230
3	10900	7500	6000	5150	4600	4150	3820	3550	3300	3120	2810	2550	2180
3-1/2	16000	11000	8900	7600	6750	6200	5650	5250	4950	4650	4150	3800	3200
4	22500	15500	12400	10600	9300	8500	7900	7300	6800	6400	5700	5200	4400

Size (in.)	250	300	350	400	450	500	550	600
1/2	30	27	25	23	22	21	20	19
3/4	63	57	52	48	45	43	41	39
1	118	108	100	92	86	81	77	74
1-1/4	238	215	200	185	172	162	155	150
1-1/2	350	320	295	275	255	240	230	220
2	690	625	570	535	500	470	450	430
2-1/2	1100	1000	920	850	800	760	720	690
3	1930	1750	1600	1500	1400	1320	1250	1200
3-1/2	2860	2600	2400	2200	2100	2000	1900	1800
4	3950	3600	3250	3050	2850	2700	2570	2450

Pipe carrying natural gas of .65 specific gravity. Size of gas piping - maximum delivery in cfh for gas pressures from 4 ounces (7") to 0.9 lb.

Based on ½ inch water column pressure drop.

Table 8-15. Pipe Sizes for Pipe Carrying Natural Gas

Consumer no.	cfh	Length of pipe (ft)	Sections included	Pipe size (in.)
1	100	600	A	1-¼
	250 + 100	600	C	2
	200+250+100	600	E	2-½
	1500 + 200 + 250 + 100	600	G	4
2	250	550*	B	2
3	200	300	D	1-¼
4	1500	350	F	3

*Length 550 to section 2 is 150 + 100 + 200 + 100 = 550 ft. G + E + C + B.
The same type of calculation is used for consumers 3 and 4.

Table 8-16. Pipe Sizes for Each Pipe Section

NOTES

[1] Normal hot water supply temperature is considered 140°F. Normal incoming cold water is considered 40°F (usually the temperature of the incoming water is a little higher than that). This is the reason why the calculations are usually based on a temperature difference of 100°F.

[2] The actual sum of this calculation is 9960, but it is rounded to 10,000.

[3] For actual consumption, multiply the amount resulted by this demand coefficient.

[4] Water temperature difference of 100°F is based on 140°F - 40°F, which is the exiting water temperature minus entering water temperature.

[5] In these examples, once exact calculations are determined the figures are rounded up to a higher number. The number rounding is necessary because water heater tanks are manufactured in certain incremental sizes.

[6] Considering total hot water demand of 90 gph and actual demand of 60 gph, the simultaneity coefficient is 0.66 (60 gph ÷ 90 gph).

Chapter Nine

Drainage System

The sanitary drainage system collects the liquid and solid wastes from toilets, as well as waste water from lavatories, sinks, bathtubs, floor drains, etc., and conveys them by gravity through sloped horizontal pipes and vertical stacks to the building sewer and then outside to a street sewer. It is prohibited to discharge raw sewage on the ground surface, in waterways (unless previously treated), or in the storm drainage system (unless the municipal drainage system is a common, treated, storm and sanitary sewer).

In general, an underground waste pipe from a sanitary system is located below the frost line in the yard. In a public system, it will discharge into a sanitary drain pipe that ends up in a municipal sewage treatment plant. If there is no public sewer system in the area, a private sewage treatment system is used, which involves a local septic tank and a leaching (tile) field. A sewer system discharging into the city sewer pipe, as well as a private sewage disposal and treatment system, must conform to the applicable codes, rules, and regulations of the local authority having jurisdiction. There are two special plumbing fixtures that are not connected to the sewer system: an emergency shower and an emergency eye-wash.

The sanitary system must be designed, constructed, and maintained to guard against fouling the environment; clogging the pipes; and depositing suspended solids that are floating in the sewage. These three items can be accomplished by designing the system with large enough pipeways; adequate liquid velocities to prevent deposits; and enough cleanouts for proper maintenance.

To prevent the deterioration of a sanitary drainage system, the following rules must be followed:

- No explosive mixtures shall be discharged into the sewer.
- Sewer system must not be interconnected with the storm drainage system (unless the code allows it).
- Wastes with temperatures over 140°F (such as blowdown from boilers) shall not be discharged directly into the sewer pipes, because these pipes are not designed for expansion and contraction. This type of hot waste has to be cooled first in retention tanks.

Materials that can safely be poured down the drain include disinfectants, expired medicine, rust remover, hair relaxer, water-based glue, drain cleaner, aluminum cleaner, window cleaner, antifreeze, and windshield washer solution. Materials that cannot be poured down the drain (must be sent to a landfill) include latex paint, fertilizer, oven cleaner, solidified nail polish, and shoe polish. Hazardous wastes that should go to a licensed hazardous waste contractor (some are recyclable) include battery acid, gasoline, varnish, paint thinner, kerosene, car wax with solvent, turpentine, motor oil, automatic transmission fluid, bug spray, diesel fuel, and mercury batteries.

Traps

The trap's function is to maintain a water seal, which prevents sewer gases from returning into the area where the plumbing fixture is located. If the trap is not vented, siphonage might occur. This situation involves the water seal being sucked into the drain, allowing sewer gases to return into the area where the plumbing fixture is located.

The drainage system originates at every fixture outlet. The first item downstream from the outlet at each fixture is the trap. As a rule of thumb, the trap is located a maximum of 2 feet below each fixture that is connected to the sewer. Plumbing fixtures that discharge waste indirectly through drains being farther away then 4 ft from the fixture outlet must have a water seal trap. If the plumbing installation for a kitchen includes only a two or three compartment sink, one single trap shall be provided for the assembly. In this case no sink should be equipped with a garbage disposal.

Traps that do not meet special criteria are prohibited. To be acceptable for installation, a trap must have the following characteristics:

- It must be self cleaning.
- It must have a smooth interior.
- It should have no movable parts.
- The seal size shall be between minimum 2 in. and maximum 4 in.
- It shall have an accessible cleanout or shall be easily removable.

Traps that include movable parts can cause wastes to accumulate, which may ultimately render the trap unusable or ineffective. Crown vent traps may also induce the growth of slime at the point of connection, which may render them inoperative. Any item that could cause clogging in a sewer pipe must be eliminated. A trap with a smooth interior helps prevent waste accumulation and enhances the "self-cleaning" characteristics.

Easy access to the plumbing fixture trap is necessary so that in case of clogging, the stoppage can be easily removed. A cleanout at the bottom of the trap is convenient for trap cleanup. If the trap does not include a cleanout, it is necessary to include an easy way to remove the trap itself for cleaning.

It is not necessary to keep up old traditions of unnecessarily oversizing a system. For example, if the code and local authorities allow it, use a 3" trap and drain pipe from each toilet to help save money, Figure 9-1a; Figure 9-1b shows the oversized pipe arrangement. If the application is for a large system (e.g., mall, high rise, or department store) in which many fixtures are installed, the savings can be substantial. In addition, the newest WC flush discharges 1.5 to 1.6 gal of water compared with the present 3.5 gal. However, to install a 3" drain pipe from a WC, approval is required from the authority having jurisdiction.

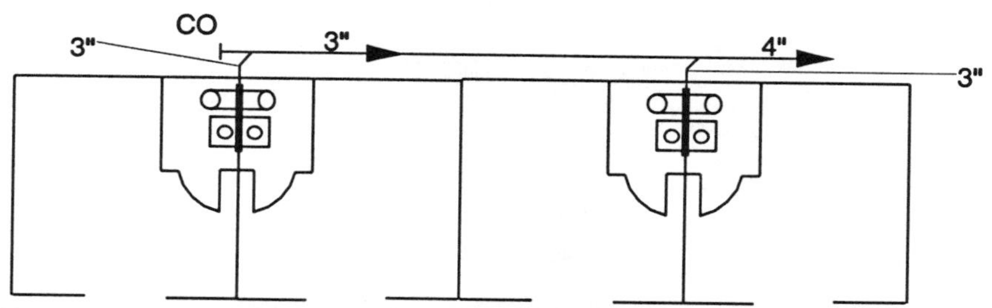

a) For a back-to-back store toilet arrangement, the horizontal branch may be 3 inches since it only receives the discharge of 2 water closets and 2 lavatories (This is acceptable if code allows).

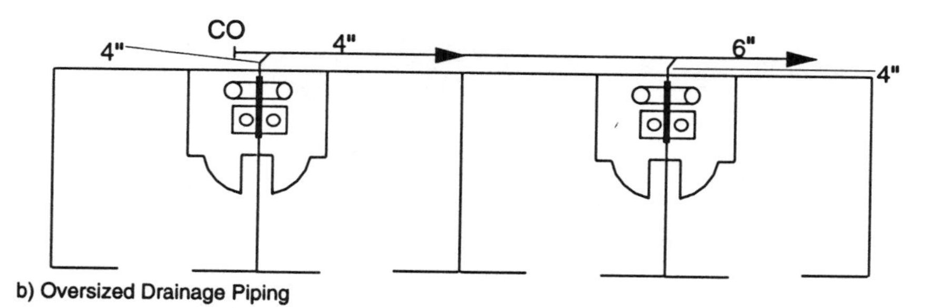

b) Oversized Drainage Piping

Figure 9-1. Economical Toilet Pipe Size

There are two basic types of traps: a *P-trap*, Figure 9-2, and an *S-trap*, Figure 9-3. The P-trap, which in fact is a half S-trap, is the one in use today in plumbing installations. It allows for easy air pressure equilibrium as well as adequate venting. The S traps are now prohibited in most states due to their configuration, which consists of one extra bend, making them more prone to clogging. The S trap was part of the early original toilet design.

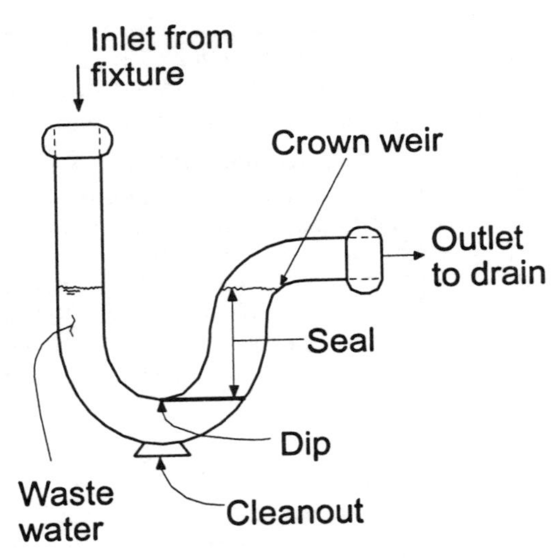

Figure 9-2. P-Trap

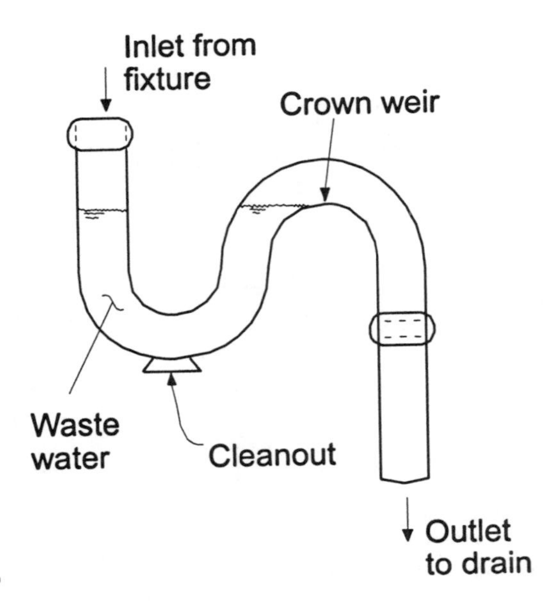

Figure 9-3. S-Trap

The vent pipe (system and location with respect to the trap is detailed separately) is located close to the fixture's drain trap and is interconnected with the drainage system. It is prohibited to connect a vent pipe closer to the crown than two pipe (trap) diameters, because if the vent pipe is located too close to the trap, either the seal has the tendency to evaporate, or the vent tends to clog.

There is a length limitation between the fixture outlet and the trap weir, Figure 9-4. This is because the atmospheric pressure downstream of the trap must equal the atmospheric pressure at the outlet. A column of water higher than 2 ft disturbs this equilibrium.

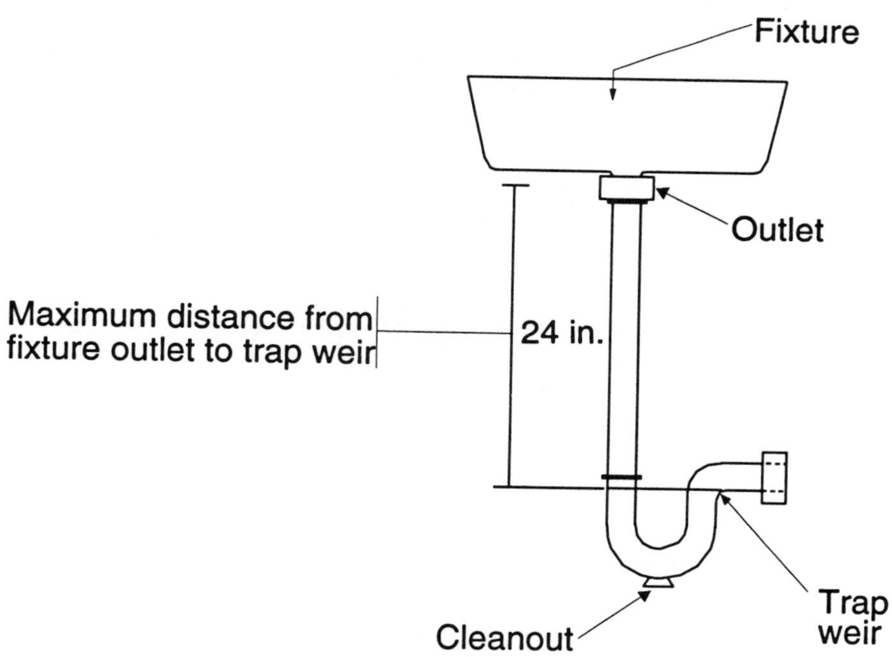

Figure 9-4. Length Limitation

Interceptors

Wastes may include products that will clog the sewer, so these materials must be screened out. To do this, specialized traps or interceptors are included in the drainage system. These devices separate, retain, or trap certain materials normally suspended in the waste stream. After separation, the effluent is usually discharged into the regular sanitary sewer system. The installation of interceptors is mandatory for special waste and required by most state codes. Some of these interceptors and their applications are as follows:

Interceptors of solids:

- Grease interceptors (restaurants, cafeterias, snack bars, and candle factories). Expected to trap cheeses, fats, grease, paraffin, and wax.
- Plaster or barium traps (radiological offices)
- Glass grindings (sand paper manufacturing)
- Sand and gravel traps (garages, parking areas, ramps)
- Hair and lint traps (beauty salons, barber shops)
- Precious metals (specialized manufacturing processes)

Interceptors of objectionable liquids:

- Oil traps or oil interceptors (auto work station, vehicle garages)
- Liquids lighter than water interceptors, which are separated in volatile waste separators (benzene, gasoline, kerosene, lubricating oils, naphtha, etc.)

Interceptors are very useful in a sewer system, because they protect the installation from clogging, and they separate environmentally-harmful substances from sewer waste. The sanitary drainage piping helps to convey the waste from the fixtures through these specialty traps as applicable.

There are a few general requirements for interceptors:

- Install the required interceptors based on the State Plumbing Code and in accordance with the authority having jurisdiction.
- Specific traps shall be provided where they are able to remove objectionable substances (solids or liquids) from the wastes.
- Traps or interceptors shall be installed only in accordance with their application.
- Trap type and size shall be in strict accordance with the quantity of waste expected and trap manufacturer's instructions.
- Trap shall be provided with access for maintenance and cleaning.

Sewage Backup

Liquid waste usually flows in drain pipes due to the force of gravity. In some buildings, it may be possible for the building sewer system to discharge at a lower elevation than the street sewer. Therefore, the waste must either be pumped out from a special container or ejected by compressed air to the higher discharge point. If the same building also has areas located above the street sewer elevation, they will normally be drained by gravity. This reduces power consumption for the sewage pumps or pneumatic ejectors. In such a case, two separate sewer lines shall be installed: one by gravity from the upper floors, and one pumped for the fixtures located at or below the street sewer level.

If the pumped waste is connected to the gravity sewer line rather than separately connected into the street sewer line, the possibility of sewage backup exists. Sewage backup is never desirable and in certain locations, it is absolutely prohibited. These locations include areas in which sterile goods are prepared. In these cases, waste lines from fixtures handling such products must have indirect waste pipes connected to the sewer through air gaps.

Buildings taller than 10 stories should direct the drainage to a sump, and then pump or eject the drainage out. Otherwise, the solids in the sanitary drainage might produce pipe or joint damage due to the velocity and acceleration during the vertical fall. Figure 9-5 shows the relationship between the number of drainage fixture units and the expected flow in gpm.

Sumps and Ejectors

As just stated, when a building's sewer line is below the street sewer line, the building's sanitary waste must accumulate in a sump. With the help of one

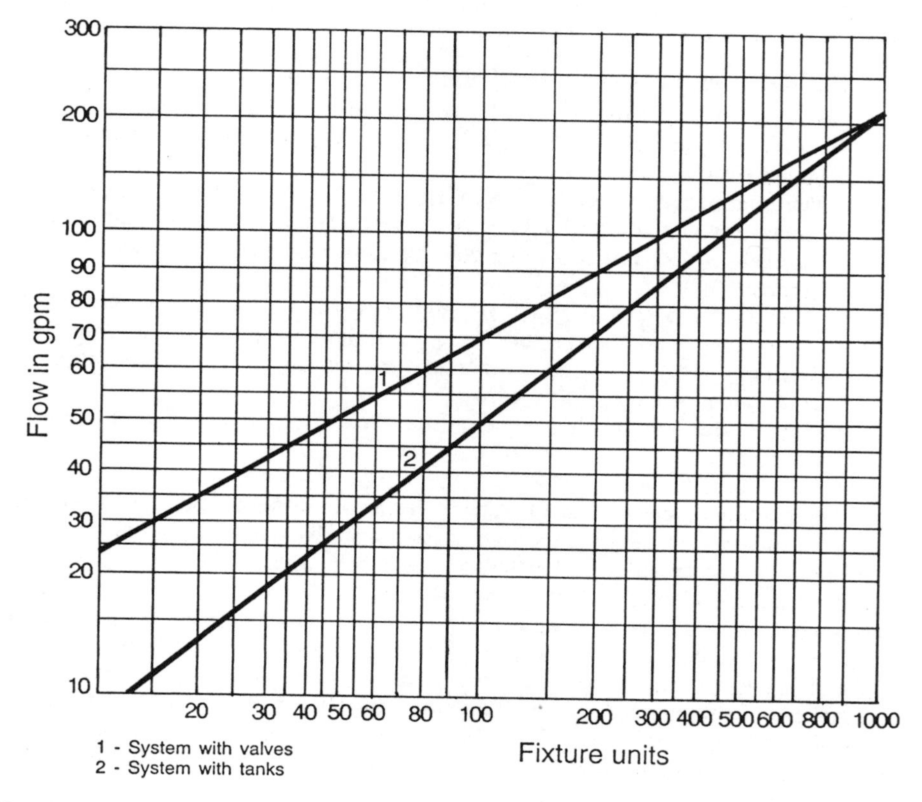

Figure 9-5. Drainage Conversion of Fixture Units to GPM

or two electric sump pumps, the contents must be pumped out (ejected) from the sump. This equipment assembly is called a sanitary lift station.

The electric, motor-driven pump may be a submersible, a vertical lift type, or a self-priming type. Table 9-1 shows receiving tank dimensions, and Figure 9-6 illustrates operating dimensions.

An alternative to the sump and pump is a pneumatic ejector. The pneumatic sewage ejector has no rotating parts and consists of an air compressor that injects air into a totally enclosed receiving tank. The pressure of the air forces the contents through a discharge pipe.

The liquid waste circulating in a drainage system with a pneumatic ejector is not exposed to the atmosphere, so no strong odors are expected. However, a sanitary lift station that is either equipped with pumps or a pneumatic ejector should be located at the lowest elevation of the building in a separate, well-ventilated room.

Both systems (the electric pump and pneumatic ejector) require calculations to determine the amount of sewage and the actual sizing of the ejectors. Figure 9-5 illustrated a curve, which could be used for these calculations.

The following rules apply to sump pumps and pneumatic ejectors:

- The lift station sump (with electric pump) must be ventilated.
- Pumps should not start more often than every three to five minutes and operate for ten minutes.
- Ventilation system for the room shall be based on the equipment heat release plus an additional 10% to 15% (safety) or a certain number of air changes as determined by the code or previous experience, whichever is greater.
- Ejector system (pumps) shall be able to pass 2" to 2-1/2" diameter solids.
- Pneumatic system is sealed except for the incoming gravity pipe, the discharge (lift) pipe, and the compressed air supply pipe. The collecting tank is not vented in this case.
- Tank dimensions shall be determined by the equipment manufacturer.

The pneumatic ejector has two check valves to ensure correct operation. One valve on the sewage inlet pipe allows waste to be discharged only one way into the tank. The other valve is located on the ejector discharge pipe and ensures the contents are ejected out only one way. When the ejector fills to the predetermined level it allows the compressed air inside, which causes contents to be expelled. A special valve allows the tank overpressure to be relieved if necessary. This air must be evacuated separately into the atmosphere. A 3" pipe with a cleanout is recommended for this purpose. The air pressure furnished for ejection of contents must be equal to the friction developed in the discharge pipe plus the static pressure difference between the tank and the gravity sewer line in which the contents are discharged.

Pump capacity (gpm)	One pump (simplex) basin diameter (in.)	Two pumps (duplex) basin diameter (in.)
50 to 125	30.0	36
125 to 200	36 to 42	42 to 48
200 to 300	36 to 42	48 to 60
350 to 500	36 to 48	48 to 60
larger	consult manufacturer	

(a)

Tank diameter (in.)	Capacity per ft depth (gal)
18	14.0
24	24.0
30	38.0
36	53.0
42	77.0
48	95.0
60	150.0
72	212.0

(b)

Table 9-1. Receptor Diameter vs Pump Capacity

To properly calculate ejector size, consider the following points:

- Determine the number of ejections (discharges) per hour. This number will determine:
  (a) size of retaining tank.
  (b) capacity of pump(s) and the head or flow amount expelled during a lift.
- Select the type of operation and the type of equipment to do the job (determine the solid sizes that will pass through the system)
- Select the type of controls for smooth and safe operation.

Sanitary Drainage Pipe Selection and Installation

Drainage pipes must be selected to produce a minimum liquid velocity of 2 fps (feet per second) but usually no more than 4 fps. This range of velocity for sewage ensures a scouring effect without pipe erosion. While drainage velocity in a vertical pipe (stack) may momentarily reach 10 to 15 fps, this only happens

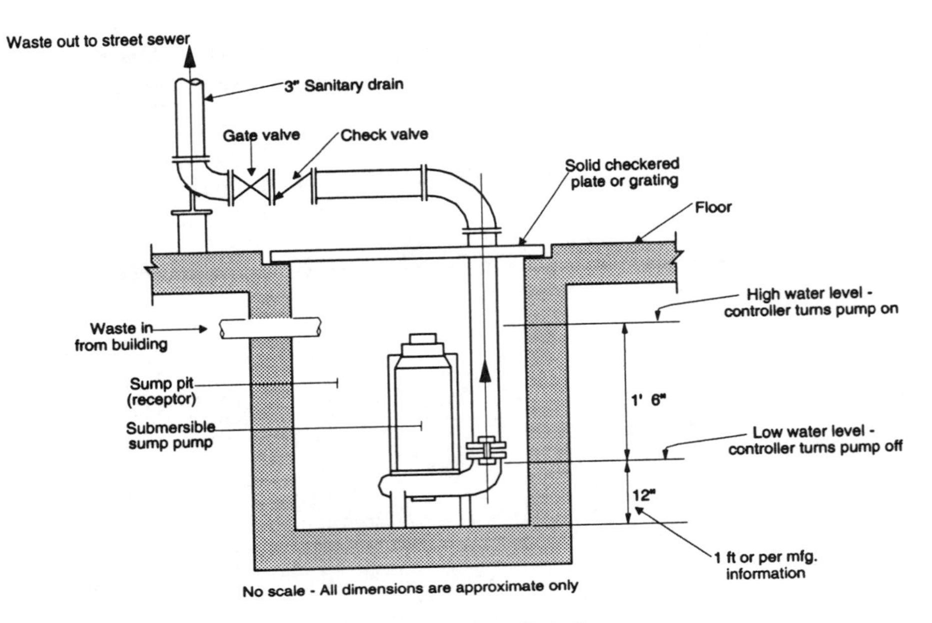

Figure 9-6. Simplex Sump Pump — Piping Detail

for very short periods. To obtain the normal velocity of 2 to 4 fps in a horizontal gravity flow system, recommended pipe slopes are:

Pipes 3" and smaller	1/4" per ft
4" to 6"	1/8" per ft
8" and larger	1/16" per ft

Horizontal drainage piping must be installed in a uniform alignment and slope to produce the desired velocity. To make sure this occurs, the following rules should be applied when installing drainage pipes:

- Drainage pipe must be installed with the hub in the direction of the flow. A tee branch discharging into the sewer lines shall be located in the upper part of the pipe and shall be angled in the direction of the flow, Figure 9-7.

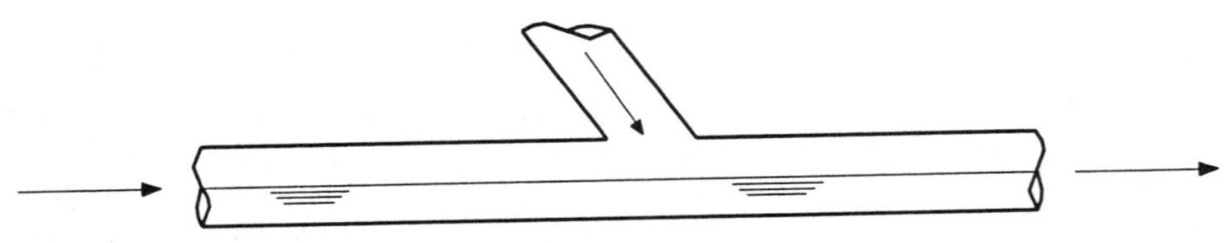

Figure 9-7. Hub in Direction of Flow

- Drainage pipes must not be installed in stairwells, elevator shafts, in front of any building opening, or in any place where they may pass above areas where food products are prepared or stored.
- Drainage pipes must be watertight and gastight.
- Maintenance must be performed based on a maintenance plan (see Chapter 12) to prevent clogging that may occur when drains become lined with deposits.
- Vertical drainage pipes are designed differently based on the type of fixture served (e.g., the **soil stack** receives waste from toilets, and the **waste stack** receives waste liquids from lavatories, bathtubs, sinks, etc.); therefore, horizontal branches must connect to each fixture and then discharge into the vertical stack(s).
- Vertical stacks shall run as straight as possible with minimum offsets.
- Drainage pipes originating in rooms with groups of showers must be sized for 100% usage rather than waste water fixture units.
- Drainage waste originating in laboratories must be treated and/or neutralized before being discharged into the sewer system. Piping from such systems before neutralization must be certified by the piping manufacturer to be acid resistant.
- Drain pipe must be sized for maximum expected flow (use drainage pipe size as indicated by state plumbing code).
- Horizontal and vertical drainage pipes must be firmly supported with hangers. Recommended hanger location and distance are listed in Table 9-2.

Pipe material	Vertical	Horizontal
Cast iron	At base & each story	5 to 10 ft
Steel (screwed conn.)	Every other floor max 25 ft	10 ft intervals
Copper tube	Every other floor max 10 ft	10 ft intervals
Lead	At 4 ft intervals	One continous support
Plastic	At 4 ft intervals	At 4 ft intervals
Glass	Every floor	At 8 ft intervals

Table 9-2. Distance and Location of Pipe Hangers

FORMULAS

There are several different formulas used to calculate the flow in a pipe. They may be used in calculating the flow and velocity in a drain pipe. They are used to develop various tables.

The general fluid flow formula is:

$$Q = AV$$

where:
- Q = flow in cubic ft per second
- A = cross section of pipe in ft^2
- V = velocity in pipe in ft/sec

For example, Table 9-6 and Figure 9-5 were prepared based on *Manning's Gravity Flow Formula:*

$$V = \frac{1.486}{n} r^{2/3} s^{1/2}$$

where:
- V = velocity in feet per second (fps)
- r = hydraulic radius in feet[1]
- s = slope of pipe in ft/ft
- n = roughness coefficient (depending on the piping material)

The following table indicates the roughness coefficient for steel pipe:

Pipe size (in.)	n
1½	0.012
2 - 3	0.013
4	0.014
5 - 6	0.015
8 and larger	0.016

As a rule, sanitary drains are normally designed to flow half full to a maximum of three-quarters full. Table 9-3 lists the values for $R^{2/3}$ for half-full flow for various pipe sizes.

Pipe size (in.)	R=D/4 (ft)	$R^{2/3}$	A (ft^2)
1 ¼	0.0290	0.09440	0.00520
1 ½	0.0335	0.10390	0.00706
2	0.0418	0.12040	0.01090
2 ½	0.0520	0.13930	0.01704
3	0.0625	0.15750	0.02455
4	0.0833	0.19070	0.04365
5	0.1040	0.22120	0.06820
6	0.1250	0.25000	0.09820
8	0.1670	0.30330	0.17460
10	0.2080	0.35100	0.27270
12	0.2500	0.39700	0.39270
15	0.3125	0.46050	0.61350

R = Depth of flow

A = Cross-section of flow area

Table 9-3. Values for Half-Full Flow

Calculations can be made to determine how much flow (in gpm) can be drained from a certain fixture. To do that, it is necessary to measure the water line height above the drain. The mathematical formula that gives this relationship is the following:

$$g = 13.17d^2h^{1/2}$$

where:
- g = flow in gpm
- d = diameter of outlet in inches
- h = vertical height of water above the outlet in ft
- 13.17 = experimental coefficient

For example, consider the sanitary flow from a bathtub in which d = 1/2" and h = 1 ft:

$$g = (13.17)\ (0.5^2)\ (1^{1/2}) = 3.3 \text{ gpm}$$

As soon as the water level drops, the flow decreases accordingly.

WASTE WATER FIXTURE UNIT

A waste water fixture unit (FU) is the number of gallons of water (liquid) a fixture will drain in one minute. To size a drainage system, use drainage fixture units and slopes as listed in the applicable state plumbing code.

It needs to be understood that the fixture unit values for drainage are different than the fixture unit values for water supply. One reason is that the rate of discharge is different than for the water supply rate. Another reason is that the water supply flows in pressurized pipes that are running full, while drainage flows by gravity with pipes running only partially full. Table 9-4 lists some fixture unit values.

Example 9-1. Size all branches and interconnecting sections for the sanitary system in the three-story building shown in Figure 9-8. The vertical pipes are stacks, which collect wastes from fixtures installed on upper floors.

Solution 9-1. Figure 9-8 shows a horizontal piping plan (branches) in the basement, which collects the waste from six vertical stacks. The fixture connections and their location above the basement is not shown, only listed. Sizing the pipes is done based on tables. Start this problem at the point that is farthest away from the sewer line exit point. The fixture units can be obtained from Table 9-4, horizontal branch sizes from Table 9-5, maximum number of fixture units from Table 9-6, and vertical pipe sizes for stacks from Table 9-7.

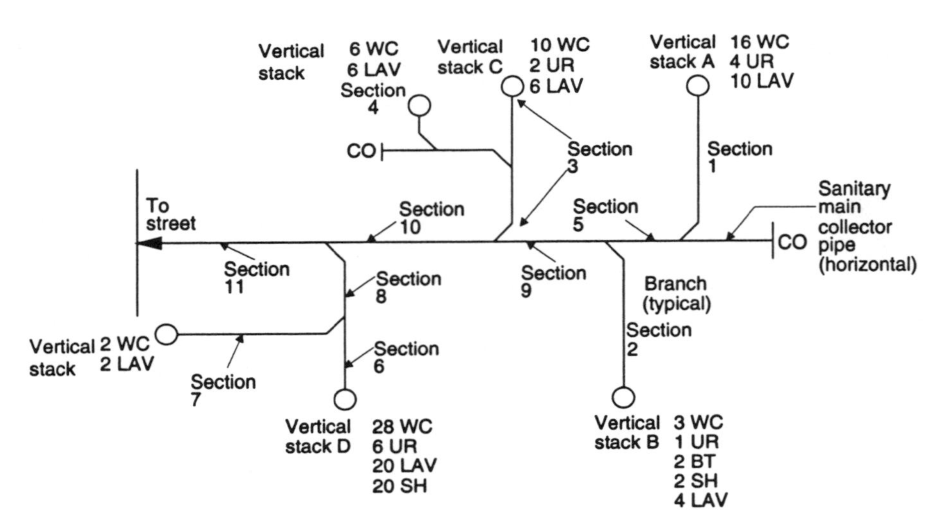

Figure 9-8. Sanitary Drainage Plan

Fixture or group	Pipe size (in.)	Fixture units
Residential:		
Automatic clothes washer, domestic	2	3
Bathroom group consisting of a water closet, lavatory, and bathtub or shower stall:		
Flushometer valve closet		8
Tank-type closet		6
Bathtub (with or without overhead shower)	1½	2
Bidet	1¼	2
Dishwasher, domestic	1½	2
Floor drain	2	3
Food-waste grinder, domestic	1½	2
Kitchen sink, domestic	1½	2
Kitchen sink, domestic, with dishwasher and food-waste grinder	2	2
Kitchen sink and wash (laundry) tray with single 1-½ in. trap	1½	2
Lavatory, common (small)	1¼	1
Lavatory, common (large)	1¼	2
Laundry tray (1 or 2 compartments)	1½	2
Shower stall, single head	2	2
Sink, bar, private	1½	1
Public toilet rooms (commercial, office):		
Urinal, pedestal, siphon-jet blowout	3	8
Urinal, stall, washout	2	4
Urinal, wall	1½	4
Water closet, flushometer valve	3	8
Industrial:		
Interceptors for grease, oil, solids, etc.	2	3
Interceptors for sand, auto wash, etc.	3	6
Lavatory, multiple type (wash fountain or wash sink)	1½	2
Showers, gang (one unit per head)	2	2

Table 9-4. Sanitary Drainage Fixture Unit Values (continued on next page)

Fixture or group	Pipe size (in.)	Fixture units
Commercial:		
Dishwasher, commercial	2	2
Food-waste grinder, commercial	2	3
Receptors (floor sinks), indirect waste receptors for refrigerators, coffee urn, water stations, etc.	1½	1
Receptors, indirect waste receptors for commercial sinks, dishwashers, airwashers, etc.	2	3
Sink, bar, commercial	1½	2
Sink, commercial, with food-waste grinder	2	3
Sink, commercial (pot, scullery, or similar type)	2	4
Sink (flushing-rim type, flush valve supplied)	3	6
Sink (service type with trap standard)	3	3
Washing machines, commercial	2	3
Medical:		
Dental unit or cuspidor	1¼	1
Dental lavatory	1¼	1
Lavatory (surgeon's, barber shop, beauty parlor)	1½	2
Sink (surgeon's)	1½	3
Miscellaneous:		
Drinking fountain	1¼	½
Mobile home park traps (one for each trailer)	3	6
Trap size 1-¼ in. or less	1¼	1
Trap size 1-½ in.	1¼	2
Trap size 2 in.	2	3
Trap size 2-½ in.	2¼	4
Trap size 3 in.	3	5
Trap size 4 in.	4	6

Table 9-4. Continued from previous page

Branch diameter (in.)	Total load FU
1½	3
2	6
2½	12
3	20*
4	160
5	360
6	620
8	1400

*No more than two water closets permited.

Table 9-5. Maximum Loads for Horizontal Sanitary Branches

Diameter of pipe (in.)	Pipe slope in./ft			
	1/16 in.	1/8 in.	1/4 in.	1/2 in.
2	-	-	21	26
2.5	-	-	24	31
3	-	20*	27*	36*
4	-	180	216	250
5	-	390	480	575
6	-	700	840	1000
8	1400	1600	1920	2300
10	2500	2900	3500	4200
12	3900	4600	5600	6700

*No more than two water closets.

Source: IAMPO (International Association of Plumbing and Mechanical Officials)

Table 9-6. Maximum Number of Fixture Units

Stack diameter (in.)	Stacks three stories or less in height FU	Stacks more than three stories in height FU	Total discharge into one branch interval FU
2	10	24	6
2 ½	20	42	9
3	30*	60*	16**
4	240	500	90
5	540	1100	200
6	960	1900	350
8	2200	3600	600
10	3800	5600	1000
12	6000	8400	1500

*No more than six water closets permitted.

**No more than two water closets permitted.

Table 9-7. Maximum Loads for Sanitary Stacks

Sizing Vertical Stack A and Horizontal Section 1

16 WC x 8 FU (Flushometer valve)	=	128 FU
4 UR x 4 FU (Wall hung)	=	16 FU
10 LAV x 2 FU (Large)	=	20 FU
Total	=	164 FU

A pipe size of 4" will work for vertical Stack A (per Table 9-7), because the building in question is three stories or less in height, and 4" pipe carries up to 240 FU.

For horizontal Section 1, the horizontal slope is 1/4" per ft. The pipe size needed is 4", because this size carries up to 216 FU (per Table 9-6).

Sizing Vertical Stack B and Horizontal Sections 2 and 9

3 WC x 8 FU	=	24 FU
1 UR x 4 FU	=	4 FU
2 BT x 2 FU (Residential)	=	4 FU
2 SH x 2 FU (Residential)	=	4 FU
4 LAV x 1 FU (Small)	=	4 FU
Total	=	40 FU

A pipe size of 4" will work for vertical Stack B (per Table 9-7), because the building in question is three stories or less in height, and 4" pipe carries up to 240 FU.

For horizontal Section 2, the horizontal slope is 1/4" per ft. The pipe size needed is 4", because this size carries up to 216 FU.

For horizontal Section 9, the horizontal slope is 1/4" per ft. The total FU discharge from vertical Stacks A and B is 204 FU (164 FU + 40 FU). Therefore, a pipe size of 4" can still be used here.

Sizing Vertical Stack C and Horizontal Sections 3, 4, and 10

10 WC x 8 FU	=	80 FU
2 UR x 4 FU	=	8 FU
6 LAV x 2 FU	=	12 FU
Total	=	100 FU

A pipe size of 4" will work for vertical Stack C (per Table 9-7), because the building in question is three stories or less in height, and 4" pipe carries up to 240 FU.

For horizontal Section 3, the horizontal slope is 1/4" per ft. The pipe size needed is 4", because this size carries up to 216 FU.

For horizontal Section 4, the horizontal slope is 1/4" per ft. The pipe size needed is 4", because this size carries up to 216 FU:

6 WC x 8	=	48 FU
6 LAV x 2	=	12 FU
Total	=	60 FU

Even though Section 3 is downstream of Section 4, 4" pipe can still be used. This is because the total fixture units from Sections 3 and 4 only add up to 160, and 4" inch pipe can carry up to 216 FU.

For horizontal Section 10, the horizontal slope is 1/4" per ft. To determine the total FU sum for Section 10, add Sections 9 and 3 together for a total of 364 FU (204 FU + 160 FU). The pipe size needed is 5", because this size can carry up to 480 FU. While 5" diameter pipe is available, it may be difficult to obtain in some areas of the country. For this reason the selection will be 6" pipe, which carries up to 840 FU (per Table 9-6).

Sizing Vertical Stack D and Horizontal Sections 6, 7, and 8

28 WC x 8 FU	=	224 FU
6 UR x 4 FU	=	24 FU
20 LAV x 1 FU	=	20 FU
20 SH x 2 FU	=	40 FU
Total	=	308 FU

A pipe size of 5" will work for vertical Stack D (per Table 9-7), because the building in question is three stories or less in height, and 5" pipe carries up to 540 FU (per Table 9-7). However, for the reasons outlined earlier, the selection will be a 6" pipe which accepts up to 960 FU.

For horizontal Section 6, the horizontal slope is 1/4" per ft. The pipe size needed is 5", because this size carries up to 480 FU. Again, however, 6" pipe will be used. Another reason why 6" pipe will be used is because the pipe size from the vertical to horizontal should not be decreased.

For horizontal Section 7, the horizontal slope is 1/4" per ft. It seems that a 2-1/2" pipe would work based on the total sum of 20 FU:

2 WC x 8 FU	=	16 FU
2 LAV x 2 FU	=	4 FU
Total	=	20 FU

However, there is a discharge pipe from the toilets, and whenever there are WCs discharging into a branch or stack, the minimum drain pipe size is 4" in diameter. Therefore, 4" pipe must be used in this case.

For horizontal Section 8, the horizontal slope is 1/4" per ft. To determine the total FU sum for Section 8, add Sections 6 and 7 together for a total of 328 FU (308 FU + 20 FU). The pipe size needed is 6", because pipe reduction is not acceptable.

Sizing Horizontal Section 11

Horizontal Section 11 is the sum of Sections 10 and 11, or 692 FU (364 FU + 328 FU). Both Sections 8 and 10 are served by a 6" pipe. Since a 6" pipe can accept up to 840 FU, Section 11 will also be a 6" pipe.

Alternative Solution to Example 9-1. Another possible way to size the system is to determine the drainage system flow using the FU (fixture units). Add these together and read the corresponding number of gpm from a curve very similar to the Hunter's curve. From there, it is possible to look at a table to select a pipe size to run half full.

When sizing sanitary drains, the design must consider that larger-than-recommended pipes will not improve the drainage. In fact, negative results might occur, because in larger pipe the velocity is lower, the water depth is less, and the depositing of solids will occur.

The drainage system in a high-rise building poses specific problems. Garbage disposals, cooking that involves a great deal of greasy food, and the use of dishwashing detergents impose a special load on the drainage stack. One basic recommendation is to isolate the first floor from the rest of the building; in other words, separate the drainage from the first floor. This will offset problems for the first floor since it is the base of the stack. If grease and other clogging materials are included in the normal drainage, more frequent rodding of the drainage pipe and an additional number of cleanouts are required.

CLEANOUTS

Cleanouts are provided in a drainage piping system to permit access for cleaning and opening stoppages within pipes. Horizontal pipes within a building must be provided with cleanouts every 50 feet.

A cleanout is really just a pipe plug that is threaded into a ferrule (a metal sleeve internally threaded that is installed at the end or as a Y-branch in a pipe). To serve its purpose of maintenance and elimination of stoppages in a sanitary sewer, a cleanout must be watertight and gastight. Figure 9-9 shows an end of pipe cleanout as well as a floor cleanout connected into a sewer pipe.

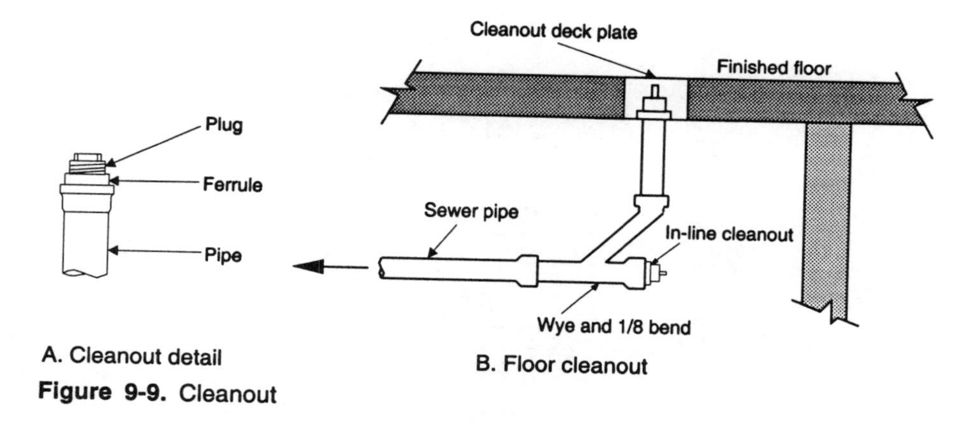

A. Cleanout detail

B. Floor cleanout

Figure 9-9. Cleanout

There are several rules that apply to location and position of cleanouts. A cleanout:

- must be installed if there is any change in direction from the straight line with an angle greater than 45°.
- must be installed if there is any change in direction with an angle of 90°.
- shall open in a direction opposite to the flow of liquid.
- for an interceptor shall be outside of the interceptor.
- shall be installed flush with the floor or under a cover, flush with the floor, and easily accessible.
- shall be the same size as the pipe it serves, up to 4" diameter. Pipes larger than 4" may use cleanouts of 4" diameter.
- must be an accessible manhole in an outdoor yard. Manholes must be installed every 100 ft on a straight line or at every 90° turn.
- must be provided inside the building at the sewer exit of the building.
- shall be installed at the base of each stack.

In summary, properly installed cleanouts and proper maintenance at regular intervals will ensure a long and adequate life of a drainage system.

Drainage Connections

To facilitate the understanding of drainage connections, Figure 9-10 illustrates a very simple plan for back-to-back toilets and lavatories, and Figure 9-11 shows the respective waste water riser diagram.

Figure 9-12 shows a waste and vent isometric diagram for a toilet room that includes a WC, LAV, and floor drain.

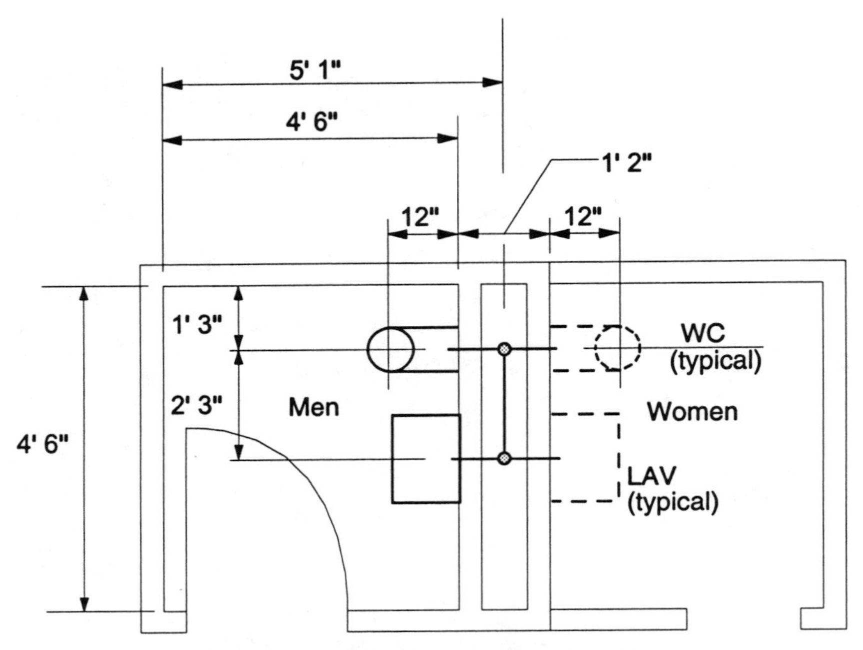

Figure 9-10. Drainage from Two Toilet Rooms

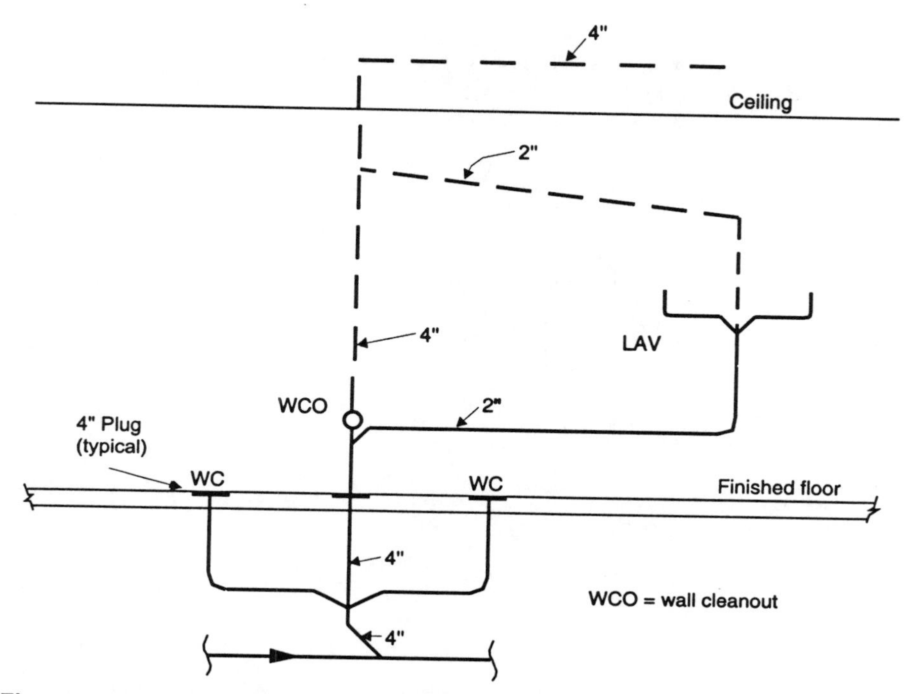

Figure 9-11. Drainage and Vent Riser Diagram

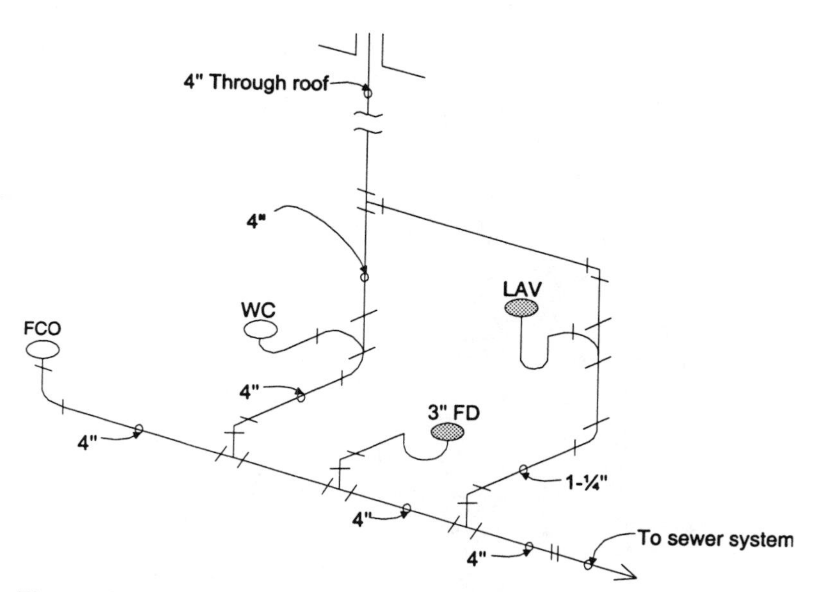

Figure 9-12. Waste and Vent Isometric Diagram

NOTES

[1] The hydraulic radius is the area of flow divided by the wetted perimeter. It can also be expressed as $r = a \div p$.

Chapter Ten

Vent System

The vent system is an integral part of the plumbing/drainage system. Its application was first proposed at a conference of plumbing masters and journeymen, which took place in New York City in 1874. At this conference, the vent pipe theory was presented as a way to protect the water seal in the (plumbing) fixture's trap. This theory was tested, proven correct, and quickly spread throughout the U.S. as well as abroad.

Vent pipes equalize pressure on both sides of the trap. The trap inlet is normally open to atmospheric pressure. Connecting the trap outlet with a pipe reaching up through the roof allows the atmospheric pressure to act downstream of the trap. This arrangement produces an equilibrium of pressure on both sides of the seal and consequently prevents siphonage.

The primary function of the water seal in the trap is to prevent the return of foul sewer odor into the fixture's area location. While vent pipes prevent the loss of water seal through siphonage, they also help sewer gases travel upward to be exhausted above the roof level. Since sewer gases are warmer than their surroundings, they rise naturally. The vent system is sized to maintain or limit the air pressure variation in all fixture drains (not to exceed 1-in. water gauge).

The goal of the vent system is to provide sufficient air passage in order to disrupt any siphonage action that might occur at the trap. The vent system also helps circulate air, which keeps anaerobic bacteria from growing and producing slime. If slime (which is produced at a low rate) is not scoured by the water flow, it might block the liquid flow through the pipe.

Vent Stacks

The terminals of vent stacks on the roof should not be located closer than ten feet from any door, window, or air intake or less than two feet above the same. If the roof has other uses (e.g., habitable penthouse), the vent pipe should be installed five feet above the roof level of the penthouse. The top of the vent pipe should be turned down (gooseneck) with a bird screen protecting its end. However, most vent ends are unprotected and are installed straight up from the roof.

To be effective, the vent pipe at the fixture must not be installed any closer than two pipe diameters from the trap crown but not farther away than shown in Table 10-1. The two reasons for not installing the vent too close are: 1) if the vent is too close to the trap, the seal tends to evaporate; and 2) slime grows much faster and tends to clog the vent pipe.

Diameter drain pipe (in.)	Distance of vent from trap crown (ft)
1¼	2½
1½	3½
2	5
3	6
4	10

Table 10-1. Relationship of Drain Pipe and Vent

If the distance between stacks conforms with a maximum acceptable length, (15-20 ft) stack vents and waste vents may be connected into a common header extended through the roof. This arrangement is helpful if only one roof penetration is allowed. Table 10-2 illustrates recommended vent sizes and lengths.

Vent Sizing

In general, the minimum size for any vent pipe is 1-1/4". To determine the correct vent pipe size, multiply the pipe diameter to which it is connected by 0.5. An exception to this rule is a WC vent, which may be a minimum of 1-1/2" for a 4" waste pipe. A vent that is the same size as the waste pipe will be no problem, but it will be more expensive because of a larger-than-normal pipe diameter.

Size of soil or waste stack (in.)	Fixture units connected	Diameter of vent required (in.)								
		1¼	1½	2	2½	3	4	5	6	8
		Maximum length of vent (ft)								
1¼	2	30	-	-	-	-	-	-	-	-
1½	8	50	150	-	-	-	-	-	-	-
2	10	30	100	-	-	-	-	-	-	-
2	12	30	75	200	-	-	-	-	-	-
2	20	26	50	150	-	-	-	-	-	-
2½	42	-	30	100	300	-	-	-	-	-
3	10	-	30	100	200	600	-	-	-	-
3	30	-	-	60	200	500	-	-	-	-
3	60	-	-	50	80	400	-	-	-	-
4	100	-	-	35	100	260	1000	-	-	-
4	200	-	-	30	90	250	900	-	-	-
4	500	-	-	20	70	180	700	-	-	-
5	200	-	-	-	35	80	350	1000	-	-
5	500	-	-	-	30	70	300	900	-	-
5	1100	-	-	-	20	50	200	700	-	-
6	350	-	-	-	25	50	200	400	1300	-
6	620	-	-	-	15	30	125	300	1100	-
6	960	-	-	-	-	24	100	250	1000	-
6	1900	-	-	-	-	20	70	200	700	-
8	600	-	-	-	-	-	50	150	500	1300
8	1400	-	-	-	-	-	40	100	400	1200
8	2200	-	-	-	-	-	30	80	350	1100
8	3600	-	-	-	-	-	25	60	250	800
10	1000	-	-	-	-	-	-	75	125	1000
10	2500	-	-	-	-	-	-	50	100	500
10	3800	-	-	-	-	-	-	30	80	350
10	5600	-	-	-	-	-	-	25	60	250

Note: Twenty percent of the total shown may be installed in a horizontal position.

Table 10-2. Recommended Size and Length of Vents

To calculate the maximum allowable length of a vent pipe, the following formula is used:

$$L = \frac{2226d^5}{(f)(g)}$$

where:
- 2226 = experimental coefficient
- L = length of vent pipe (ft)
- d = pipe diameter in inches
- f = gas friction coefficient
- g = gas flow in cfm (cubic feet per minute)

Taking into consideration all of the factors included in this formula, it lends itself to the following result:

L = 2/3 of the equivalent length of the basic circuit

This is the maximum length of the longest vent leg. If the length of one leg is longer, two or more vent pipes must be provided through the roof. For example:

Basic circuit = 200 ft
Equivalent length using 50% fittings
200 x 1.5 = 300 ft equivalent length

Maximum admissible vent length is 2/3 x 300 = 200 feet.

Figure 10-1 illustrates two waste-vent riser diagrams and is included for the reader's convenience.

DEFINITIONS

The following are definitions of different types of vents. These definitions might vary based an on individual designer's method and experience.

Circuit Vent - A branch vent that serves two or more traps, Figure 10-2. It extends from the last fixture connection to a vent stack. It is usually allowed for venting floor-mounted fixtures such as WCs, shower stalls, urinals, and floor drains. The vent must serve between two and eight fixtures connected to a branch. The vent is located between the two most remote fixtures.

Common Vent - A vent that connects two or more fixtures with the single vent line serving those fixtures, Figure 10-3.

Continuous Vent - A vertical vent that is a continuation of the waste vent to which it is connected, Figure 10-4.

Individual Vent or Revent - A vent that connects directly to only one fixture and extends to either a branch vent or stack, Figure 10-5.

Loop Vent - A vent branch pipe that serves two or more traps and extends in front of the last fixture connection to a stack vent, Figure 10-6.

Main Vent - A principal building vent that remains undiminished in size from the connection with the drainage system to its terminal.

Relief Vent - An auxiliary vent that connects the vent stack to the soil or waste stack in multi-story buildings and equalizes pressure between them. Such a connection will occur at offsets and at set vertical intervals determined by code.

Side Vent - A vent pipe that connects to a drain pipe through a fitting with an angle that is vertically no less than 45°, Figure 10-7.

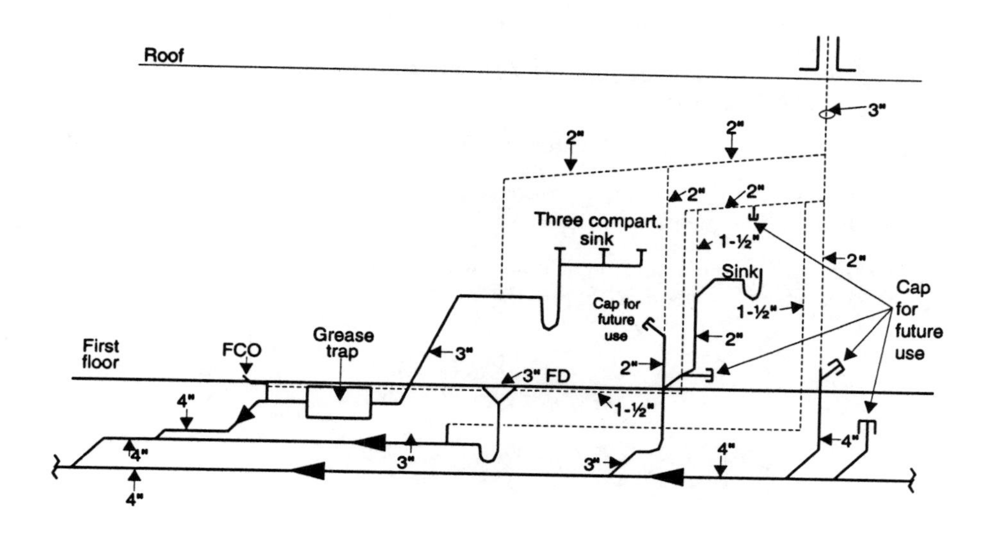

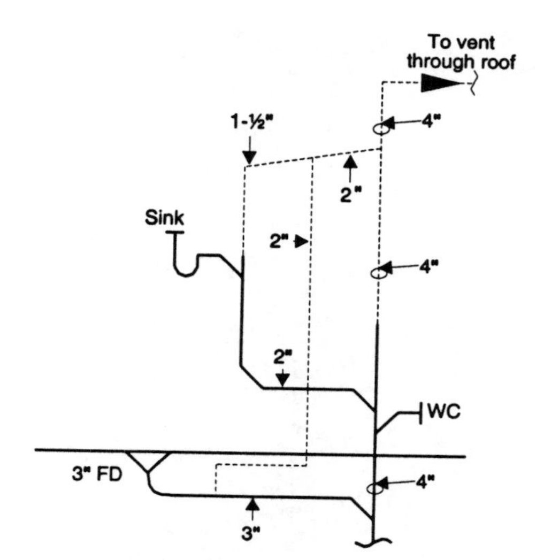

Figure 10-1. Sanitary Riser Diagrams

Stack Vent - A vent pipe that is an extension of a soil or waste stack. It is above the highest horizontal connection to that stack. It is also the name of a venting method using the stack vent for a branch vent connection.

Suds Venting - A method of venting where there is a suds pressure zone.

Trap Arm - The portion of one drain pipe between the trap and the vent.

Vent Stack - A vertical pipe that extends one or more stories and terminates in the outside air. It allows air circulation into and out of the drainage system.

Vent Terminal - The open air location where the end of the vent stack is placed.

Wet Vent - A vent into which a fixture other than a water closet can discharge. Figure 10-8 shows positions A and B.

Yoke Vent - A vent pipe connected to a soil or waste stack that continues upward to the vent stack connection, Figure 10-9.

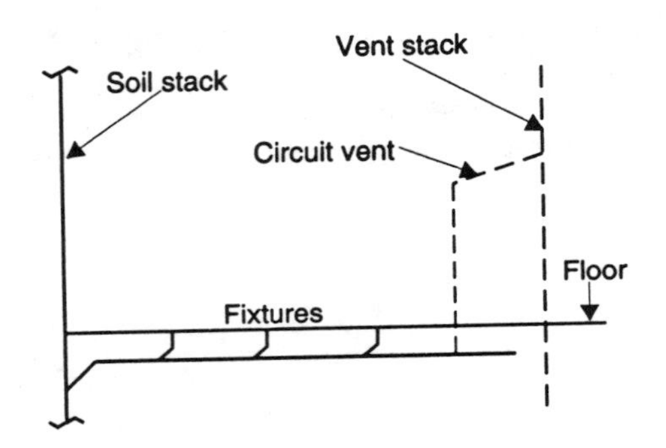

Figure 10-2. Circuit Vent

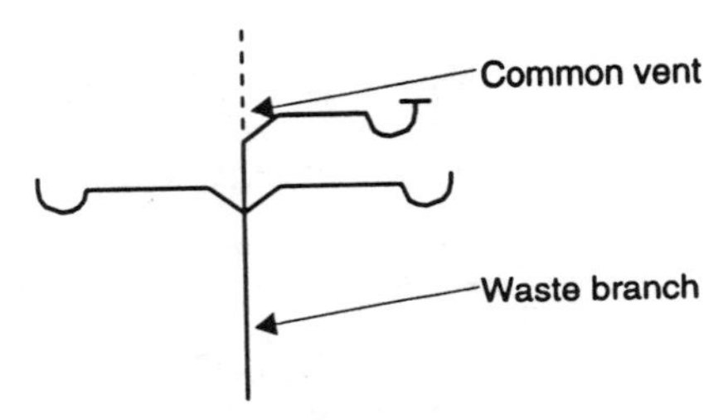

Figure 10-3. Common Vent

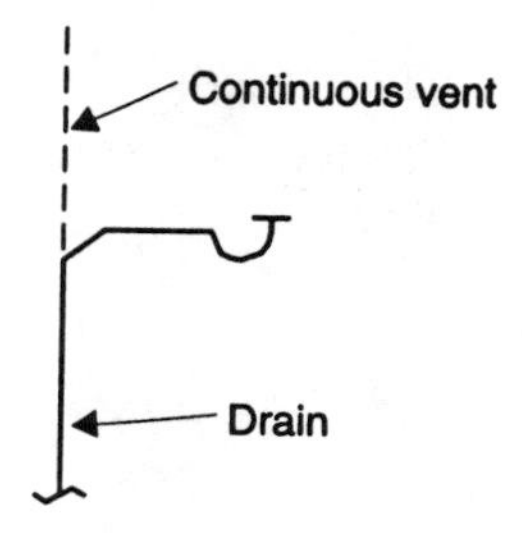

Figure 10-4. Continuous Vent

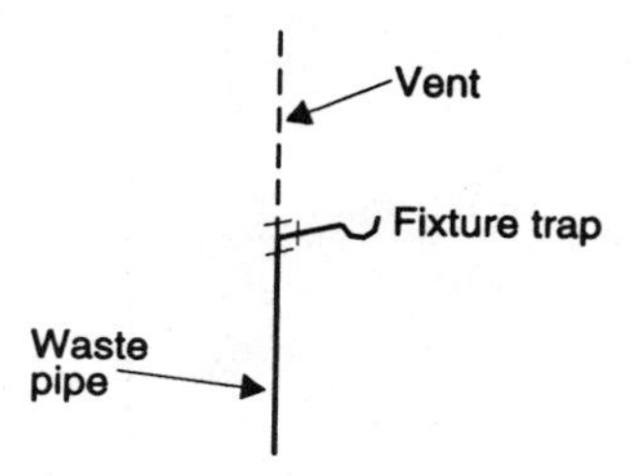

Figure 10-5. Individual Vent

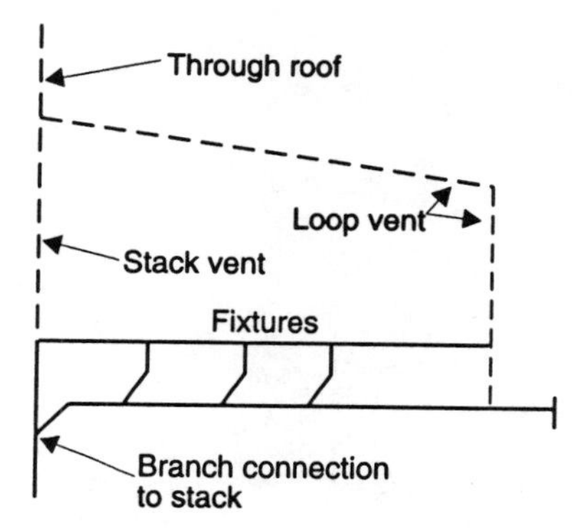

Figure 10-6. Loop Vent

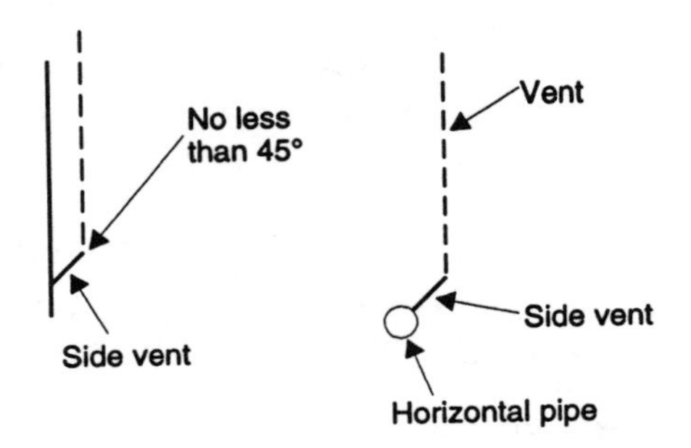

Figure 10-7. Side Vent

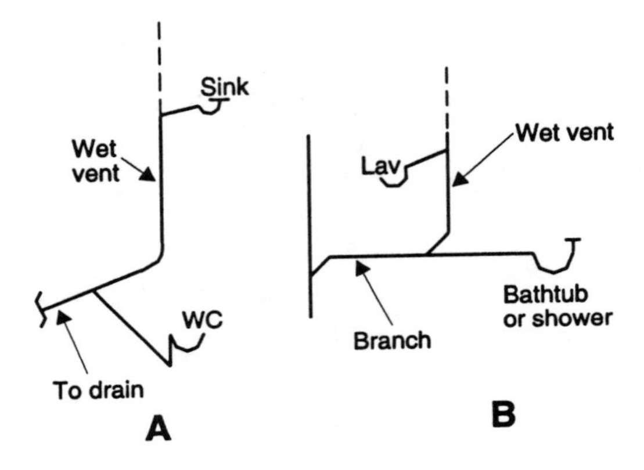

Figure 10-8. Wet Vent

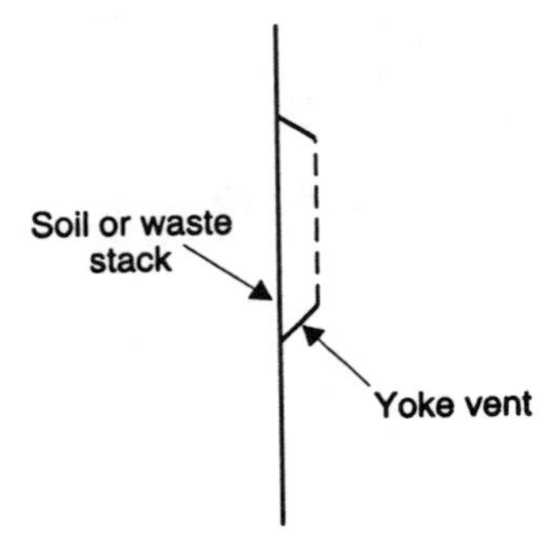

Figure 10-9. Yoke Vent

Chapter Eleven

Private Sewage Treatment Systems

A private sewage treatment system is used when there is no municipal sewer available. It usually serves a limited number of people. The private sewage treatment system includes a septic tank and a leaching area. The septic tank, which is the principal piece of equipment in this system, is a holding tank equipped with baffles. These baffles guide the slow-moving liquid to deposit the solids. The deposited matter is biodegraded with the help of bacteria. A simple schematic, septic tank-leaching area arrangement system is shown in Figure 11-1.

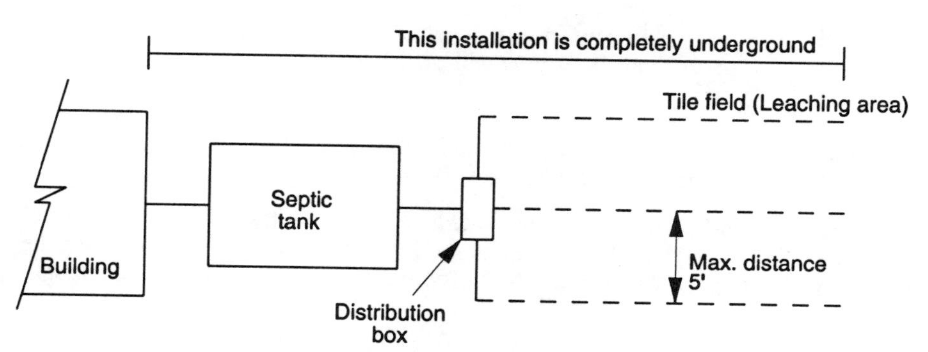

Figure 11-1. Septic Tank - Leaching Area Arrangement

Septic tanks may be constructed of the following materials:

- Poured concrete
- Precast concrete
- Cement blocks
- Steel-lined with cement
- Coated steel

LEACHING AREA

Downstream from a septic tank is a distribution box followed by a leaching area, usually a tile field. To determine the site for the tile field, percolation tests are required. These tests will determine how quickly liquid (poured into a small test hole) will seep into the ground after saturation. To install a leaching area in a particular location, tests must show that water percolates into water-saturated ground in a predetermined amount of time. The ground under the leaching area acts as a natural filter. When the liquid percolates through the ground and reaches the water table below, it is considered *clean water*.

If percolation does not occur within the prescribed time limit and the spot selected for the field cannot be changed, it is still possible to use it as a leaching area. However, in this case, a man-made filter, usually a sand bed, must be installed under the tiles. This type of installation increases the total cost, but it will make the installation feasible.

The calculation for the percolation test is based on the following formula:

$$C = t + \frac{6.24}{29}$$

where: t = time in minutes required for the water level to fall 1" after saturation

C = percolation coefficient measured in sq ft/gal

6.24 and 29 = practical coefficients

For the area to be acceptable as a tile field, the result of this equation should be 0.285 after 2 minutes and 1.25 after 30 minutes.

The tile field is not made up of ceramic tiles. Instead, the tiles are a special type of perforated pipes. These pipes are laid with perforations face down on the ground and are installed with a small distance (1" to 2") between each length of pipe, Figure 11-2.

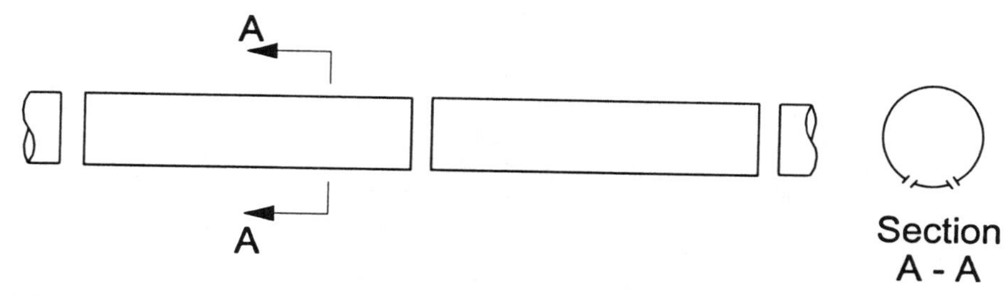

Figure 11-2. Tile Arrangement

The following must be considered when installing the leaching area:

- Locate field at a minimum of 100 feet away from any ground or underground water supply.
- Locate field at a minimum of 10 feet away from any structure.
- Area should have a slight downstream slope.
- Tile field width is recommended not to exceed 60 ft (100 ft maximum in special circumstances).
- Tiles shall be sloped 2" to 4" per 100 ft, and the field slope shall not exceed 6"/100 ft.
- Clean graded gravel, broken bricks, or washed rocks should be installed around the tiles.
- Entire field must be installed below the frost line or be covered with biodegradable material (hay) for protection against freezing.
- Tiles must be installed with a space between them.
- Distribution box (located downstream of the septic tank and ahead of the tile field) must have all outlets at the same elevation.

For information, nonmetallic, perforated tile pipes are 12-ft long and have a standard diameter of 4". Tiles may be clay bell, and spigot sewer pipes may each be 2- to 3-ft long.

A dosing tank must be installed downstream of the septic tank and ahead of the distribution box in case the tile field is not large enough to accommodate the inflow. The dosing tank must have a capacity of 70% of the tile field. This tank helps store the overflow before slowly distributing it to the tile field.

A tile field is not the only method available for a leaching area. Another method involves allowing the liquid waste to seep directly into the ground along some specially prepared and covered trenches with dividers, Figure 11-3. This system is less expensive to install, but the seepage into the ground is less uniform.

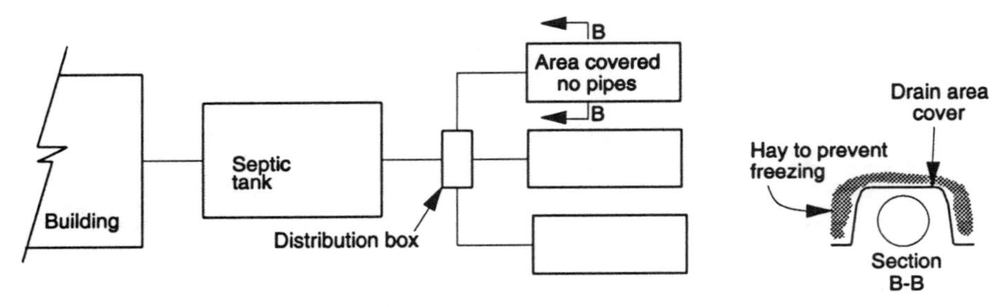

Figure 11-3. Septic Tank and Leaching Area

MAINTENANCE

A well-maintained, septic tank drainage system has an expected operating life of 20 to 30 years. There are certain maintenance procedures and observations that will help ensure good system operation. For example:

- Septic tank must be pumped (emptied) by specialists at the rate recommended by the system manufacturer or the installing contractor.
- Leaching area must be kept free of bushes and trees.
- No structure (e.g., decks) must be built above the leaching area.
- Effluent sent down the drain should not contain:
  (a) cooking oil and/or fats
  (b) food scraps
  (c) plastic material or tin foil
  (d) non-biodegradable objects such as cigarette butts or tampons
  (e) concentrated or aggressive chemicals (they may kill the needed bacteria growing in the tank)
- Use non-colored toilet paper.
- Keep the conveying tank supply pipes conventionally clean (rod through once a year).
- Conserve water, as this will help to prevent the septic tank and the system from overloading.

Chapter Twelve

Inspection and Maintenance

After a plumbing installation is completed by a contractor, the owner's representative must visit the site and perform an installation inspection **before the system is covered**. The inspection ensures everything is installed in accordance with the project documents, which consist of the drawings and specifications. Based on the uncovered system, the inspector must then prepare a list of deficiencies, called a *punch list*, which must be corrected by the contractor/installer. A punch list consists of the following inspection details and the deficiencies encountered:

- Is the system properly installed?
- What is the quality of the workmanship, piping, fittings, fixtures, and equipment?
- Are there any leaks?
- Does the equipment function correctly?
- Is the hot water temperature set per specification?
- Is the relief valve functioning?
- Is the pipe slope correct?
- Are the pipe sizes, insulation, cleanouts, interceptors, and accessibility correct?

The initial inspection includes checking the operation of each fixture and making sure it is solidly installed. The first inspection is usually scheduled when the system is uncovered, and a second inspection is performed when the installation is completed. The initial inspection is necessary, because a completed plumbing system is not visible in its entirety. Pipes, certain valves, fittings, connections, etc., are normally hidden in pipe chases behind fixtures. Pipe chases are usually designed as part of a room's walls or partitions, and they must be provided with access doors at cleanouts, valve locations, gauges, thermometers, etc., for inspection, adjustment, or maintenance.

MAINTENANCE

After the plumbing system is installed, checked again, and final corrections are made, the system is taken over by the owner. At this point, the maintenance of the system begins.

The equipment reliability depends on adequate and sustained maintenance. Without proper care, the equipment is bound to fail at some point. To perform the required maintenance, the employed mechanics must be familiar with the installed equipment. This, in turn, requires manufacturer installation and maintenance instructions to be available at the site.

A definition of scheduled maintenance is given by the National Fire Protection Association (NFPA) Standard No. 10 and is definitely applicable to the plumbing system as well. The standard states: "A thorough system check, intended to give maximum assurance that the system will operate effectively and safely. It includes any necessary repair or replacement of components."

In spite of the fact that the plumbing system is not as complicated as other building services, a building would be rendered inoperable if the water supply or sanitary system were interrupted. Due to a lack of adequate maintenance, such a situation could occur.

Maintenance activities can be divided into several categories:

- Inspecting
- Testing
- Cleaning
- Preventive maintenance
- Repair and replacement

Inspecting

Inspection schedules are usually based on manufacturer recommendations for the particular equipment and must be conducted to identify early warning signs of failure. It is important that the maintenance instructions for various pieces of equipment are integrated in the general maintenance area.

The maintenance inspection must include any exposed parts, piping, valves, backflow preventers, hangers and supports, pumps, fixtures, etc., which must be observed for leaks, discoloration, rust, or incorrect operating position. The functional equipment should be checked for operation within the manufacturer's range.

Testing

All equipment testing must include performance and safety checks. Testing must be performed periodically as determined by the person in charge and based on practical experience and/or manufacturer recommendations. This ensures the equipment meets specification requirements.

Cleaning

A regular, supervised cleaning program is required and should include all plumbing fixtures installed throughout the premises of the respective commercial, industrial, or institutional facility. Maintenance personnel should perform basic janitorial duties within each building, including cleaning floors, washing windows, removing refuse, and servicing rest rooms.

Unclean fixtures may contaminate users and spread disease. The market is overflowing with good quality cleaning products of a specific nature, such as WC bowl cleaner, tile cleaner, etc. Fixtures are smooth and easy to clean, so regular cleaning gives good results. Cleaning **should be done on a regular basis** and based on a precise schedule with the right materials and tools.

Preventive Maintenance

All equipment and drain lines must be scheduled for preventive maintenance. Regular inspection and a scheduled preventive maintenance program must be implemented. Preventive maintenance should prevent unplanned and undesired breakdowns. At the opposite end of the spectrum is breakdown maintenance, which is nothing more than running the equipment until it breaks.

Replacing washers in dripping faucets is part of an educated and intelligent way to save money and contribute to an overall water management program. Washers, which are made from rubber or special man-made synthetic materials, age with use. Gaskets and washers crack and break and must be changed at scheduled intervals with correct replacement parts and the right tools.

An important part of safety maintenance is the operation of the pressure and temperature relief valve on water heaters. As soon as a valve shows signs of leakage or being stuck, it requires immediate replacement.

Repair and Replacement

As the installation ages, there will be a need at some point for both the repair and replacement of equipment. It will be necessary to store and maintain spare parts for interim repairs. In certain cases, old and inefficient equipment might be scheduled to be replaced with a more economical, new, and better type.

Often unseen and undetected, corrosion in piping and fittings might occur. When any sign of corrosion becomes apparent, immediate action is required. For example, vent pipes become clogged, traps are siphoned out, and foul odors return to the fixture area.

When occupants in residential buildings, multi-purpose high rises, commercial complexes, or institutional/industrial establishments complain about fixture problems, such as leaky faucets, stoppages, broken parts, etc., they must be serviced promptly. The maintenance personnel team is usually composed of qualified, well-trained people who can accomplish this kind of work quickly and efficiently.

All fixtures must be provided with accessible shut-off valves on the water supply lines (cold and hot) to allow for quick and efficient repairs. If fixtures do not have these devices, the person in charge must schedule installation of such valves.

Piping materials for water supply and vents in plumbing systems may be galvanized steel pipes, copper tubing, or plastic pipes. Each type is different, and the details of their repair and/or replacement follow.

Galvanized Steel Pipe — Galvanized steel pipe is strong, durable, and relatively inexpensive. It is usually available in standard 21-ft lengths, but it is also available in a large variety of precut, prethreaded sizes or may be cut to order. If pipe is cut to order, it can be threaded at the same time.

When connecting pipes, fittings, or valves, apply joint-sealing compound or plastic joint-sealing tape to the outside threads. Tighten joints with a wrench to avoid straining in the rest of the pipes. Do not overtighten joints, and never screw a pipe more than one turn after the last thread is no longer visible.

If a break occurs in a long stretch of pipe, money can be saved by replacing only the damaged section. Cut out the damaged piece, then connect a new pipe section with a coupling and a union fitting.

Copper Tubing — Copper pipe or tubing is lighter and more resistant to corrosion than steel pipe. However, the cost is nearly twice as much per foot as galvanized steel pipe. There are two kinds of copper tubing: flexible (soft temper) and rigid.

Copper tubing is connected only by soldering. If soldering a connection to a fitting that already has other tubing soldered to it, wrap the finished joints with wet rags to protect them from heat. ***Note: Before using a torch, drain all water from pipes.***

Flexible copper tubing will accept only compression fittings or flare fittings. To make a compression fitting, slide on the nut and then tighten the nut to the fitting. The soft metal of the compression ring will be pressed against the fitting and pipe to form a watertight, long-lasting seal.

The flexible type copper pipes can usually be shaped by hand. However, a special bending tool is available from plumbing material suppliers.

Plastic Pipe — Plastic pipe is light, inexpensive, easy to work with, and corrosion resistant. As mentioned before, it has two disadvantages: 1) it expands more than metallic pipes when hot water is flowing; and 2) it is less resistant to applied heat.

Plastic piping is available in either a flexible or rigid form. There are three types of rigid plastic pipes: PVC, ABS, and CPVC. PVC and ABS are suitable for various cold water applications. CPVC, which is rated at 100 psi at 180°F, is recommended for hot water.

Rigid plastic pipes are connected by a solvent cement. The flexible types are joined with insert fittings. Adapters are available for connecting plastic pipe to metal pipe. **Do not mix different types of plastics in the same system.**

Before installing a hot water pipe, check the relief valve setting of the water heater. It should be maintained at or below 100 psi at a maximum 180°F. Connect the pipe to the heater with a heat-dissipating fitting.

Other Maintenance

After years of service, sewer pipes become clogged from sewage containing grease, dirt, organic/inorganic material, etc., which form undesirable deposits in the pipes. To keep from reaching a crisis point, the waste lines should be cleaned per a scheduled maintenance program. When preventive maintenance is **not** performed, repair becomes necessary due to stoppage or breakage. The repair should be done correctly, using new materials, the right tools and parts, and by personnel specially trained for the kind of work needed.

When discussing this problem, a case comes to mind concerning an apartment building that consisted of three apartments. The building had three levels: basement, first, and second floors.

The plumbing fixtures in the basement apartment were located at the same elevation as the combined (storm and sanitary) street sewer. To prevent backflow during heavy rains, the sewer line was provided with its own accessible clapper (check valve).

The occupants on the second floor smoked cigarettes and had the bad habit of throwing the cigarette butts in the toilet. The butts that did not disintegrate got stuck in the clapper, which then did not close when it rained. This produced sewage backflow and overflowed the basement toilet. A simple sign above the toilet that read, "Do not throw cigarette butts in the toilet" solved the problem.

If a fixture is cracked, it must be scheduled for replacement. Improved materials come on the market constantly, and if they prove to be better, they should be used. For example, Teflon tape is now used to makes tight connections between screwed pipes or pipes and fittings. Another example is the 1.6 gallon per flush toilet and the 2 gpm showerhead. Examples such as these are numerous.

Maintenance Frequency

Here are a few items that should be included in regular maintenance programs and their recommended frequencies:

- Once every two months open and close each fixture's stop valve twice. Replace or repair the leaky ones.
- Once every six months purge the water heater.
- Every month operate each safety valve within the system (mainly on the water heaters) to make sure they are operating properly. Replace the defective ones.
- At least once a month check all gauges in the system. Make adjustments as needed.

It must be emphasized that in various parts of the country, due to the water quality variations (hard water, salty water, water containing fine grit or sand, etc.), the maintenance schedule may vary. Plans have to be adopted based on local conditions.

Part of the plumbing maintenance program should include periodic cleaning of roof drains. Leaves and other light debris carried by the wind land on the roof and soon accumulate at the roof drains, blocking them. Cleaning the drains in the fall (mainly in the northern part of the U.S.), twice a month during September through December, and once a month during the rest of the year, will ensure normal storm water drainage.

In commercial and industrial buildings, heating, ventilating, and air conditioning (HVAC) equipment drains should be included in the regular maintenance program. Condensate water drips and drains away from the equipment on the roof or in mechanical rooms where the equipment is installed. The drains should always be open to the atmosphere (not connected directly to the sewer in order to avoid potential contamination).

Chapter Thirteen

Storm Drainage

The storm drainage consists of two independent and separate systems: the roof drainage and the yard drainage. The roof drainage is usually the area of activity for the plumbing technician, while civil personnel usually take care of the yard drainage.

RAINFALL

Prior to beginning storm drainage design, it is necessary to consider rainfall intensity (measured in inches/hr), frequency (measured in years), and duration (measured in minutes). All of this data is available free from the U.S. Weather Bureau.

Rainfall intensity indicates how many inches of water fell during a particular rainfall. If the storm lasted only 30 minutes and the accumulation was 1.6 inches, the rainfall intensity was 3.2 inches/hr (1.6 inches ÷ 0.5 hr).

Rainfall frequency is also called a return period. Based on statistics for a particular location, it indicates how many years have passed since a storm of a certain intensity has occurred. For example, assume that a 4-inch rainfall intensity happens approximately once in a 25-year interval. The return period is 25 years. The return period is given in years of 5, 10, 25, 50, and 100 years. The larger the return period, the greater the storm intensity. For each type of building, various specialized design manuals recommend the return period to be used in the design and calculation.

Rainfall (storm) duration is measured in minutes. This information is also based on statistics and indicates that a storm of a certain intensity lasted 5, 10, 15, 20, 30, or 60 minutes or more. Normally, the shorter the duration the heavier the intensity.

Appendix D shows a rainfall intensity map and a list of rainfall intensities for various cities in the U.S. This data used to be updated every ten years. Due to their general nature, rainfall intensity maps are not very accurate. For a particular location and project, rainfall data for storm drainage design should be obtained from the closest U.S. Weather Bureau. ***Note: the tables included in this book are for information only. Do not use them for an actual design.***

It is important to mention that since 1960, the rainfall pattern has been changing more rapidly. Wet patterns for certain areas drastically modify the existing statistical numbers. The Weather Bureau has available records for specific locations:

- State of Washington, May 1992: the driest in history
- State of California, February 1992: heaviest rains since 1938
- Southern U.S. to Georgia, Spring 1992: twice as much rain as usual
- State of Colorado: April 1992, dry and hot; June 1992, snow

Such unusual patterns have shown up in Mexico, South America, and Europe.

Roof Drainage Design

At the beginning of this century, roof drainage systems consisted of ornate metallic (copper, which oxidized over time and became green) or masonry outdoor perimeter gutters with metallic downspouts to help spill storm water on the ground, sidewalk, etc. At the point where the gutters changed direction from horizontal to vertical, collector boxes were installed. Simpler versions of this type are still used for residential homes. The big disadvantage of such a system is that in the northern part of the U.S., freezing problems develop. To prevent freezing, various types of relief systems are used with fair to good results.

The roof or deck drainage design must comply with the applicable plumbing code. However, pipe sizes that are larger than the minimum required by code may be used to accommodate unexpected downpours. This is considered a safer design. Code pipe sizes already have built-in safety factors, however, the design engineer usually tends to greatly oversize the storm drain pipes and increase the number of roof drains. These additional safety factors increase the installation costs. The recommendation is a minimum of two drains per roof and a 10% to 25% overdesign. Additional safety factors applied to a design will cost too much money.

One type of storm drainage system has the roofs or decks being constructed so that the water drains off as soon as it is collected. This means that the roof area must be sloped or pitched toward the drains. If the roof has a parapet around its perimeter, the roof drains may be smaller in size. An alternative to this design allows the water to pond on the roof and drain slower. This system is called a *controlled flow type*. To use such a design, the roof supports must be calculated for the additional water weight (62.3 lb/ft^3). Again, the design (structure) must be safe but not largely oversized.

To establish the amount of rainfall (in gallons) collected from a roof, it is necessary to calculate the roof's area. This area should include the total horizontal (projected) area plus 50% of any vertical wall area located above the roof. If the roof slope is steep, the roof area from which the water is collected must be increased based on the following information:

Slope (per ft)	Coefficient of roof area increase
Flat to 3"	1.0
4" to 5"	1.5
6" to 8"	1.10
9" to 11"	1.20
12"	1.30

The indoor leaders and downspouts must run down undiminished and be insulated to avoid condensation. For the design of the storm drainage piping system, water velocity in the horizontal pipes must be selected at 3 to 5 fps.

Commercial, high-rise, and industrial buildings collect storm water from the roof at low points where the roof drains are located. In turn, the roof drains are connected to indoor horizontal leaders and then to indoor vertical downspouts. Downstream, these are connected underground to a municipal storm drainage or to a combined sewer. If the building is not located in an area where there is a municipal storm sewer, then the storm water may be directed to a pond, river, or man-made drywell. Storm water may be discharged without treatment into a natural body of water; however, this must be discussed with the plumbing inspector and/or other authorities having jurisdiction.

The roof drainage system design should consider the following roof construction details:

Type of roof	Minimum slope (per ft)	Preferred slope (per ft)
Flat roofs with gravel surface	1/8"	1/4"
Concrete decks	1/8"	1/8"

Roof drains with gratings must be installed at a maximum of 2" below the finished roof level. Modern buildings have drains and indoor downspouts. The downspouts must not be installed in unheated areas.

Very steep roofs, which include the majority of suburban residential buildings, have outside perimeter gutters and outdoor downspouts leading to drywells. To prevent freeze-up, low intensity electric wires are installed in the vicinity of the gutters.

Figure 13-1 indicates half-round gutter sizes based on rainfall intensity and roof area. This chart may also be used for a building with a parapet and half-round gutters.

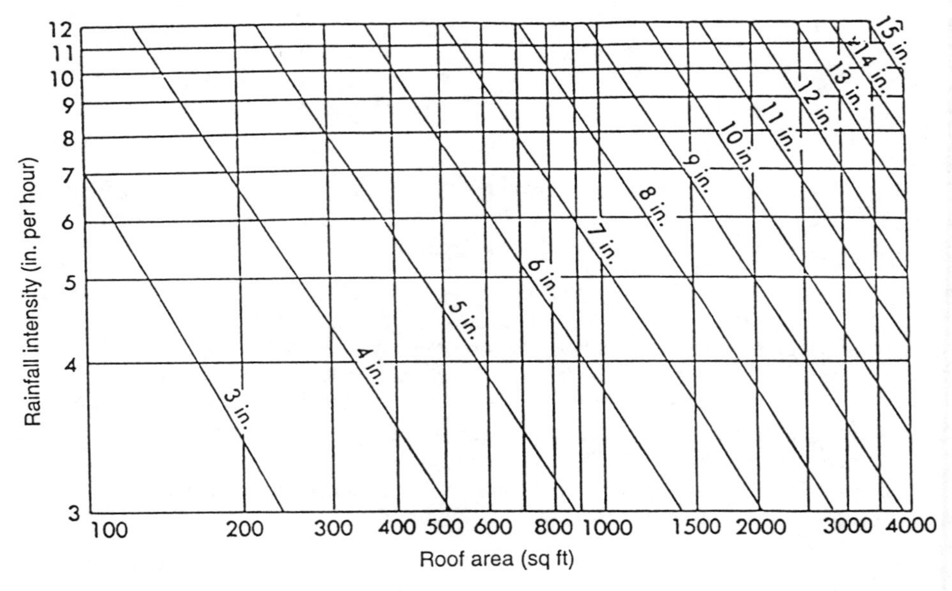

Figure 13-1. Half-Round Gutter Sizes

The minimum roof drain diameter is 3". Below are the elements needed to determine roof drain size:

- Area to be drained
- Rainfall intensity (rate) for the area
- Pitch or slope of the roof
- Nature of roof material

The recommended piping materials for storm drainage are yoloy (alloy mixture of steel and nickel); cast iron soil pipe; wrought iron; and carbon steel with wrought welding fittings or mitered elbows. Piping materials must also be acceptable to the local authority having jurisdiction. It is necessary to find out the local regulations before the design is developed. This is accomplished during a preliminary discussion with the authority having jurisdiction. More details about the storm drains and piping materials are included in the specification at the end of this book.

After the basic elements are established, including the roof's high and low points for the location of roof drains, it is possible to perform the roof drainage calculations and select the correct pipe size. The calculation includes the total amount of water (precipitation) expected to fall on a particular roof during a certain time period. The actual calculation is of a general nature; in practice, roof areas and pipe size tables from the plumbing code are used.

Roof Drains

The roof drain is the main fixture in a roof drainage system. To be acceptable, the roof drain must meet a number of criteria, including the following:

- Acceptable dome shape, preferably with low profile
- Non-rusting dome or flat grate material
- Effective debris penetration prevention
- Overflow drainage in case of debris build-up
- Gravel stop
- Seepage control channel

Depending on the application, the variety of roof drains available is very large and encompasses shape, discharge pipe position, material, size, and type of connection. Care should be exercised when making the selection so that it will be the correct fixture for the application. Visiting various projects to observe similar installations and discussing the operation of drains already installed may produce better results for new and/or future applications.

When connected to a combined sewer, storm drains must be trapped. No traps are required for a storm sewer connected into a storm system carrying only storm water.

For certain types of construction, the location and number of roof drains must be established in close cooperation with the architect and structural engineer. A roof drain must be placed at each low point or in each collecting valley. As a rule, all roofs must have a minimum of two roof drains to ensure proper drainage (in case one becomes clogged). A general recommendation is that the roof area assigned to one drain should not exceed 5000 sq ft. The size should be based on local conditions such as rainfall intensity. The applicable code governs the roof area, drain size, and all of the other design details.

The maximum distance between roof drains along the same valley should not exceed 50 ft. The maximum distance between two drains should not exceed 200 ft. Based on these dimensions, the roof area discharging into roof drains is a maximum of 10,000 sq ft (200 ft x 50 ft). This concurs with the minimum of two drains per roof. In practice, every roof drain should serve, as mentioned, a roof area of approximately 5000 sq ft. This includes a sizable safety factor. For a safer design, the calculation should consider smaller roof areas per drain and the areas allowable by code.

Example 13-1. Based on Figure 13-2, how many drains are required for this structure, and what size must they be? For this building's location, the rainfall intensity is 1.0 in./hr and rainfall return is 100 years.

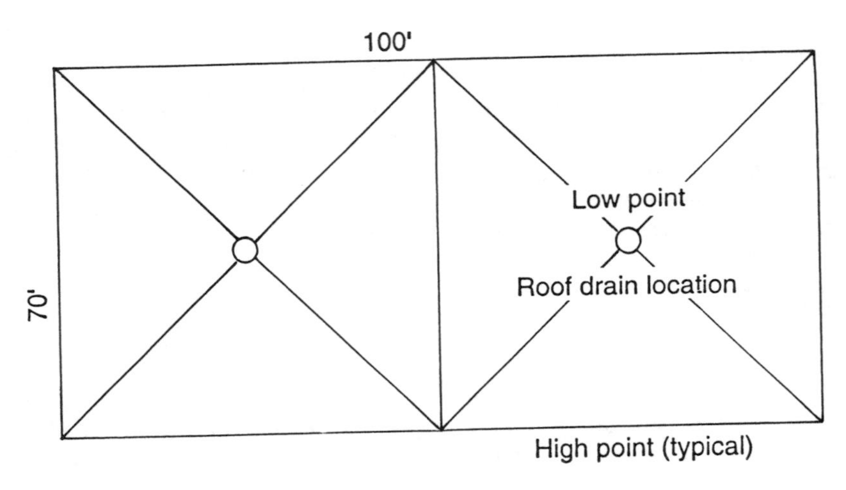

Figure 13-2. Sizing a Drain

Solution 13-1. Based on Figure 13-2, it is possible to determine that the roof area is 7000 sq ft (100' x 70').

The BOCA code states that for a 1.0 in./hr rainfall intensity and 100-year return, one 3" diameter roof drain installed in a horizontal roof at a slope of 1/8" will drain 3288 sq ft. For a 7000 sq ft roof area, two 3" drains would not be satisfactory (3288 sq ft x 2 = 6576 sq ft). There are two solutions: either select a slope of 1/4" per ft for the roof, which in this case can carry the water collected from 4640 sq ft to one 3" drain, or maintain the 1/8" per ft slope and provide two 4" drains. However, Table 13-1 shows more conservative figures, which are recommended for this design. For vertical leaders, sizes shown in Table 13-2 may be used.

Diameter of drain (in.)	Maximum projected roof area for drains of various slopes sq ft at: 1/8 in. slope	1/4 in. slope	1/2 in. slope
3	-	1160	1644
4	1880	2650	3760
5	3340	4720	6680
6	5350	7550	10,700
8	11,500	16,300	23,000
10	20,700	29,200	41,400
12	33,300	47,000	66,600
15	59,500	84,000	119,000

*Based on a maximum rainfall of 4 inches per hour.

Table 13-1. Size of Horizontal Storm Drains

Maximum projected roof area (ft^2)	Diameter of leaders (in.)	Maximum projected roof area (ft^2)	Diameter of leaders (in.)
720	2	4600	4
1300	2½	8650	5
2200	3	13,500	6
		29,000	8

Table 13-2. Size of Vertical Leaders

Some designers include a second *overflow* drainage in the roof drainage system. This safety storm system is installed at a slightly higher elevation (usually 6" to 8") than the normal system. This overflow usually discharges on the ground surface.

Yard Drainage Design

For the yard drainage design, *runoff*[1] must be determined. Storm water runoff is the portion of water that is not lost by infiltration into the soil or through evaporation. Due to many unknown variables, it is very difficult to come up with an accurate estimate of the amount of water that reaches the drainage system. All portions of the storm drainage system must be designed to handle the peak flow anticipated under design conditions.

Infiltration is the natural movement of water through and below the ground surface. The infiltration rate depends on the existing quantity of water in the pores and the filling of pores with surface soil or other materials. For such reasons, infiltration rates are usually determined from experimental data. For example, the infiltration rate in bare soils (under average summer conditions) and after being subjected to a rainfall that lasts one hour, is approximately 0.1 in./hr for clay soils and 1.0 in./hr for sandy soil. Infiltration rates after 12 hours of rainfall may be as little as 20% of the rate at the rainfall's start. Ground covered with grass increases the infiltration rate by up to 7%.

The yard drainage formula that yields the amount of water collected from an area is called the rational formula. This formula is:

$$Q = ciA$$

where:
- Q = flow in cfs (cubic ft per second)
- c = runoff coefficient
- i = rainfall intensity (in./hr)
- A = area to be drained (in acres)

As a reminder:

1 square mile = 640 acres
1 acre = 4840 sq yards = 43,560 sq ft
1 cfs = 448 gpm
gpm = gallons per minute

For Q measured in gpm the formula becomes:

$$Q = 0.0104\, ciA$$

where:
- A = area in sq ft
- 0.0104 = conversion factor
- c and i = same as above

The storm-water runoff coefficient (c) is the actual amount of water that reaches a channel or collecting manhole. It represents a percentage of the rainfall calculated for the respective area. This coefficient depends on the nature of the terrain (e.g., rough, cultivated, paved, sandy, etc.), evaporation, and terrain slope. From a paved area, there is more water collected in the storm sewer than from rough land, where a good portion of it infiltrates into the ground.

Practical and approximate runoff coefficients (c) are:

Pavement — 0.70 to 0.90
Grass — Average 0.30
Cultivated — 0.20 to 0.60
Roofs — 1.0

Another element of the yard drainage design includes manhole locations, which must be at:

- every change in the pipe direction.
- every change in the pipe size.
- every substantial (drop) in the pipe elevation.
- every major connection.
- every 300 ft maximum distance along a straight pipe line.
- a recommended maximum manhole depth of 15 ft.

The minimum metallic pipe size to be installed in the yard is 10" diameter. Larger pipes are installed as a result of calculations based on the area to be drained.

NOTES

[1] Runoff is the actual amount of water that reaches the drainage system. Ultimately, it is the base for the yard storm drainage sizing criteria.

References

American Society of Heating, Refrigerating and Air Conditioning Engineers, Inc. (ASHRAE), *Handbook 1987*, Chapter 54.

American Society of Plumbing Engineers (ASPE) Data Books:

- No. 3 - *Cold Water System* 1988
- No. 6 - *Fuel Gas Piping* 1988
- No. 8 - *Water Closet Types* 1991
- No. 13 - *Hangers and Supports* 1992
- No. 21 - *Formulas, Symbols and Terminology* 1989

Church, James C., P.E., *Practical Plumbing Design Guide*, McGraw-Hill, 1979.

Crocker, Sabin, M.E., *Piping Handbook*, McGraw-Hill, 1973.

Harris, Cyril M., PhD., *Handbook of Utilities for Buildings Planning, Design and Installation*, McGraw-Hill, 1990.

Heald, C.C. (editor), *Cameron Hydraulic Data*, Ingersoll-Rand Company, 1988.

Hornung, William J., *Plumbers and Pipe Fitters Handbook*, Prentice-Hall, Inc., 1984.

Neilsan, Louis S., P.E., *Standard Plumbing Engineering Design*, McGraw-Hill, 1981.

Plomberie, Michel Matana, *Canalisations • Sanitaires • Gaz • Eau Chaude • Bruits,* Syros - Alternatives, Paris, France.

Steele, Alfred, P.E., *Engineered Plumbing Design*, Plumbing, Heating, Piping, Chicago, IL, 1982.

Appendix A

Symbols and Abbreviations

Symbols are a graphic representation of a written description. They convey information and simplify the drawings. This appendix consists of the symbols and abbreviations commonly found on plumbing drawings. Some of the items shown here may not be represented by both a symbol and an abbreviation.

Name	Abbreviation	Symbol
LAVATORY	LAV	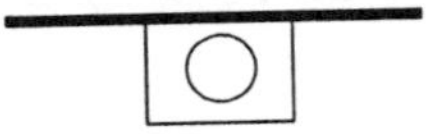
WATER COOLER, WALL MOUNTED	WCL	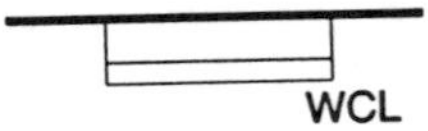 WCL
URINAL, WALL HUNG	UR	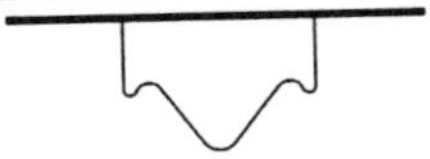
URINAL, STALL	UR	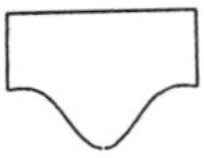
WATER CLOSET, FLOOR	WC	
WATER CLOSET, WALL HUNG	WC	
SERVICE SINK	SS	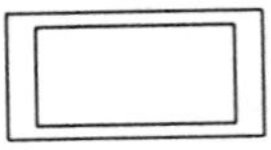
CIRCULAR WASH FOUNTAIN	WF	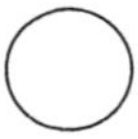
SEMI-CIRCULAR WASH FOUNTAIN	WF	
FLOOR DRAIN	FD	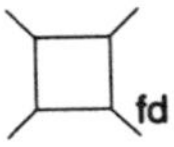 fd
ROOF DRAIN	RD	
SHOWER HEAD	SH HD	

Name	Abbreviation	Symbol
HOSE BIBB	HB	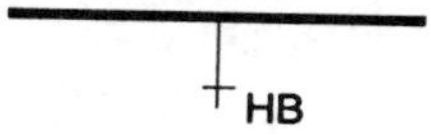 HB
MANHOLE (IDENTIFY BY NUMBER)	MH	
INLET BASIN (IDENTIFY BY NUMBER)	IB	
CATCH BASIN (IDENTIFY BY NUMBER)	CB	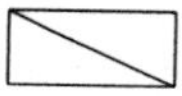
WALL HYDRANT OR SIAMESE		
LIMITS OF WORK (TRADE, CONTRACTS, ETC.)		
WALL CASTING		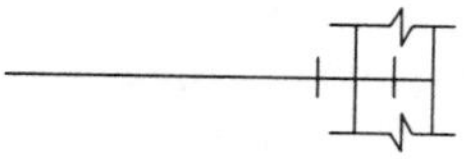
GATE VALVE	GV	
GLOBE VALVE	GL.V	
PLUG VALVE	PV	
CHECK VALVE	CK.V	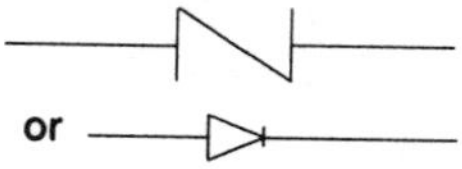 or
BUTTERFLY VALVE	BV	

Name	Abbreviation	Symbol
PRESSURE REDUCING VALVE	PRV	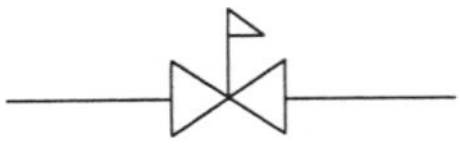
HOSE GATE	HG	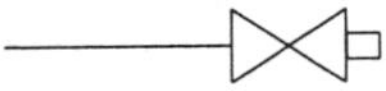
SHOCK ABSORBER	SA	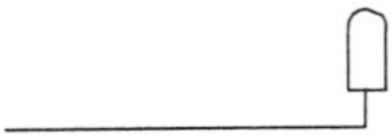
RELIEF OR SAFETY VALVE		
ANGLE GATE VALVE		
PLUG COCK		
THREE-WAY VALVE		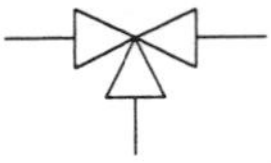
MOTOR OPERATED VALVE		 M
DIAPHRAGM OPERATED VALVE		
REDUCING VALVE (SELF ACTUATED)		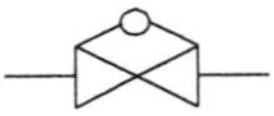
REDUCING VALVE (EXTERNAL PILOT CONNECTION)		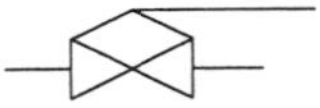
ECCENTRIC REDUCER		
CONCENTRIC REDUCER		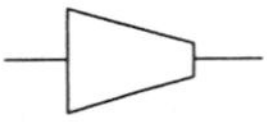

Name	Abbreviation	Symbol
STRAINER		
INDICATING THERMOMETER		
SOLENOID VALVE		
PRESSURE GAUGE		
POST INDICATOR GATE VALVE		
DIRECTION OF FLOW		
DOWNWARD PITCH		
UNION		
GAS COCK		
SERVICE WATER SYSTEM	SWS	—— SWS ——
GAS LINE	G	—— G ——
WASTE LINE	W	—— W ——
SOIL LINE	S	—— S ——

Name	Abbreviation	Symbol
RAIN WATER LEADER	RWL	——RWL——
ACID RESISTANT WASTE	AW	——AW——
CHEMICAL RESISTANT WASTE	CRW	——CRW——
DRAIN LINE	D	——D——
VENT LINE	V	-------V-------
ACID RESISTANT VENT	AV	------ AV------
SANITARY SEWER	SS	——SS——
INDIRECT WASTE	IW	——IW——
STORM DRAINAGE	SD	——SD——
INDIRECT VENT	IV	—— IV ——
TEMPERED WATER, POTABLE	T	—— T ——
FIRE LINE	F	—— F ——
COMPRESSED AIR LINE	A	—— A ——

Name	Abbreviation	Symbol
VACUUM CLEANING LINE	V	——— V ———
SUMP PUMP DISCHARGE	SPD	——— SPD ———
CLEANOUT	CO	
WALL CLEANOUT	WCO	———|WCO
FLOOR CLEANOUT	FCO	
COLD WATER	CW	——— - ——— - ——— - ———
HOT WATER	HW	——— - - ——— - - ——— - -

Name	Abbreviation
FLUSH VALVE	FV
AIR CHAMBER	AC
VACUUM CLEANING INLET	VCI
FLUSH OUT VALVE	FOV
DOWN SPOUT	DS
OPEN END DRAIN	OED
CONDENSATE DRAIN	CD
UNIT HEATER	UH
GAS WATER HEATER	GWH
ABOVE CEILING	ABC
AT CEILING	ATC
ABOVE FINISH FLOOR	AFF
UNDER FLOOR	UF

Name	Abbreviation
ACCESS PANEL	AP
CUBIC FEET PER HOUR	CFH
GENERAL CONTRACTOR	GC
PLUMBING CONTRACTOR	PC
DRAIN	DR
WATER HEATER AND NUMBER	WH-1
SUMP PUMP AND NUMBER	SP-1
COLD WATER RISER	CWR
HOT WATER RISER	HWR
VENT RISER	VR
VENT THROUGH ROOF	VTR
VENT STACK	VS

Appendix B

This appendix contains the following information:

Part One: Net Positive Suction Head (N.P.S.H.)

Part Two: Friction of Water

- Copper and Brass
- New Steel Pipe
- Asphalt-dipped Cast Iron

Part Three: Friction Losses in Pipe Fittings

Part Four: Corresponding Pressure Table

PART ONE: NET POSITIVE SUCTION HEAD

Pump Details

The Net Positive Suction Head is defined as the total suction head measured in feet of liquid (in absolute pressure value at the pump centerline or what is called the impeller eye) from which is deducted the absolute vapor pressure (measured also in feet of liquid) of the liquid being pumped. This value has to be positive.

Based on the suction type, there are two formulas:

1. Suction lift or the supply level being below the pump centerline.
   $NPSH = h_a - h_{vpa} - h_{st} - h_{fs}$
2. Flooded suction or the supply is above the pump centerline.
   $NPSH = h_a - h_{vpa} + h_{st} - h_{fs}$

where:
- h_a = The absolute pressure (measured in feet of liquid) on the surface of the liquid supply level. (This is the atmospheric pressure for an open tank or sump or the absolute pressure existing in a closed tank.)
- h_{vpa} = The head in feet corresponding to the vapor pressure of the liquid at the temperature at which it is being pumped.
- h_{st} = The static height in feet that the liquid supply level is above or below the pump centerline.
- h_{fs} = All suction line head friction losses (measured in feet) including entrance losses, friction losses through pipe, valves, and fittings, etc.

For an example, see the figure on the next page.

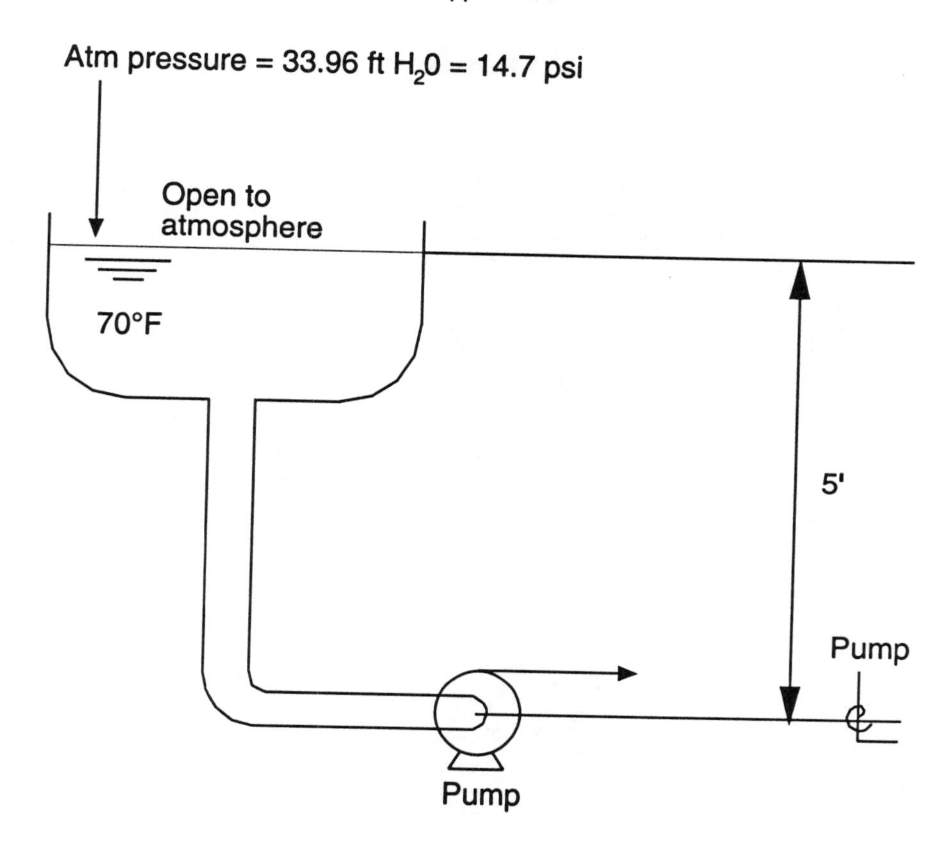

System is flooded (supply above pump centerline).

Liquid water is at 70°F.

Location is at sea level. (It is known that the atm pressure varies with altitude.)

NPSH = 33.96 (ft of water) - 0.79 ft vapor pressure + 5 static - 2.40 head loss = 35.77 ft

Net Positive Suction Head available is 35.77 feet.

PART TWO: FRICTION OF WATER

Friction of Water

(Based on Darcy's Formula)

Copper Tubing—*S.P.S. Copper and Brass Pipe

⅜ Inch

Flow — US gal per min	Type K tubing .402" inside dia .049" wall thk		Type L tubing .430" inside dia .035" wall thk		Type M tubing .450" inside dia .025" wall thk		*Pipe .494" inside dia .0905" wall thk		Flow — US gal per min
	Velocity ft/sec	Head loss ft/100 ft	Velocity ft/sec	Head loss ft/100 ft	Velocity ft/sec	Head loss ft/100 ft	Velocity ft/sec	Head loss ft/100 ft	
0.2	0.51	**0.66**	0.44	**0.48**	0.40	**0.39**	0.34	**0.26**	0.2
0.4	1.01	**2.15**	0.88	**1.57**	0.81	**1.27**	0.67	**0.82**	0.4
0.6	1.52	**4.29**	1.33	**3.12**	1.21	**2.52**	1.00	**1.63**	0.6
0.8	2.02	**7.02**	1.77	**5.11**	1.61	**4.12**	1.34	**2.66**	0.8
1	2.52	**10.32**	2.20	**7.50**	2.01	**6.05**	1.68	**3.89**	1
1½	3.78	**20.86**	3.30	**15.15**	3.02	**12.21**	2.51	**7.84**	1½
2	5.04	**34.48**	4.40	**20.03**	4.02	**20.16**	3.35	**12.94**	2
2½	6.30	**51.03**	5.50	**37.01**	5.03	**29.80**	4.19	**19.11**	2½
3	7.55	**70.38**	6.60	**51.02**	6.04	**41.07**	5.02	**26.32**	3
3½	8.82	**92.44**	7.70	**66.98**	7.04	**53.90**	5.86	**34.52**	3½
4	10.1	**117.1**	8.80	**84.85**	8.05	**68.26**	6.70	**43.70**	4
4½	11.4	**144.4**	9.90	**104.6**	9.05	**84.11**	7.53	**53.82**	4½
5	12.6	**174.3**	11.0	**126.1**	10.05	**101.4**	8.36	**64.87**	5

Note: No allowance has been made for age, difference in diameter, or any abnormal condition of interior surface. Any factor of safety must be estimated from the local conditions and the requirements of each particular installation. It is recommended that for most commercial design purposes a safety factor of 15 to 20% be added to the values in the tables

Reprinted with permission from Cameron Hydraulic Data book.

Appendix B

Friction of Water

(Based on Darcy's Formula)

Copper Tubing—* S.P.S. Copper and Brass Pipe

½ Inch

Flow — US gal per min	Type K tubing .527" inside dia .049" wall thk		Type L tubing .545" inside dia .040" wall thk		Type M tubing .569" inside dia .028" wall thk		*Pipe .625" inside dia .1075" wall thk		Flow — US gal per min
	Velocity ft/sec	**Head loss ft/100 ft**	Velocity ft/sec	**Head loss ft/100 ft**	Velocity ft/sec	**Head loss ft/100 ft**	Velocity ft/sec	**Head loss ft/100 ft**	
½	0.74	**0.88**	0.69	**0.75**	0.63	**0.62**	0.52	**0.40**	½
1	1.47	**2.87**	1.38	**2.45**	1.26	**2.00**	1.04	**1.28**	1
1½	2.20	**5.77**	2.06	**4.93**	1.90	**4.02**	1.57	**2.58**	1½
2	2.94	**9.52**	2.75	**8.11**	2.53	**6.61**	2.09	**4.24**	2
2½	3.67	**14.05**	3.44	**11.98**	3.16	**9.76**	2.61	**6.25**	2½
3	4.40	**19.34**	4.12	**16.48**	3.79	**13.42**	3.13	**8.59**	3
3½	5.14	**25.36**	4.81	**21.61**	4.42	**17.59**	3.66	**11.25**	3½
4	5.87	**32.09**	5.50	**27.33**	5.05	**22.25**	4.18	**14.22**	4
4½	6.61	**39.51**	6.19	**33.65**	5.68	**27.39**	4.70	**17.50**	4½
5	7.35	**47.61**	6.87	**40.52**	6.31	**32.99**	5.22	**21.07**	5
6	8.81	**65.79**	8.25	**56.02**	7.59	**45.57**	6.26	**29.09**	6
7	10.3	**86.57**	9.62	**73.69**	8.84	**59.93**	7.31	**38.23**	7
8	11.8	**109.9**	11.0	**93.50**	10.1	**76.03**	8.35	**48.47**	8
9	13.2	**135.6**	12.4	**115.4**	11.4	**93.82**	9.40	**59.79**	9
10	14.7	**163.8**	13.8	**139.4**	12.6	**113.3**	10.4	**72.16**	10

⅝ Inch

Flow — US gal per min	Type K tubing .652" inside dia .049" wall thk		Type L tubing .666" inside dia .042" wall thk		Type M tubing .690" inside dia .030" wall thk		*Pipe		Flow — US gal per min
	Velocity ft/sec	**Head loss ft/100 ft**	Velocity ft/sec	**Head loss ft/100 ft**	Velocity ft/sec	**Head loss ft/100 ft**			
½	0.48	**0.31**	0.46	**0.29**	0.43	**0.24**			½
1	0.96	**1.05**	0.92	**0.95**	0.86	**0.76**			1
1½	1.44	**2.11**	1.38	**1.91**	1.29	**1.53**			1½
2	1.92	**3.47**	1.84	**3.14**	1.72	**2.51**			2
2½	2.40	**5.11**	2.30	**4.62**	2.14	**3.68**			2½
3	2.88	**7.02**	2.75	**6.35**	2.57	**5.07**			3
3½	3.36	**9.20**	3.21	**8.32**	3.00	**6.64**			3½
4	3.84	**11.63**	3.67	**10.51**	3.43	**8.40**			4
4½	4.32	**14.30**	4.13	**12.93**	3.86	**10.35**			4½
5	4.80	**17.22**	4.59	**15.56**	4.29	**12.49**			5
6	5.75	**23.76**	5.51	**21.47**	5.15	**17.21**			6
7	6.71	**31.22**	6.42	**28.21**	6.00	**22.58**			7
8	7.67	**39.58**	7.35	**35.75**	6.85	**28.54**			8
9	8.64	**48.81**	8.25	**44.09**	7.71	**35.35**			9
10	9.60	**58.90**	9.18	**53.19**	8.57	**42.48**			10
11	10.6	**69.83**	10.1	**63.06**	9.43	**50.47**			11
12	11.5	**81.59**	11.0	**73.67**	10.3	**59.1**			12
13	12.5	**94.18**	11.9	**85.03**	11.2	**68.8**			13

Note: No allowance has been made for age, difference in diameter, or any abnormal condition of interior surface. Any factor of safety must be estimated from the local conditions and the requirements of each particular installation. It is recommended that for most commercial design purposes a safety factor of 15 to 20% be added to the values in the tables

Friction of Water

(Based on Darcy's Formula)

Copper Tubing—* S.P.S. Copper and Brass Pipe

¾ Inch

Flow — U S gal per min	Type K tubing .745" inside dia .065" wall thk		Type L tubing .785" inside dia .045" wall thk		Type M tubing .811" inside dia .032" wall thk		*Pipe .822" inside dia .114" wall thk		Flow — U S gal per min
	Velocity ft/sec	Head loss ft/100 ft	Velocity ft/sec	Head loss ft/100 ft	Velocity ft/sec	Head loss ft/100 ft	Velocity ft/sec	Head loss ft/100 ft	
1	0.74	**0.56**	0.66	**0.44**	0.62	**0.38**	0.60	**0.35**	1
2	1.47	**1.84**	1.33	**1.44**	1.24	**1.23**	1.21	**1.16**	2
3	2.21	**3.73**	1.99	**2.91**	1.86	**2.49**	1.81	**2.34**	3
4	2.94	**6.16**	2.65	**4.81**	2.48	**4.12**	2.42	**3.86**	4
5	3.67	**9.12**	3.31	**7.11**	3.10	**6.09**	3.02	**5.71**	5
6	4.41	**12.57**	3.98	**9.80**	3.72	**8.39**	3.62	**7.86**	6
7	5.14	**16.51**	4.64	**12.86**	4.34	**11.01**	4.23	**10.32**	7
8	5.88	**20.91**	5.30	**16.28**	4.96	**13.94**	4.83	**13.07**	8
9	6.61	**25.77**	5.96	**20.06**	5.59	**17.17**	5.44	**16.10**	9
10	7.35	**31.08**	6.62	**24.19**	6.20	**20.70**	6.04	**19.41**	10
11	8.09	**36.83**	7.29	**28.66**	6.82	**24.52**	6.64	**22.99**	11
12	8.83	**43.01**	7.95	**33.47**	7.44	**28.63**	7.25	**26.84**	12
13	9.56	**49.62**	8.61	**38.61**	8.06	**33.02**	7.85	**30.96**	13
14	10.3	**56.66**	9.27	**44.07**	8.68	**37.69**	8.45	**35.33**	14
15	11.0	**64.11**	9.94	**49.86**	9.30	**42.64**	9.05	**39.97**	15
16	11.8	**71.97**	10.6	**55.97**	9.92	**47.86**	9.65	**44.86**	16
17	12.5	**80.24**	11.25	**62.39**	10.55	**53.35**	10.25	**50.00**	17
18	13.2	**88.92**	11.92	**69.13**	11.17	**59.10**	10.85	**55.40**	18

1 Inch

Flow — U S gal per min	Type K tubing .995" inside dia .065" wall thk		Type L tubing 1.025" inside dia .050" wall thk		Type M tubing 1.055" inside dia .035" wall thk		*Pipe 1.062" inside dia .1265" wall thk		Flow — U S gal per min
	Velocity ft/sec	Head loss ft/100 ft	Velocity ft/sec	Head loss ft/100 ft	Velocity ft/sec	Head loss ft/100 ft	Velocity ft/sec	Head loss ft/100 ft	
2	0.82	**0.47**	0.78	**0.41**	0.73	**0.36**	0.72	**0.35**	2
3	1.24	**0.95**	1.17	**0.82**	1.10	**0.72**	1.08	**0.70**	3
4	1.65	**1.56**	1.56	**1.35**	1.47	**1.18**	1.45	**1.14**	4
5	2.06	**2.30**	1.95	**2.00**	1.83	**1.74**	1.81	**1.69**	5
6	2.48	**3.17**	2.34	**2.75**	2.20	**2.40**	2.17	**2.32**	6
7	2.89	**4.15**	2.72	**3.60**	2.56	**3.14**	2.53	**3.04**	7
8	3.30	**5.25**	3.11	**4.56**	2.93	**3.97**	2.89	**3.85**	8
9	3.71	**6.47**	3.50	**5.61**	3.30	**4.89**	3.25	**4.74**	9
10	4.12	**7.79**	3.89	**6.76**	3.66	**5.89**	3.61	**5.71**	10
12	4.95	**10.76**	4.67	**9.33**	4.40	**8.13**	4.34	**7.88**	12
14	5.77	**14.15**	5.45	**12.27**	5.13	**10.69**	5.05	**10.36**	14
16	6.60	**17.94**	6.22	**15.56**	5.86	**13.55**	5.78	**13.13**	16
18	7.42	**22.14**	7.00	**19.20**	6.60	**16.72**	6.50	**16.20**	18
20	8.24	**26.73**	7.78	**23.18**	7.33	**20.18**	7.22	**19.55**	20
25	10.30	**39.87**	9.74	**34.56**	9.16	**30.09**	9.03	**29.15**	25
30	12.37	**55.33**	11.68	**47.96**	11.00	**41.74**	10.84	**40.43**	30
35	14.42	**73.06**	13.61	**63.31**	12.82	**55.09**	12.65	**53.37**	35
40	16.50	**93.00**	15.55	**80.58**	14.66	**70.11**	14.45	**67.90**	40
45	18.55	**115.1**	17.50	**99.72**	16.50	**86.75**	16.25	**84.02**	45
50	20.60	**139.4**	19.45	**120.7**	18.32	**105.0**	18.05	**101.7**	50

Note: No allowance has been made for age, difference in diameter, or any abnormal condition of interior surface. Any factor of safety must be estimated from the local conditions and the requirements of each particular installation. It is recommended that for most commercial design purposes a safety factor of 15 to 20% be added to the values in the tables

Reprinted with permission from Cameron Hydraulic Data book.

Friction of Water

(Based on Darcy's Formula)

Copper Tubing—*S.P.S. Copper and Brass Pipe

1¼ Inch

Flow — US gal per min	Type K tubing 1.245" inside dia .065" wall thk Velocity ft/sec	Type K tubing Head loss ft/100 ft	Type L tubing 1.265" inside dia .055" wall thk Velocity ft/sec	Type L tubing Head loss ft/100 ft	Type M tubing 1.291" inside dia .042" wall thk Velocity ft/sec	Type M tubing Head loss ft/100 ft	*Pipe 1.368" inside dia .146" wall thk Velocity ft/sec	*Pipe Head loss ft/100 ft	Flow — US gal per min
5	1.31	**0.79**	1.28	**0.74**	1.22	**0.67**	1.09	**0.51**	5
6	1.58	**1.09**	1.53	**1.01**	1.47	**0.92**	1.31	**0.70**	6
7	1.84	**1.43**	1.79	**1.32**	1.71	**1.20**	1.53	**0.91**	7
8	2.11	**1.81**	2.04	**1.67**	1.96	**1.52**	1.75	**1.15**	8
9	2.37	**2.22**	2.30	**2.06**	2.20	**1.87**	1.96	**1.42**	9
10	2.63	**2.67**	2.55	**2.48**	2.45	**2.25**	2.18	**1.71**	10
12	3.16	**3.69**	3.06	**3.42**	2.93	**3.10**	2.62	**2.35**	12
15	3.95	**5.47**	3.83	**5.07**	3.66	**4.60**	3.27	**3.49**	15
20	5.26	**9.13**	5.10	**8.46**	4.89	**7.67**	4.36	**5.81**	20
25	6.58	**13.59**	6.38	**12.59**	6.11	**11.42**	5.46	**8.65**	25
30	7.90	**18.83**	7.65	**17.44**	7.33	**15.82**	6.55	**11.98**	30
35	9.21	**24.83**	8.94	**23.00**	8.55	**20.86**	7.65	**15.79**	35
40	10.5	**31.57**	10.2	**29.24**	9.77	**26.51**	8.74	**20.06**	40
45	11.8	**38.03**	11.5	**36.15**	11.0	**32.77**	9.83	**24.80**	45
50	13.2	**47.20**	12.8	**43.71**	12.2	**39.63**	10.9	**29.98**	50
60	15.8	**65.65**	15.3	**60.78**	14.7	**55.10**	13.1	**41.66**	60
70	18.4	**86.82**	17.9	**80.38**	17.1	**72.86**	15.3	**55.07**	70
80	21.1	**110.7**	20.4	**102.5**	19.6	**92.85**	17.5	**70.16**	80
90	23.7	**137.2**	23.0	**127.0**	22.0	**115.1**	19.6	**86.91**	90
100	26.3	**166.3**	25.5	**153.9**	24.4	**139.4**	21.8	**105.3**	100

1½ Inch

Flow — US gal per min	Type K tubing 1.481" inside dia .072" wall thk Velocity ft/sec	Type K tubing Head loss ft/100 ft	Type L tubing 1.505" inside dia .060" wall thk Velocity ft/sec	Type L tubing Head loss ft/100 ft	Type M tubing 1.527" inside dia .049" wall thk Velocity ft/sec	Type M tubing Head loss ft/100 ft	*Pipe 1.600" inside dia .150" wall thk Velocity ft/sec	*Pipe Head loss ft/100 ft	Flow — US gal per min
8	1.49	**0.79**	1.44	**0.73**	1.40	**0.68**	1.27	**0.55**	8
9	1.67	**0.97**	1.62	**0.90**	1.57	**0.84**	1.43	**0.67**	9
10	1.86	**1.17**	1.80	**1.08**	1.75	**1.01**	1.59	**0.81**	10
12	2.23	**1.61**	2.16	**1.49**	2.10	**1.39**	1.91	**1.12**	12
15	2.79	**2.39**	2.70	**2.21**	2.63	**2.07**	2.39	**1.65**	15
20	3.72	**3.98**	3.60	**3.68**	3.50	**3.44**	3.19	**2.75**	20
25	4.65	**5.91**	4.51	**5.48**	4.38	**5.11**	3.98	**4.09**	25
30	5.58	**8.19**	5.41	**7.58**	5.25	**7.07**	4.78	**5.65**	30
35	6.51	**10.79**	6.31	**9.99**	6.13	**9.31**	5.58	**7.45**	35
40	7.44	**13.70**	7.21	**12.68**	7.00	**11.83**	6.37	**9.45**	40
45	8.37	**16.93**	8.11	**15.67**	7.88	**14.61**	7.16	**11.68**	45
50	9.30	**20.46**	9.01	**18.94**	8.76	**17.66**	7.96	**14.11**	50
60	11.2	**28.42**	10.8	**26.30**	10.5	**24.53**	9.56	**19.59**	60
70	13.0	**37.55**	12.6	**34.74**	12.3	**32.40**	11.2	**25.87**	70
80	14.9	**47.82**	14.4	**44.24**	14.0	**41.25**	12.8	**32.93**	80
90	16.7	**59.21**	16.2	**54.78**	15.8	**51.07**	14.4	**40.76**	90
100	18.6	**71.70**	18.0	**66.34**	17.5	**61.84**	15.9	**49.34**	100
110	20.5	**85.29**	19.8	**78.90**	19.3	**73.55**	17.5	**58.67**	110
120	22.3	**99.95**	21.6	**92.46**	21.0	**86.18**	19.1	**68.74**	120
130	24.2	**115.7**	23.4	**107.0**	22.8	**99.73**	20.7	**79.53**	130

Note: No allowance has been made for age, difference in diameter, or any abnormal condition of interior surface. Any factor of safety must be estimated from the local conditions and the requirements of each particular installation. It is recommended that for most commercial design purposes a safety factor of 15 to 20% be added to the values in the tables

Friction of Water
(Based on Darcy's Formula)

Copper Tubing—*S.P.S. Copper and Brass Pipe

2 Inch

Flow — US gal per min	Type K tubing 1.959" inside dia .083" wall thk		Type L tubing 1.985" inside dia .070" wall thk		Type M tubing 2.009" inside dia .058" wall thk		*Pipe 2.062" inside dia .1565" wall thk		Flow — US gal per min
	Velocity ft/sec	**Head loss ft/100 ft**	Velocity ft/sec	**Head loss ft/100 ft**	Velocity ft/sec	**Head loss ft/100 ft**	Velocity ft/sec	**Head loss ft/100 ft**	
10	1.07	**0.31**	1.04	**0.29**	1.01	**0.27**	.96	**0.24**	10
12	1.28	**0.43**	1.24	**0.40**	1.21	**0.38**	1.15	**0.33**	12
14	1.49	**0.56**	1.45	**0.52**	1.42	**0.50**	1.34	**0.44**	14
16	1.70	**0.71**	1.66	**0.66**	1.62	**0.63**	1.53	**0.55**	16
18	1.92	**0.87**	1.87	**0.82**	1.82	**0.77**	1.72	**0.68**	18
20	2.13	**1.05**	2.07	**0.98**	2.02	**0.93**	1.92	**0.82**	20
25	2.66	**1.55**	2.59	**1.46**	2.53	**1.38**	2.39	**1.22**	25
30	3.19	**2.15**	3.11	**2.01**	3.03	**1.90**	2.87	**1.68**	30
35	3.73	**2.82**	3.62	**2.65**	3.54	**2.50**	3.35	**2.21**	35
40	4.26	**3.58**	4.14	**3.36**	4.05	**3.17**	3.83	**2.80**	40
45	4.79	**4.42**	4.66	**4.15**	4.55	**3.92**	4.30	**3.46**	45
50	5.32	**5.34**	5.17	**5.01**	5.05	**4.73**	4.80	**4.17**	50
60	6.39	**7.40**	6.21	**6.95**	6.06	**6.56**	5.75	**5.79**	60
70	7.45	**9.76**	7.25	**9.16**	7.07	**8.65**	6.70	**7.63**	70
80	8.52	**12.42**	8.28	**11.65**	8.09	**11.00**	7.65	**9.70**	80
90	9.58	**15.36**	9.31	**14.41**	9.10	**13.60**	8.61	**12.00**	90
100	10.65	**18.58**	10.4	**17.43**	10.1	**16.45**	9.57	**14.51**	100
110	11.71	**22.07**	11.4	**20.71**	11.1	**19.55**	10.5	**17.24**	110
120	12.78	**25.84**	12.4	**24.25**	12.1	**22.88**	11.5	**20.18**	120
130	13.85	**29.88**	13.4	**28.04**	13.1	**26.45**	12.5	**23.33**	130
140	14.9	**34.18**	14.5	**32.07**	14.2	**30.26**	13.4	**26.69**	140
150	16.0	**38.75**	15.5	**36.36**	15.2	**34.30**	14.4	**30.25**	150
160	17.0	**43.58**	16.5	**40.89**	16.2	**38.58**	15.3	**34.01**	160
170	18.1	**48.67**	17.6	**45.66**	17.2	**43.08**	16.3	**37.98**	170
180	19.2	**54.01**	18.6	**50.67**	18.2	**47.81**	17.2	**42.15**	180
190	20.2	**59.61**	19.6	**55.92**	19.2	**52.76**	18.2	**46.51**	190
200	21.3	**65.46**	20.7	**61.41**	20.2	**57.94**	19.2	**51.07**	200
210	22.4	**71.57**	21.7	**67.14**	21.2	**63.34**	20.1	**55.83**	210
220	23.4	**77.93**	22.8	**73.10**	22.2	**68.96**	21.0	**60.78**	220
230	24.5	**84.53**	23.8	**79.29**	23.2	**74.80**	22.0	**65.93**	230
240	25.6	**91.38**	24.8	**85.72**	24.3	**80.86**	23.0	**71.26**	240
250	26.6	**98.43**	25.9	**92.37**	25.3	**87.14**	23.9	**76.79**	250
260	27.7	**105.8**	26.9	**99.26**	26.3	**93.63**	24.9	**82.51**	260
270	28.8	**113.4**	27.9	**106.4**	27.3	**100.3**	25.8	**88.42**	270
280	29.8	**121.3**	29.0	**113.7**	28.3	**107.3**	26.8	**94.52**	280
290	30.9	**129.3**	30.0	**121.3**	29.4	**114.4**	27.8	**100.8**	290
300	32.0	**137.6**	31.1	**129.1**	30.4	**121.8**	28.7	**107.3**	300

Note: No allowance has been made for age, difference in diameter, or any abnormal condition of interior surface. Any factor of safety must be estimated from the local conditions and the requirements of each particular installation. It is recommended that for most commercial design purposes a safety factor of 15 to 20% be added to the values in the tables

Friction of Water

(Based on Darcy's Formula)

Copper Tubing—*S.P.S. Copper and Brass Pipe

2½ Inch

Flow — U S gal per min	Type K tubing 2.435" inside dia .095" wall thk		Type L tubing 2.465" inside dia .080" wall thk		Type M tubing 2.495" inside dia .065" wall thk		*Pipe 2.500" inside dia .1875" wall thk		Flow — U S gal per min
	Velocity ft/sec	**Head loss ft/100 ft**	Velocity ft/sec	**Head loss ft/100 ft**	Velocity ft/sec	**Head loss ft/100 ft**	Velocity ft/sec	**Head loss ft/100 ft**	
20	1.38	**0.37**	1.34	**0.35**	1.31	**0.33**	1.31	**0.33**	20
25	1.72	**0.55**	1.68	**0.52**	1.64	**0.49**	1.63	**0.49**	25
30	2.07	**0.76**	2.02	**0.72**	1.97	**0.68**	1.96	**0.67**	30
35	2.41	**1.00**	2.35	**0.94**	2.30	**0.89**	2.29	**0.88**	35
40	2.76	**1.26**	2.69	**1.19**	2.62	**1.13**	2.61	**1.12**	40
45	3.10	**1.56**	3.02	**1.47**	2.95	**1.39**	2.94	**1.38**	45
50	3.45	**1.88**	3.36	**1.77**	3.28	**1.68**	3.26	**1.66**	50
60	4.14	**2.61**	4.03	**2.46**	3.93	**2.32**	3.92	**2.30**	60
70	4.82	**3.43**	4.70	**3.24**	4.59	**3.06**	4.57	**3.03**	70
80	5.51	**4.36**	5.37	**4.12**	5.25	**3.88**	5.22	**3.85**	80
90	6.20	**5.39**	6.04	**5.08**	5.90	**4.80**	5.88	**4.75**	90
100	6.89	**6.52**	6.71	**6.15**	6.55	**5.80**	6.53	**5.74**	100
110	7.58	**7.74**	7.38	**7.30**	7.21	**6.89**	7.19	**6.82**	110
120	8.27	**9.06**	8.05	**8.54**	7.86	**8.05**	7.84	**7.98**	120
130	8.96	**10.46**	8.73	**9.87**	8.52	**9.31**	8.49	**9.22**	130
140	9.65	**11.97**	9.40	**11.28**	9.18	**10.64**	9.14	**10.54**	140
150	10.35	**13.56**	10.1	**12.78**	9.83	**12.06**	9.79	**11.94**	150
160	11.0	**15.24**	10.8	**14.36**	10.5	**13.55**	10.45	**13.42**	160
170	11.7	**17.01**	11.4	**16.03**	11.1	**15.12**	11.1	**14.98**	170
180	12.4	**18.87**	12.1	**17.79**	11.8	**16.78**	11.8	**16.61**	180
190	13.1	**20.81**	12.8	**19.62**	12.5	**18.51**	12.4	**18.33**	190
200	13.8	**22.85**	13.4	**21.54**	13.1	**20.31**	13.1	**20.12**	200
220	15.2	**27.18**	14.8	**25.61**	14.4	**24.16**	14.4	**23.93**	220
240	16.5	**31.84**	16.1	**30.01**	15.7	**28.31**	15.7	**28.03**	240
260	17.9	**36.85**	17.5	**34.73**	17.1	**32.75**	17.0	**32.44**	260
280	19.3	**42.19**	18.8	**39.76**	18.4	**37.50**	18.3	**37.13**	280
300	20.7	**47.86**	20.1	**45.10**	19.7	**42.53**	19.6	**42.12**	300
320	22.1	**53.86**	21.5	**50.75**	21.0	**47.86**	20.9	**47.40**	320
340	23.4	**60.18**	22.8	**56.71**	22.3	**53.48**	22.2	**52.96**	340
360	24.8	**66.83**	24.2	**62.97**	23.6	**59.38**	23.5	**58.81**	300
380	26.2	**73.80**	25.5	**69.54**	24.9	**65.57**	24.8	**64.94**	380
400	27.6	**81.09**	26.9	**76.41**	26.2	**72.04**	26.1	**71.35**	400
420	29.0	**88.70**	28.2	**83.57**	27.5	**78.80**	27.4	**78.04**	420
440	30.3	**96.62**	29.5	**91.04**	28.8	**85.83**	28.7	**85.00**	440
460	31.7	**104.9**	30.9	**98.80**	30.2	**93.15**	30.0	**92.24**	460
480	33.1	**113.4**	32.2	**106.8**	31.5	**100.7**	31.4	**99.76**	480
500	34.5	**122.3**	33.6	**115.2**	32.8	**108.6**	32.6	**107.5**	500

Note: No allowance has been made for age, difference in diameter, or any abnormal condition of interior surface. Any factor of safety must be estimated from the local conditions and the requirements of each particular installation. It is recommended that for most commercial design purposes a safety factor of 15 to 20% be added to the values in the tables

Friction of Water

(Based on Darcy's Formula)

Copper Tubing—*S.P.S. Copper and Brass Pipe

3 Inch

Flow — US gal per min	Type K tubing 2.907" inside dia .109" wall thk		Type L tubing 2.945" inside dia .090" wall thk		Type M tubing 2.981" inside dia .072" wall thk		*Pipe 3.062" inside dia .219" wall thk		Flow — US gal per min
	Velocity ft/sec	**Head loss ft/100 ft**	Velocity ft/sec	**Head loss ft/100 ft**	Velocity ft/sec	**Head loss ft/100 ft**	Velocity ft/sec	**Head loss ft/100 ft**	
20	0.96	**0.16**	0.94	**0.15**	0.92	**0.14**	0.87	**0.13**	20
30	1.45	**0.33**	1.41	**0.31**	1.37	**0.29**	1.30	**0.25**	30
40	1.93	**0.54**	1.88	**0.51**	1.83	**0.48**	1.74	**0.42**	40
50	2.41	**0.81**	2.35	**0.76**	2.29	**0.72**	2.17	**0.63**	50
60	2.89	**1.12**	2.82	**1.05**	2.75	**0.99**	2.61	**0.87**	60
70	3.38	**1.47**	3.29	**1.38**	3.20	**1.30**	3.04	**1.15**	70
80	3.86	**1.87**	3.76	**1.75**	3.66	**1.65**	3.48	**1.45**	80
90	4.34	**2.30**	4.23	**2.16**	4.12	**2.04**	3.91	**1.80**	90
100	4.82	**2.78**	4.70	**2.61**	4.59	**2.47**	4.35	**2.17**	100
110	5.30	**3.30**	5.17	**3.10**	5.05	**2.93**	4.79	**2.57**	110
120	5.79	**3.86**	5.64	**3.63**	5.50	**3.42**	5.21	**3.01**	120
130	6.27	**4.46**	6.11	**4.19**	5.95	**3.95**	5.65	**3.47**	130
140	6.75	**5.10**	6.58	**4.79**	6.41	**4.52**	6.09	**3.97**	140
150	7.24	**5.77**	7.05	**5.42**	6.87	**5.12**	6.52	**4.50**	150
160	7.72	**6.49**	7.52	**6.09**	7.34	**5.75**	6.95	**5.05**	160
170	8.20	**7.24**	7.99	**6.80**	7.79	**6.41**	7.39	**5.64**	170
180	8.69	**8.03**	8.46	**7.54**	8.25	**7.11**	7.82	**6.25**	180
190	9.16	**8.85**	8.93	**8.32**	8.70	**7.84**	8.25	**6.89**	190
200	9.64	**9.71**	9.40	**9.13**	9.16	**8.61**	8.70	**7.56**	200
220	10.6	**11.55**	10.3	**10.85**	10.1	**10.23**	9.56	**8.99**	220
240	11.6	**13.52**	11.3	**12.70**	11.0	**11.98**	10.4	**10.52**	240
260	12.6	**15.64**	12.2	**14.69**	11.9	**13.85**	11.3	**12.17**	260
280	13.5	**17.90**	13.2	**16.81**	12.8	**15.85**	12.2	**13.93**	280
300	14.5	**20.30**	14.1	**19.06**	13.7	**17.97**	13.0	**15.79**	300
320	15.4	**22.83**	15.0	**21.44**	14.7	**20.22**	13.9	**17.76**	320
340	16.4	**25.50**	16.0	**23.95**	15.6	**22.58**	14.8	**19.83**	340
360	17.4	**28.30**	16.9	**26.58**	16.5	**25.06**	15.7	**22.01**	360
380	18.3	**31.24**	17.9	**29.34**	17.4	**27.66**	16.5	**24.29**	380
400	19.3	**34.32**	18.8	**32.22**	18.3	**30.38**	17.4	**26.68**	400
450	21.7	**42.58**	21.2	**39.98**	20.6	**37.69**	19.6	**33.09**	450
500	24.1	**51.65**	23.5	**48.50**	22.9	**45.72**	21.7	**40.14**	500
550	26.6	**61.54**	25.8	**57.77**	25.2	**54.46**	23.9	**47.81**	550
600	29.0	**72.22**	28.2	**67.80**	27.5	**63.91**	26.1	**56.10**	600
650	31.4	**83.69**	30.6	**78.56**	29.8	**74.05**	28.2	**65.00**	650
700	33.8	**95.95**	32.9	**90.06**	32.1	**84.89**	30.4	**74.50**	700
750	36.2	**109.0**	35.2	**102.3**	34.4	**96.41**	32.6	**84.61**	750
800	38.6	**122.8**	37.6	**115.3**	36.6	**108.6**	34.8	**95.31**	800

Note: No allowance has been made for age, difference in diameter, or any abnormal condition of interior surface. Any factor of safety must be estimated from the local conditions and the requirements of each particular installation. It is recommended that for most commercial design purposes a safety factor of 15 to 20% be added to the values in the tables

Reprinted with permission from Cameron Hydraulic Data book.

Friction of Water New Steel Pipe
(Based on Darcy's Formula)

¼ Inch

Flow US gal per min	Standard wt steel—sch 40			Extra strong steel—sch 80		
	0.364" inside dia			0.302" inside dia		
	Velocity ft per sec	Velocity head-ft	**Head loss ft per 100 ft**	Velocity ft per sec	Velocity head-ft	**Head loss ft per 100 ft**
0.4	1.23	0.024	**3.7**	1.79	0.05	**9.18**
0.6	1.85	0.053	**7.6**	2.69	0.11	**19.0**
0.8	2.47	0.095	**12.7**	3.59	0.20	**32.3**
1.0	3.08	0.148	**19.1**	4.48	0.31	**48.8**
1.2	3.70	0.213	**26.7**	5.38	0.45	**68.6**
1.4	4.32	0.290	**35.6**	6.27	0.61	**91.7**
1.6	4.93	0.378	**45.6**	7.17	0.80	**118.1**
1.8	5.55	0.479	**56.9**	8.07	1.01	**147.7**
2.0	6.17	0.591	**69.4**	8.96	1.25	**180.7**
2.4	7.40	0.850	**98.1**	10.75	1.79	**256**
2.8	8.63	1.157	**132**	12.54	2.44	**345**

⅜ Inch

Flow US gal per min	Standard wt steel—sch 40			Extra strong steel—sch 80		
	0.493" inside dia			0.423" inside dia		
	Velocity ft per sec	Velocity head-ft	**Head loss ft per 100 ft**	Velocity ft per sec	Velocity head-ft	**Head loss ft per 100 ft**
0.5	0.84	0.011	**1.26**	1.14	0.02	**2.63**
1.0	1.68	0.044	**4.26**	2.28	0.08	**9.05**
1.5	2.52	0.099	**8.85**	3.43	0.18	**19.0**
2.0	3.36	0.176	**15.0**	4.57	0.32	**32.4**
2.5	4.20	0.274	**22.7**	5.71	0.51	**49.3**
3.0	5.04	0.395	**32.0**	6.85	0.73	**69.6**
3.5	5.88	0.538	**42.7**	8.00	0.99	**93.3**
4.0	6.72	0.702	**55.0**	9.14	1.30	**120**
5.0	8.40	1.097	**84.2**	11.4	2.0	**185**
6.0	10.08	1.58	**119**	13.7	2.9	**263**

Note: No allowance has been made for age, difference in diameter, or any abnormal condition of interior surface. Any factor of safety must be estimated from the local conditions and the requirements of each particular installation. It is recommended that for most commercial design purposes a safety factor of 15 to 20% be added to the values in the tables

Friction of Water New Steel Pipe

(Based on Darcy's Formula)

½ Inch

Flow US gal per min	Standard wt steel—sch 40			Extra strong steel—sch 80			Schedule 160		
	.622" inside dia			.546" inside dia			.464" inside dia		
	Velocity ft per sec	Velocity head ft	**Head loss ft per 100 ft**	Velocity ft per sec	Velocity head ft	**Head loss ft per 100 ft**	Velocity ft per sec	Velocity head ft	**Head loss ft per 100 ft**
0.7	0.739	.008	**0.74**	.96	.01	**1.39**			
1.0	1.056	.017	**1.86**	1.37	.03	**2.58**	1.90	.056	**1.68**
1.5	1.58	.039	**2.82**	2.06	.07	**5.34**	2.85	.126	**5.73**
2.0	2.11	.069	**4.73**	2.74	.12	**9.02**	3.80	.224	**12.0**
2.5	2.64	.108	**7.10**	3.43	.18	**13.6**	4.74	.349	**20.3**
3.0	3.17	.156	**9.94**	4.11	.26	**19.1**	5.69	.503	**30.8**
3.5	3.70	.212	**13.2**	4.80	.36	**25.5**	6.64	.684	**43.5**
4.0	4.22	.277	**17.0**	5.48	.47	**32.7**	7.59	.894	**58.2**
4.5	4.75	.351	**21.1**	6.17	.59	**40.9**	8.54	1.13	**75.0**
5.0	5.28	.433	**25.8**	6.86	.73	**50.0**	9.49	1.40	**94.0**
5.5	5.81	.524	**30.9**	7.54	.88	**59.9**	10.44	1.69	**115**
6.0	6.34	.624	**36.4**	8.23	1.05	**70.7**	11.38	2.01	**138**
6.5	6.86	.732	**42.4**	8.91	1.23	**82.4**	12.33	2.36	**163**
7.0	7.39	.849	**48.8**	9.60	1.43	**95.0**	13.28	2.74	**190**
7.5	7.92	.975	**55.6**	10.3	1.6	**109**	14.23	3.14	**220**
8.0	8.45	1.109	**63.0**	11.0	1.9	**123**			
8.5	8.98	1.25	**70.7**	11.6	2.1	**138**			
9.0	9.50	1.40	**78.9**	12.3	2.4	**154**			
9.5	10.03	1.56	**87.6**	13.0	2.6	**171**			
10	10.56	1.73	**96.6**	13.7	2.9	**189**			

¾ Inch

Flow US gal per min	Standard wt steel—sch 40			Extra strong steel—sch 80			Steel—schedule 160		
	.824" inside dia			.742" inside dia			.612" inside dia		
	Velocity ft per sec	Velocity head ft	**Head loss ft per 100 ft**	Velocity ft per sec	Velocity head ft	**Head loss ft per 100 ft**	Velocity ft per sec	Velocity head ft	**Head loss ft per 100 ft**
1.5	0.90	.013	**0.72**	1.11	.02	**1.19**	1.64	.042	**3.05**
2.0	1.20	.023	**1.19**	1.48	.03	**1.99**	2.18	.074	**5.12**
2.5	1.50	.035	**1.78**	1.86	.05	**2.97**	2.73	.115	**7.70**
3.0	1.81	.051	**2.47**	2.23	.08	**4.14**	3.27	.166	**10.8**
3.5	2.11	.069	**3.26**	2.60	.11	**5.48**	3.82	.226	**14.3**
4.0	2.41	.090	**4.16**	2.97	.14	**7.01**	4.36	.295	**18.4**
4.5	2.71	.114	**5.17**	3.34	.17	**8.72**	4.91	.374	**22.9**
5.0	3.01	.141	**6.28**	3.71	.21	**10.6**	5.45	.462	**28.0**
6	3.61	.203	**8.80**	4.45	.31	**14.9**	6.54	.665	**39.5**
7	4.21	.276	**11.7**	5.20	.42	**19.9**	7.64	.905	**53.0**
8	4.81	.360	**15.1**	5.94	.55	**25.6**	8.73	1.18	**68.4**
9	5.42	.456	**18.8**	6.68	.69	**32.1**	9.82	1.50	**85.8**
10	6.02	.563	**23.0**	7.42	.86	**39.2**	10.91	1.85	**105**
11	6.62	.681	**27.6**	8.17	1.04	**47.0**	12.00	2.23	**126**
12	7.22	.722	**32.5**	8.91	1.23	**55.5**	13.09	2.66	**149**
13	7.82	.951	**37.9**	9.63	1.44	**64.8**	14.18	3.13	**175**
14	8.42	1.103	**43.7**	10.4	1.7	**74.7**	15.27	3.62	**202**
16	9.63	1.44	**56.4**	11.9	2.2	**96.7**	17.45	4.73	**261**
18	10.8	1.82	**70.8**	13.4	2.8	**121**			
20	12.0	2.25	**86.8**	14.8	3.4	**149**			

Note: No allowance has been made for age, difference in diameter, or any abnormal condition of interior surface. Any factor of safety must be estimated from the local conditions and the requirements of each particular installation. It is recommended that for most commercial design purposes a safety factor of 15 to 20% be added to the values in the tables

Reprinted with permission from Cameron Hydraulic Data book.

Friction of Water New Steel Pipe

(Based on Darcy's Formula)

1 Inch

Flow U S gal per min	Standard wt steel—sch 40			Extra strong steel—sch 80			Schedule 160 steel		
	1.049" inside dia			.957" inside dia			.815" inside dia		
	Velocity ft per sec	Velocity head ft	**Head loss ft per 100 ft**	Velocity ft per sec	Velocity head ft	**Head loss ft per 100 ft**	Velocity ft per sec	Velocity head ft	**Head loss ft per 100 ft**
2	0.74	.009	**.385**	.89	.01	**.599**	1.23	.023	**1.26**
3	1.11	.019	**.787**	1.34	.03	**1.19**	1.85	.053	**2.60**
4	1.48	.034	**1.270**	1.79	.05	**1.99**	2.46	.094	**4.40**
5	1.86	.054	**1.90**	2.23	.08	**2.99**	3.08	.147	**6.63**
6	2.23	.077	**2.65**	2.68	.11	**4.17**	3.69	.211	**9.30**
8	2.97	.137	**4.50**	3.57	.20	**7.11**	4.92	.376	**15.9**
10	3.71	.214	**6.81**	4.46	.31	**10.8**	6.15	.587	**24.3**
12	4.45	.308	**9.58**	5.36	.45	**15.2**	7.38	.845	**34.4**
14	5.20	.420	**12.8**	6.25	.61	**20.4**	8.61	1.15	**46.2**
16	5.94	.548	**16.5**	7.14	.79	**26.3**	9.84	1.50	**59.7**
18	6.68	.694	**20.6**	8.03	1.00	**32.9**	11.07	1.90	**74.9**
20	7.42	.857	**25.2**	8.92	1.24	**40.3**	12.30	2.35	**91.8**
22	8.17	1.036	**30.3**	9.82	1.50	**48.4**	13.53	2.84	**110**
24	8.91	1.23	**35.8**	10.7	1.8	**57.2**	14.76	3.38	**131**
26	9.65	1.45	**41.7**	11.6	2.1	**66.8**	15.99	3.97	**153**
28	10.39	1.68	**48.1**	12.5	2.4	**77.1**			
30	11.1	1.93	**55.0**	13.4	2.8	**88.2**			
35	13.0	2.62	**74.1**	15.6	3.8	**119**			
40	14.8	3.43	**96.1**	17.9	5.0	**154**			
45	16.7	4.33	**121**	20.1	6.3	**194**			

1¼ Inch

Flow U S gal per min	Standard wt steel—sch 40			Extra strong steel—sch 80			Schedule 160—steel		
	1.380" inside dia			1.278" inside dia			1.160" inside dia		
	Velocity ft per sec	Velocity head ft	**Head loss ft per 100 ft**	Velocity ft per sec	Velocity head ft	**Head loss ft per 100 ft**	Velocity ft per sec	Velocity head ft	**Head loss ft per 100 ft**
4	.858	.011	**.35**	1.00	.015	**.51**	1.21	.023	**.806**
5	1.073	.018	**.52**	1.25	.024	**.75**	1.52	.036	**1.20**
6	1.29	.026	**.72**	1.50	.034	**1.04**	1.82	.051	**1.61**
7	1.50	.035	**.95**	1.75	.048	**1.33**	2.13	.070	**2.14**
8	1.72	.046	**1.20**	2.00	.062	**1.69**	2.43	.092	**2.73**
10	2.15	.072	**1.74**	2.50	.097	**2.55**	3.04	.143	**4.12**
12	2.57	.103	**2.45**	3.00	.140	**3.57**	3.64	.206	**5.78**
14	3.00	.140	**3.24**	3.50	.190	**4.75**	4.25	.280	**7.72**
16	3.43	.183	**4.15**	4.00	.249	**6.10**	4.86	.366	**9.92**
18	3.86	.232	**5.17**	4.50	.315	**7.61**	5.46	.463	**12.4**
20	4.29	.286	**6.31**	5.00	.388	**9.28**	6.07	.572	**15.1**
25	5.36	.431	**9.61**	6.25	.607	**14.2**	7.59	.894	**23.2**
30	6.44	.644	**13.6**	7.50	.874	**20.1**	9.11	1.29	**32.9**
35	7.51	.876	**18.2**	8.75	1.19	**27.0**	10.63	1.75	**44.2**
40	8.58	1.14	**23.5**	10.0	1.55	**34.9**	12.14	2.29	**57.3**
50	10.7	1.79	**36.2**	12.5	2.43	**53.7**	15.18	3.58	**88.3**
60	12.9	2.57	**51.5**	15.0	3.50	**76.5**	18.22	5.15	**126**
70	15.0	3.50	**69.5**	17.5	4.76	**103**	21.25	7.01	**170**
80	17.2	4.53	**90.2**	20.0	6.21	**134**	24.29	9.16	**221**
90	19.3	5.79	**114**	22.5	7.86	**168**	27.32	11.59	**279**

Note: No allowance has been made for age, difference in diameter, or any abnormal condition of interior surface. Any factor of safety must be estimated from the local conditions and the requirements of each particular installation. It is recommended that for most commercial design purposes a safety factor of 15 to 20% be added to the values in the tables

Reprinted with permission from Cameron Hydraulic Data book.

Friction of Water New Steel Pipe

(Based on Darcy's Formula)

1½ Inch

Flow U S gal per min	Standard wt steel—sch 40			Extra strong steel—sch 80			Schedule 160—steel		
	1.610" inside dia			1.500" inside dia			1.338" inside dia		
	Velocity ft per sec	Velocity head ft	**Head loss ft per 100 ft**	Velocity ft per sec	Velocity head ft	**Head loss ft per 100 ft**	Velocity ft per sec	Velocity head ft	**Head loss ft per 100 ft**
4	.63	.006	**.166**	.73	.01	**.233**	.913	.013	**.404**
5	.79	.010	**.246**	.91	.01	**.346**	1.14	.020	**.601**
6	.95	.014	**.340**	1.09	.02	**.478**	1.37	.029	**.832**
7	1.10	.019	**.447**	1.27	.03	**.630**	1.60	.040	**1.10**
8	1.26	.025	**.567**	1.45	.03	**.800**	1.83	.052	**1.35**
9	1.42	.031	**.701**	1.63	.04	**.990**	2.05	.065	**1.67**
10	1.58	.039	**.848**	1.82	.05	**1.20**	2.28	.081	**2.03**
12	1.89	.056	**1.18**	2.18	.07	**1.61**	2.74	.116	**2.84**
14	2.21	.076	**1.51**	2.54	.10	**2.14**	3.20	.158	**3.78**
16	2.52	.099	**1.93**	2.90	.13	**2.74**	3.65	.207	**4.85**
18	2.84	.125	**2.40**	3.27	.17	**3.41**	4.11	.262	**6.04**
20	3.15	.154	**2.92**	3.63	.20	**4.15**	4.56	.323	**7.36**
22	3.47	.187	**3.48**	3.99	.25	**4.96**	5.02	.391	**8.81**
24	3.78	.222	**4.10**	4.36	.30	**5.84**	5.48	.465	**10.4**
26	4.10	.261	**4.76**	4.72	.35	**6.80**	5.93	.546	**12.1**
28	4.41	.303	**5.47**	5.08	.40	**7.82**	6.39	.634	**13.9**
30	4.73	.347	**6.23**	5.45	.46	**8.91**	6.85	.727	**15.9**
32	5.04	.395	**7.04**	5.81	.52	**10.1**	7.30	.828	**18.0**
34	5.36	.446	**7.90**	6.17	.59	**11.3**	7.76	.934	**20.2**
36	5.67	.500	**8.80**	6.54	.66	**12.6**	8.22	1.05	**22.5**
38	5.99	.577	**9.76**	6.90	.74	**14.0**	8.67	1.17	**25.0**
40	6.30	.618	**10.8**	7.26	.82	**15.4**	9.13	1.29	**27.6**
42	6.62	.681	**11.8**	7.63	.90	**16.9**	9.58	1.43	**30.3**
44	6.93	.747	**12.9**	7.99	.99	**18.5**	10.04	1.57	**33.1**
46	7.25	.817	**14.0**	8.35	1.08	**20.1**	10.50	1.71	**36.1**
48	7.56	.889	**15.2**	8.72	1.18	**21.8**	10.95	1.86	**39.2**
50	7.88	.965	**16.5**	9.08	1.28	**23.6**	11.41	2.02	**42.4**
55	8.67	1.17	**19.8**	9.99	1.55	**28.4**	12.55	2.45	**51.0**
60	9.46	1.39	**23.4**	10.9	1.8	**33.6**	13.69	2.91	**60.4**
65	10.24	1.63	**27.3**	11.8	2.2	**39.2**	14.83	3.41	**70.6**
70	11.03	1.89	**31.5**	12.7	2.5	**45.3**	15.97	3.96	**81.5**
75	11.8	2.17	**36.0**	13.6	2.9	**51.8**	17.11	4.55	**93.2**
80	12.6	2.47	**40.8**	14.5	3.3	**58.7**	18.25	5.17	**106**
85	13.4	2.79	**45.9**	15.4	3.7	**66.0**	19.40	5.84	**119**
90	14.2	3.13	**51.3**	16.3	4.1	**73.8**	20.54	6.55	**133**
95	15.0	3.48	**57.0**	17.2	4.6	**82.0**	21.68	7.29	**148**
100	15.8	3.86	**63.0**	18.2	5.1	**90.7**	22.82	8.08	**164**
110	17.3	4.67	**75.8**	20.0	6.2	**109.3**	25.10	9.78	**197**
120	18.9	5.56	**89.9**	21.8	7.4	**129.6**	27.38	11.6	**234**
130	20.5	6.52	**105**	23.6	8.7	**151.6**	29.66	13.7	**274**
140	22.1	7.56	**122**	25.4	10.0	**175**			
150	23.6	8.68	**139**	27.2	11.5	**201**			
160	25.2	9.88	**158**	29.0	13.1	**228**			
170	26.8	11.15	**178**	30.9	14.8	**257**			
180	28.4	12.50	**199**	32.7	16.6	**288**			

Note: No allowance has been made for age, difference in diameter, or any abnormal condition of interior surface. Any factor of safety must be estimated from the local conditions and the requirements of each particular installation. It is recommended that for most commercial design purposes a safety factor of 15 to 20% be added to the values in the tables

Appendix B

Friction of Water New Steel Pipe

(Based on Darcy's Formula)

2 Inch

Flow US gal per min	Standard wt steel—sch 40			Extra strong steel—sch 80			Schedule 160—steel		
	2.067" inside dia			1.939" inside dia			1.687" inside dia		
	Velocity ft per sec	Velocity head ft	**Head loss ft per 100 ft**	Velocity ft per sec	Velocity head ft	**Head loss ft per 100 ft**	Velocity ft per sec	Velocity head ft	**Head loss ft per 100 ft**
5	.478	.004	**.074**	.54	.00	**.101**	.718	.008	**.197**
6	.574	.005	**.102**	.65	.01	**.139**	.861	.012	**.271**
7	.669	.007	**.134**	.76	.01	**.182**	1.01	.016	**.357**
8	.765	.009	**.170**	.87	.01	**.231**	1.15	.020	**.452**
9	.860	.012	**.209**	.98	.01	**.285**	1.29	.026	**.559**
10	.956	.014	**.252**	1.09	.02	**.343**	1.44	.032	**.675**
12	1.15	.021	**.349**	1.30	.03	**.476**	1.72	.046	**.938**
14	1.34	.028	**.461**	1.52	.04	**.629**	2.01	.063	**1.20**
16	1.53	.036	**.586**	1.74	.05	**.800**	2.30	.082	**1.53**
18	1.72	.046	**.725**	1.96	.06	**.991**	2.58	.104	**1.90**
20	1.91	.057	**.878**	2.17	.07	**1.16**	2.87	.128	**2.31**
22	2.10	.069	**1.05**	2.39	.09	**1.38**	3.16	.155	**2.76**
24	2.29	.082	**1.18**	2.61	.11	**1.62**	3.45	.184	**3.25**
26	2.49	.096	**1.37**	2.83	.12	**1.88**	3.73	.216	**3.77**
28	2.68	.111	**1.57**	3.04	.14	**2.16**	4.02	.251	**4.33**
30	2.87	.128	**1.82**	3.26	.17	**2.46**	4.31	.288	**4.93**
35	3.35	.174	**2.38**	3.80	.22	**3.28**	5.02	.392	**6.59**
40	3.82	.227	**3.06**	4.35	.29	**4.21**	5.74	.512	**8.49**
45	4.30	.288	**3.82**	4.89	.37	**5.26**	6.46	.648	**10.6**
50	4.78	.355	**4.66**	5.43	.46	**6.42**	7.18	.799	**13.0**
55	5.26	.430	**5.58**	5.98	.56	**7.70**	7.89	.967	**15.6**
60	5.74	.511	**6.58**	6.52	.66	**9.09**	8.61	1.15	**18.4**
65	6.21	.600	**7.66**	7.06	.77	**10.59**	9.33	1.35	**21.5**
70	6.69	.696	**8.82**	7.61	.90	**12.2**	10.05	1.57	**24.8**
75	7.17	.799	**10.1**	8.15	1.03	**13.9**	10.77	1.80	**28.3**
80	7.65	.909	**11.4**	8.69	1.17	**15.8**	11.48	2.05	**32.1**
85	8.13	1.03	**12.8**	9.03	1.27	**17.7**	12.20	2.31	**36.1**
90	8.60	1.15	**14.3**	9.78	1.49	**19.8**	12.92	2.59	**40.3**
95	9.08	1.28	**15.9**	10.3	1.6	**22.0**	13.64	2.89	**44.8**
100	9.56	1.42	**17.5**	10.9	1.8	**24.3**	14.35	3.20	**49.5**
110	10.52	1.72	**21.0**	12.0	2.2	**29.2**	15.79	3.87	**59.6**
120	11.5	2.05	**24.9**	13.0	2.6	**34.5**	17.22	4.61	**70.6**
130	12.4	2.40	**29.1**	14.1	3.1	**40.3**	18.66	5.40	**82.6**
140	13.4	2.78	**33.6**	15.2	3.6	**46.6**	20.10	6.27	**95.5**
150	14.3	3.20	**38.4**	16.3	4.1	**53.3**	21.53	7.20	**109**
160	15.3	3.64	**43.5**	17.4	4.7	**60.5**	22.97	8.19	**124**
170	16.3	4.11	**49.0**	18.5	5.3	**68.1**	24.40	9.24	**140**
180	17.2	4.60	**54.8**	19.6	6.0	**76.1**	25.84	10.36	**156**
190	18.2	5.13	**60.9**	20.6	6.6	**84.6**	27.27	11.54	**174**
200	19.1	5.68	**67.3**	21.7	7.3	**93.6**	28.71	12.79	**192**
220	21.0	6.88	**81.1**	23.9	8.9	**113**			
240	22.9	8.18	**96.2**	26.9	10.6	**134**			
260	24.9	9.60	**113**	28.3	12.4	**157**			
280	26.8	11.14	**130**	30.4	14.4	**181**			
300	28.7	12.8	**149**	32.6	16.5	**208**			

Note: No allowance has been made for age, difference in diameter, or any abnormal condition of interior surface. Any factor of safety must be estimated from the local conditions and the requirements of each particular installation. It is recommended that for most commercial design purposes a safety factor of 15 to 20% be added to the values in the tables

Reprinted with permission from Cameron Hydraulic Data book.

Friction of Water New Steel Pipe

(Based on Darcy's Formula)

2½ Inch

Flow	Standard wt steel—sch 40			Extra strong steel—sch 80			Schedule 160—steel		
	2.469" inside dia			2.323" inside dia			2.125" inside dia		
U S gal per min	Velocity ft per sec	Velocity head ft	**Head loss ft per 100 ft**	Velocity ft per sec	Velocity head ft	**Head loss ft per 100 ft**	Velocity ft per sec	Velocity head ft	**Head loss ft per 100 ft**
8	.536	.005	**.072**	.61	.01	**.097**	.724	.008	**.149**
10	.670	.007	**.107**	.76	.01	**.144**	.905	.013	**.221**
12	.804	.010	**.148**	.91	.01	**.199**	1.09	.018	**.305**
14	.938	.014	**.195**	1.06	.02	**.261**	1.27	.025	**.403**
16	1.07	.018	**.247**	1.21	.02	**.332**	1.45	.033	**.512**
18	1.21	.023	**.305**	1.36	.03	**.411**	1.63	.041	**.634**
20	1.34	.028	**.369**	1.51	.04	**.497**	1.81	.051	**.767**
22	1.47	.034	**.438**	1.67	.04	**.590**	1.99	.061	**.912**
24	1.61	.040	**.513**	1.82	.05	**.691**	2.17	.073	**1.03**
26	1.74	.047	**.593**	1.97	.06	**.800**	2.35	.086	**1.20**
28	1.88	.055	**.679**	2.12	.07	**.915**	2.53	.100	**1.37**
30	2.01	.063	**.770**	2.27	.08	**1.00**	2.71	.114	**1.56**
35	2.35	.086	**0.99**	2.65	.11	**1.33**	3.17	.156	**2.08**
40	2.68	.112	**1.26**	3.03	.14	**1.71**	3.62	.203	**2.66**
45	3.02	.141	**1.57**	3.41	.18	**2.13**	4.07	.257	**3.32**
50	3.35	.174	**1.91**	3.79	.22	**2.59**	4.52	.318	**4.05**
55	3.69	.211	**2.28**	4.16	.27	**3.10**	4.98	.384	**4.85**
60	4.02	.251	**2.69**	4.54	.32	**3.65**	5.43	.457	**5.72**
65	4.36	.295	**3.13**	4.92	.38	**4.25**	5.88	.537	**6.66**
70	4.69	.342	**3.60**	5.30	.44	**4.89**	6.33	.622	**7.67**
75	5.03	.393	**4.10**	5.68	.50	**5.58**	6.79	.714	**8.75**
80	5.36	.447	**4.64**	6.05	.57	**6.31**	7.24	.813	**9.90**
85	5.70	.504	**5.20**	6.43	.64	**7.08**	7.69	.918	**11.1**
90	6.03	.565	**5.80**	6.81	.72	**7.89**	8.14	1.03	**12.4**
95	6.37	.630	**6.43**	7.19	.80	**8.76**	8.59	1.15	**13.8**
100	6.70	.698	**7.09**	7.57	.89	**9.66**	9.05	1.27	**15.2**
110	7.37	.844	**8.51**	8.33	1.08	**11.6**	9.95	1.54	**18.3**
120	8.04	1.00	**10.1**	9.08	1.28	**13.7**	10.86	1.83	**21.6**
130	8.71	1.18	**11.7**	9.84	1.50	**16.0**	11.76	2.15	**25.2**
140	9.38	1.37	**13.5**	10.6	1.7	**18.5**	12.67	2.49	**29.1**
150	10.05	1.57	**15.5**	11.3	2.0	**21.1**	13.57	2.86	**33.3**
160	10.7	1.79	**17.5**	12.1	2.3	**23.9**	14.47	3.25	**37.8**
170	11.4	2.02	**19.7**	12.9	2.6	**26.9**	15.38	3.67	**42.5**
180	12.1	2.26	**22.0**	13.6	2.9	**30.1**	16.28	4.12	**47.5**
190	12.7	2.52	**24.4**	14.4	3.2	**33.4**	17.19	4.59	**52.8**
200	13.4	2.79	**27.0**	15.1	3.5	**36.9**	18.09	5.08	**58.4**
220	14.7	3.38	**32.5**	16.7	4.3	**44.4**	19.90	6.15	**70.3**
240	16.1	4.02	**38.5**	18.2	5.1	**52.7**	21.71	7.32	**83.4**
260	17.4	4.72	**45.0**	19.7	6.0	**61.6**	23.52	8.59	**97.6**
280	18.8	5.47	**52.3**	21.2	7.0	**71.2**	25.33	9.96	**113**
300	20.1	6.28	**59.6**	22.7	8.0	**81.6**	27.14	11.43	**129**
350	23.5	8.55	**80.6**	26.5	10.9	**110**	31.66	15.56	**175**
400	26.8	11.2	**105**	30.3	14.3	**144**	36.19	20.32	**228.**
450	30.2	14.1	**132**	34.1	18.1	**181**	40.71	25.72	**288**
500	33.5	17.4	**163**	37.9	22.3	**223**	45.23	31.75	**354**

Note: No allowance has been made for age, difference in diameter, or any abnormal condition of interior surface. Any factor of safety must be estimated from the local conditions and the requirements of each particular installation. It is recommended that for most commercial design purposes a safety factor of 15 to 20% be added to the values in the tables

Friction of Water

Asphalt-dipped Cast Iron and New Steel Pipe

(Based on Darcy's Formula)

3 Inch

Flow US gal per min	Asphalt-dipped cast iron			Std wt steel sch 40			Extra strong steel sch 80			Schedule 160—steel		
	3.0" inside dia			3.068" inside dia			2.900" inside dia			2.624" inside dia		
	Velocity ft per sec	Velocity head ft	Head loss ft per 100 ft	Velocity ft per sec	Velocity head ft	Head loss ft per 100 ft	Velocity ft per sec	Velocity head ft	Head loss ft per 100 ft	Velocity ft per sec	Velocity head ft	Head loss ft per 100 ft
10	.454	.00	.042	.434	.003	.038	.49	.00	.050	.593	.005	.080
15	.681	.01	.088	.651	.007	.077	.73	.01	.101	.890	.012	.164
20	.908	.01	.149	.868	.012	.129	.97	.02	.169	1.19	.022	.275
25	1.13	.02	.225	1.09	.018	.192	1.21	.02	.253	1.48	.034	.411
30	1.36	.03	.316	1.30	.026	.267	1.45	.03	.351	1.78	.049	.572
35	1.59	.04	.421	1.52	.036	.353	1.70	.04	.464	2.08	.067	.757
40	1.82	.05	.541	1.74	.047	.449	1.94	.06	.592	2.37	.087	.933
45	2.04	.06	.676	1.95	.059	.557	2.18	.07	.734	2.67	.111	1.16
50	2.27	.08	.825	2.17	.073	.676	2.43	.09	.860	2.97	.137	1.41
55	2.50	.10	.990	2.39	.089	.776	2.67	.11	1.03	3.26	.165	1.69
60	2.72	.12	1.17	2.60	.105	.912	2.91	.13	1.21	3.56	.197	1.99
65	2.95	.14	1.36	2.82	.124	1.06	3.16	.15	1.40	3.86	.231	2.31
70	3.18	.16	1.57	3.04	.143	1.22	3.40	.18	1.61	4.15	.268	2.65
75	3.40	.18	1.79	3.25	.165	1.38	3.64	.21	1.83	4.45	.307	3.02
80	3.63	.21	2.03	3.47	.187	1.56	3.88	.23	2.07	4.75	.350	3.41
85	3.86	.23	2.28	3.69	.211	1.75	4.12	.26	2.31	5.04	.395	3.83
90	4.08	.26	2.55	3.91	.237	1.95	4.37	.29	2.58	5.34	.443	4.27
95	4.31	.29	2.83	4.12	.264	2.16	4.61	.33	2.86	5.63	.493	4.73
100	4.54	.32	3.12	4.34	.293	2.37	4.85	.36	3.15	5.93	.546	5.21
110	4.99	.39	3.75	4.77	.354	2.84	5.33	.44	3.77	6.53	.661	6.25
120	5.45	.46	4.45	5.21	.421	3.35	5.81	.52	4.45	7.12	.787	7.38
130	5.90	.54	5.19	5.64	.495	3.90	6.30	.62	5.19	7.71	.923	8.61
140	6.35	.63	6.00	6.08	.574	4.50	6.79	.71	5.98	8.31	1.07	9.92
150	6.81	.72	6.87	6.51	.659	5.13	7.28	.82	6.82	8.90	1.23	11.3
160	7.26	.82	7.79	6.94	.749	5.80	7.76	.93	7.72	9.49	1.40	12.8
180	8.17	1.04	9.81	7.81	.948	7.27	8.72	1.01	9.68	10.68	1.77	16.1
200	9.08	1.28	12.1	8.68	1.17	8.90	9.70	1.46	11.86	11.87	2.19	19.8
220	9.98	1.55	14.5	9.55	1.42	10.7	10.7	1.78	14.26	13.05	2.64	23.8
240	10.9	1.84	17.3	10.4	1.69	12.7	11.6	2.07	16.88	14.24	3.15	28.2
260	11.8	2.16	20.2	11.3	1.98	14.8	12.6	2.46	19.71	15.43	3.69	32.9
280	12.7	2.51	23.4	12.2	2.29	17.1	13.6	2.88	22.77	16.61	4.28	38.0
300	13.6	2.88	26.8	13.0	2.63	19.5	14.5	3.26	26.04	17.80	4.92	43.5
320	14.5	3.28	30.4	13.9	3.00	22.1	15.5	3.77	29.53	18.99	5.59	49.4
340	15.4	3.70	34.3	14.8	3.38	24.9	16.5	4.22	33.24	20.17	6.32	55.6
360	16.3	4.15	38.4	15.6	3.79	27.8	17.5	4.73	37.16	21.36	7.08	62.2
380	17.2	4.62	42.7	16.5	4.23	30.9	18.4	5.27	41.31	22.55	7.89	69.2
400	18.2	5.12	47.3	17.4	4.68	34.2	19.4	5.81	45.67	23.73	8.74	76.5
420	19.1	5.65	52.1	18.2	5.16	37.6	20.4	6.43	50.25	24.92	9.64	84.2
440	20.0	6.20	57.1	19.1	5.67	41.2	21.4	7.13	55.05	26.11	10.58	92.2
460	20.9	6.77	62.4	20.0	6.19	44.9	22.3	7.75	60.06	27.29	11.56	101
480	21.8	7.38	67.9	20.8	6.74	48.8	23.3	8.37	65.30	28.48	12.59	109
500	22.7	8.00	73.6	21.7	7.32	52.9	24.2	9.15	70.75	29.66	13.66	119
550	25.0	9.68	88.9	23.9	8.85	63.8	26.7	11.1	85.33	32.63	16.53	143
600	27.2	11.5	106	26.0	10.5	75.7	29.1	13.1	101	35.60	19.67	170
650	29.5	13.5	124	28.2	12.4	88.6	31.6	15.5	119	38.56	23.08	199

Note: No allowance has been made for age, difference in diameter, or any abnormal condition of interior surface. Any factor of safety must be estimated from the local conditions and the requirements of each particular installation. It is recommended that for most commercial design purposes a safety factor of 15 to 20% be added to the values in the tables

Part Three: Friction Losses in Pipe Fittings

Friction of Water

Friction Loss in Pipe Fittings

Resistance coefficient K $\left(\text{use in formula } h_f = K \frac{V^2}{2g}\right)$

Note: Fittings are standard with full openings.

Fitting	L/D	Nominal pipe size ½	¾	1	1¼	1½	2	2½–3	4	6	8–10	12–16	18–24
		K value											
Gate Valves	8	0.22	0.20	0.18	0.18	0.15	0.15	0.14	0.14	0.12	0.11	0.10	0.10
Globe Valves	340	9.2	8.5	7.8	7.5	7.1	6.5	6.1	5.8	5.1	4.8	4.4	4.1
Angle Valves	55	1.48	1.38	1.27	1.21	1.16	1.05	0.99	0.94	0.83	0.77	0.72	0.66
Angle Valves	150	4.05	3.75	3.45	3.30	3.15	2.85	2.70	2.55	2.25	2.10	1.95	1.80
Ball Valves	3	0.08	0.08	0.07	0.07	0.06	0.06	0.05	0.05	0.05	0.04	0.04	0.04

Calculated from data in Crane Co. Technical Paper No. 410.

Reprinted with permission from Cameron Hydraulic Data book.

Friction of Water

Friction Losses in Pipe Fittings

Resistance coefficient K (use in formula $h_f = K \frac{V^2}{2g}$)

Note: Fittings are standard with full openings.

Fitting		L/D	Nominal pipe size ½	¾	1	1¼	1½	2	2½–3	4	6	8–10	12–16	18–24
			K value											
Butterfly Valve								0.86	0.81	0.77	0.68	0.63	0.35	**0.30**
Plug Valve straightway		18	0.49	0.45	0.41	0.40	0.38	0.34	0.32	0.31	0.27	0.25	0.23	**0.22**
Plug Valve 3-way thru-flo		30	0.81	0.75	0.69	0.66	0.63	0.57	0.54	0.51	0.45	0.42	0.39	**0.36**
Plug Valve branch-flo		90	2.43	2.25	2.07	1.98	1.89	1.71	1.62	1.53	1.35	1.26	1.17	**1.08**
Standard elbow	90°	30	0.81	0.75	0.69	0.66	0.63	0.57	0.54	0.51	0.45	0.42	0.39	**0.36**
	45°	16	0.43	0.40	0.37	0.35	0.34	0.30	0.29	0.27	0.24	0.22	0.21	**0.19**
	long radius 90°	16	0.43	0.40	0.37	0.35	0.34	0.30	0.29	0.27	0.24	0.22	0.21	**0.19**

Calculated from data in Crane Co., Technical Paper No. 410.

Friction of Water

Friction Losses in Pipe Fittings

Resistance coefficient K $\left(\text{use in formula } h_f = K\frac{V^2}{2g}\right)$

Note: Fittings are standard with full openings.

Fitting	Type of bend	L/D	Nominal pipe size: ½	¾	1	1¼	1½	2	2½–3	4	6	8–10	12–16	18–24
			K value											
Close Return Bend		50	1.35	1.25	1.15	1.10	1.05	0.95	0.90	0.85	0.75	0.70	0.65	0.60
Standard Tee	thru flo	20	0.54	0.50	0.46	0.44	0.42	0.38	0.36	0.34	0.30	0.28	0.26	0.24
	thru branch	60	1.62	1.50	1.38	1.32	1.26	1.14	1.08	1.02	0.90	0.84	0.78	0.72
90° Bends, Pipe bends, flanged elbows, butt welded elbows	r/d = 1	20	0.54	0.50	0.46	0.44	0.42	0.38	0.36	0.34	0.30	0.28	0.26	0.24
	r/d = 2	12	0.32	0.30	0.28	0.26	0.25	0.23	0.22	0.20	0.18	0.17	0.16	0.14
	r/d = 3	12	0.32	0.30	0.28	0.26	0.25	0.23	0.22	0.20	0.18	0.17	0.16	0.14
	r/d = 4	14	0.38	0.35	0.32	0.31	0.29	0.27	0.25	0.24	0.21	0.20	0.18	0.17
	r/d = 6	17	0.46	0.43	0.39	0.37	0.36	0.32	0.31	0.29	0.26	0.24	0.22	0.20
	r/d = 8	24	0.65	0.60	0.55	0.53	0.50	0.46	0.43	0.41	0.36	0.34	0.31	0.29
	r/d = 10	30	0.81	0.75	0.69	0.66	0.63	0.57	0.54	0.51	0.45	0.42	0.39	0.36
	r/d = 12	34	0.92	0.85	0.78	0.75	0.71	0.65	0.61	0.58	0.51	0.48	0.44	0.41
	r/d = 14	38	1.03	0.95	0.87	0.84	0.80	0.72	0.68	0.65	0.57	0.53	0.49	0.46
	r/d = 16	42	1.13	1.05	0.97	0.92	0.88	0.80	0.76	0.71	0.63	0.59	0.55	0.50
	r/d = 18	46	1.24	1.15	1.06	1.01	0.97	0.87	0.83	0.78	0.69	0.64	0.60	0.55
	r/d = 20	50	1.35	1.25	1.15	1.10	1.05	0.95	0.90	0.85	0.75	0.70	0.65	0.60
Mitre Bends	α = 0°	2	0.05	0.05	0.05	0.04	0.04	0.04	0.04	0.03	0.03	0.03	0.03	0.02
	α = 15°	4	0.11	0.10	0.09	0.09	0.08	0.08	0.07	0.07	0.06	0.06	0.05	
	α = 30°	8	0.22	0.20	0.18	0.18	0.17	0.15	0.14	0.14	0.12	0.11	0.10	0.10
	α = 45°	15	0.41	0.38	0.35	0.33	0.32	0.29	0.27	0.26	0.23	0.21	0.20	0.18
	α = 60°	25	0.68	0.63	0.58	0.55	0.53	0.48	0.45	0.43	0.38	0.35	0.33	0.30
	α = 75°	40	1.09	1.00	0.92	0.88	0.84	0.76	0.72	0.68	0.60	0.56	0.52	0.48
	α = 90°	60	1.62	1.50	1.38	1.32	1.26	1.14	1.08	1.02	0.90	0.84	0.78	0.72

Calculated from data in Crane Co. Technical Paper No. 410.

Reprinted with permission from Cameron Hydraulic Data book.

Friction of Water

Friction Losses in Pipe Fittings

Resistance coefficient K $\left(\text{use in formula } h_f = K\frac{V^2}{2g}\right)$

Note: Fittings are standard with full port openings.

Fitting stop-check valves	L/D	Minimum velocity for full disc lift: general ft/sec†	Minimum velocity for full disc lift: water ft/sec	Nominal pipe size: ½	¾	1	1¼	1½	2	2½–3	4	6	8–10	12–16	18–24
				K value*											
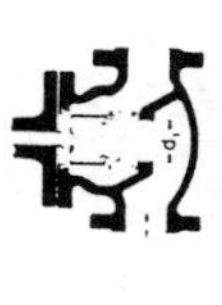	400	$55\sqrt{\bar{V}}$	6.96	10.8	10	9.2	8.8	8.4	7.5	7.2	6.8	6.0	5.6	5.2	4.8
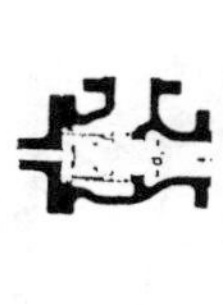	200	$75\sqrt{\bar{V}}$	9.49	5.4	5	4.6	4.4	4.2	3.8	3.6	3.4	3.0	2.8	2.6	2.4
	350	$60\sqrt{\bar{V}}$	7.59	9.5	8.8	8.1	7.7	7.4	6.7	6.3	6.0	5.3	4.9	4.6	4.2
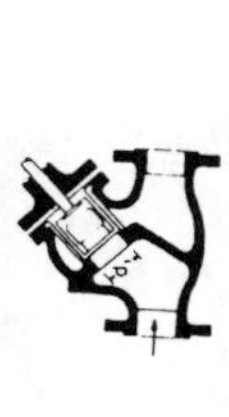	300	$60\sqrt{\bar{V}}$	7.59	8.1	7.5	6.9	6.6	6.3	5.7	5.4	5.1	4.5	4.2	3.9	3.6
	55	$140\sqrt{\bar{V}}$	17.7	1.5	1.4	1.3	1.2	1.2	1.1	1.0	.94	.83	.77	.72	.66

Calculated from data in Crane Co. Technical Paper No. 410.

* These K values for flow giving full disc lift. K values are higher for low flows giving partial disc lift.

† In these formulas, $\bar{V}$, is specific volume—ft³/lb.

Reprinted with permission from Cameron Hydraulic Data book.

Friction of Water

Friction Loss in Pipe Fittings

Resistance coefficient K $\left(\text{use in formula } h_f = K\frac{V^2}{2g}\right)$

Note: *Fittings* are standard with full port openings.

Fitting	L/D	Minimum velocity for full disc lift: general ft/sec†	Minimum velocity for full disc lift: water ft/sec	½	¾	1	1¼	1½	2	2½–3	4	6	8–10	12–16	18–24
				K value*											
Swing check valve	100	$35\sqrt{\bar{V}}$	4.43	2.7	2.5	2.3	2.2	2.1	1.9	1.8	1.7	1.5	1.4	1.3	1.2
	50	$48\sqrt{\bar{V}}$	6.08	1.4	1.3	1.2	1.1	1.1	1.0	0.9	0.9	.75	.70	.65	.6
Lift check valve	600	$40\sqrt{\bar{V}}$	5.06	16.2	15	13.8	13.2	12.6	11.4	10.8	10.2	9.0	8.4	7.8	7.2
	55	$140\sqrt{\bar{V}}$	17.7	1.5	1.4	1.3	1.2	1.2	1.1	1.0	.94	.83	.77	.72	.66
Tilting disc check valve	5°	$80\sqrt{\bar{V}}$	10.13						.76	.72	.68	.60	.56	.39	.24
	15°	$30\sqrt{\bar{V}}$	3.80						2.3	2.2	2.0	1.8	1.7	1.2	.72
Foot valve with strainer poppet disc	420	$15\sqrt{\bar{V}}$	1.90	11.3	10.5	9.7	9.3	8.8	8.0	7.6	7.1	6.3	5.9	5.5	5.0
Foot valve with strainer hinged disc	75	$35\sqrt{\bar{V}}$	4.43	2.0	1.9	1.7	1.7	1.7	1.4	1.4	1.3	1.1	1.1	1.0	.90

(Columns ½ through 18–24: Nominal pipe size.)

Calculated from data in Crane Co. Technical Paper No. 410.

* These K values for flow giving full disc lift. K values are higher for low flows giving partial disc lift.

† In these formulas, $\bar{V}$, is specific volume—ft³/lb.

Reprinted with permission from Cameron Hydraulic Data book.

Friction of Water
Friction Loss in Pipe Fittings

Resistance coefficient $\left(\text{use in formula } h_f = K \frac{V^2}{2g}\right)$

Fitting	Description	All pipe sizes K value
Pipe exit	projecting sharp edged rounded	1.0
Pipe entrance	inward projecting	0.78
Pipe entrance flush	sharp edged	0.5
	r/d = 0.02	0.28
	r/d = 0.04	0.24
	r/d = 0.06	0.15
	r/d = 0.10	0.09
	r/d = 0.15 & up	0.04

From Crane Co. Technical Paper 410.

PART FOUR: CORRESPONDING PRESSURE TABLE

Corresponding pressure in pounds per square inch (psi) to corresponding head in feet

psi	0	1	2	3	4	5	6	7	8	9
0		2.3	4.6	6.9	9.2	11.6	13.9	16.2	18.5	20.8
10	23.1	25.4	27.7	30.0	32.3	34.7	37.0	39.3	41.6	43.9
20	46.2	48.5	50.8	53.1	55.4	57.8	60.1	62.4	64.7	67.0
30	69.3	71.6	73.9	76.2	78.5	80.9	83.2	85.5	87.8	90.1
40	92.4	94.7	97.0	99.3	101.6	104.0	106.3	108.6	110.9	113.2
50	115.5	117.8	120.1	122.4	124.7	127.1	129.4	131.7	134.0	136.3
60	138.6	140.9	143.2	145.5	147.8	150.2	152.5	154.8	157.1	159.4
70	161.7	164.0	166.3	168.6	170.9	173.3	175.6	177.9	180.2	182.5
80	184.8	187.1	189.4	191.7	194.0	196.4	198.7	201.0	203.3	205.6
90	207.9	210.2	212.5	214.8	217.7	219.5	221.8	224.1	226.4	228.7
100	231.0	233.3	235.6	237.9	240.2	242.6	244.9	247.2	249.5	251.8
110	254.1	256.4	258.7	261.0	263.3	265.7	268.0	270.3	272.6	274.9
120	277.2	279.5	281.8	284.1	286.4	288.8	291.1	293.4	295.7	298.0
130	300.3	302.6	304.9	307.2	309.5	311.9	314.2	316.5	318.8	321.1
140	323.4	325.7	328.0	330.3	332.6	335.0	337.3	339.6	341.9	344.2
150	346.5	348.8	351.1	353.4	355.7	358.1	360.4	362.7	365.0	367.3
160	369.6	371.9	374.2	376.5	378.8	381.2	383.5	385.8	388.1	390.4
170	392.7	395.0	397.3	399.6	401.9	404.3	406.6	408.9	411.2	413.5
180	415.8	418.1	420.4	422.7	425.0	427.4	429.7	432.0	434.3	436.6
190	438.9	441.2	443.5	445.8	448.1	450.5	452.8	455.1	457.4	459.7
200	462.0	464.3	466.6	468.9	471.2	473.6	475.9	478.2	480.5	482.8
210	485.1	487.4	489.7	492.0	494.3	496.7	499.0	501.3	503.6	505.9
220	508.2	510.5	512.8	515.1	517.4	519.8	522.1	524.4	526.7	529.0
230	531.3	533.6	535.9	538.2	540.5	542.9	545.2	547.5	549.8	552.1
240	554.4	556.7	559.0	561.3	563.6	566.0	568.3	570.6	572.9	575.2
250	577.5	579.8	582.1	584.4	586.7	589.1	591.4	593.7	596.0	598.3

To read the table use the following example as a guide.
For 13 psi, go horizontal from 10 at the left and read under 3 vertically.
Answer is 30 ft (corresponding to 13 psi).

Appendix C

50 Cross Connection Questions, Answers, and Illustrations

Reprinted with permission from Watts Regulator.

F-50-6

50 Cross Connection Questions, Answers, & Illustrations

Relating To

Backflow Prevention Products

and

Protection of

Safe Drinking Water Supply

Prepared by

1 What is back-siphonage?

Back-siphonage is the reversal of normal flow in a system caused by a negative pressure (vacuum or partial vacuum) in the supply piping.

2 What factors can cause back-siphonage?

Back-siphonage can be created when there is stoppage of the water supply due to nearby fire-fighting, repairs or breaks in city main, etc. The effect is similar to the sipping of an ice cream soda by inhaling through a straw, which induces a flow in the opposite direction.

3 What is backpressure backflow?

Backpressure backflow is the reversal of normal flow in a system due to an increase in the downstream pressure above that of the supply pressure.

4 What factors can cause a back pressure-backflow condition?

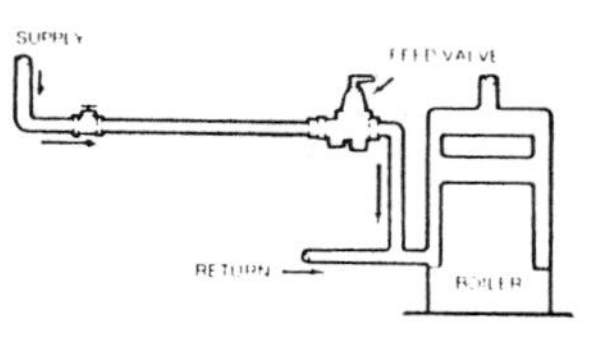

Back pressure-backflow is created whenever the downstream pressure exceeds the supply pressure which is possible in installations such as heating systems, elevated tanks, and pressure-producing systems. An example would be a hot water space-heating boiler operating under 15-20 lbs. pressure coincidental with a reduction of the city water supply below such pressure (or higher in most commercial boilers). As water tends to flow in the direction of least resistance, a back-pressure-backflow condition would be created and the contaminated boiler water would flow into the potable water supply.

5 What is a cross connection?

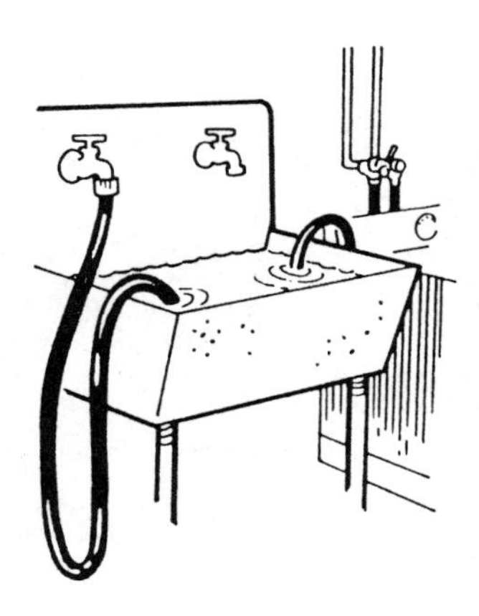

A cross connection is a direct arrangement of a piping line which allows the potable water supply to be connected to a line which contains a contaminant. An example is the common garden hose attached to a sill cock with the end of the hose lying in a cesspool. Other examples are a garden hose attached to a service sink with the end of the hose submerged in a tub full of detergent, supply lines connected to bottom-fed tanks, supply lines to boilers.

6 What is the most common form of a cross connection?

Ironically, the ordinary garden hose is the most common offender as it can be easily connected to the potable water supply and used for a variety of potentially dangerous applications.

2

7 What is potentially dangerous about an unprotected sill cock?

The purpose of a sill cock is to permit easy attachment of a hose for outside watering purposes. However, a garden hose can be extremely hazardous because they are left submerged in swimming pools, lay in elevated locations (above the sill cock) watering shrubs, chemical sprayers are attached to hoses for weed-killing, etc.; and hoses are often left laying on the ground which may be contaminated with fertilizer, cesspools, and garden chemicals.

8 What protection is required for sill cocks?

A hose bibb vacuum breaker should be installed on every sill cock to isolate garden hose applications thus protecting the potable water supply from contamination.

HOSE BIBB VACUUM BREAKER
No. 8

9 Should a hose bibb vacuum breaker be used on frost-free hydrants?

Definitely, providing the device is equipped with means to permit the line to drain after the hydrant is shut-off. A "removable" type hose bibb vacuum breaker could allow the hydrant to be drained, but the possibility exists that users might fail to remove it for draining purposes, thus defeating the benefit of the frost-proof hydrant feature. If the device is of the "Non-Removable" type, be sure it is equipped with means to drain the line to prevent winter freezing.

HOSE BIBB VACUUM BREAKER
FOR FROST-PROOF HYDRANTS
No. NF8

10 Can an atmospheric type, anti-siphon vacuum breaker be installed on a hose bibb?

Theoretically yes, but practically no. An anti-siphon vacuum breaker must be elevated above the sill cock to operate properly. This would require elevated piping up to the vacuum breaker and down to the sill cock and is normally not a feasible installation. On the other hand, a hose bibb vacuum breaker can be attached directly to the sill cock, without plumbing changes and at minor cost.

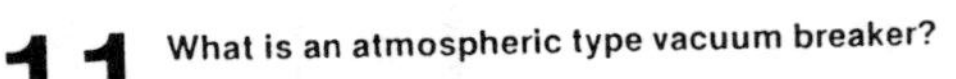

11 What is an atmospheric type vacuum breaker?

The most commonly used atmospheric type anti-siphon vacuum breakers incorporate an atmospheric vent in combination with a check valve. Its operation depends on a supply of potable water to seal off the atmospheric vent, admitting the water to downstream equipment. If a negative pressure develops in the supply line, the loss of pressure permits the check valve to drop sealing the orifice while at the same time the vent opens admitting air to the system to break the vacuum.

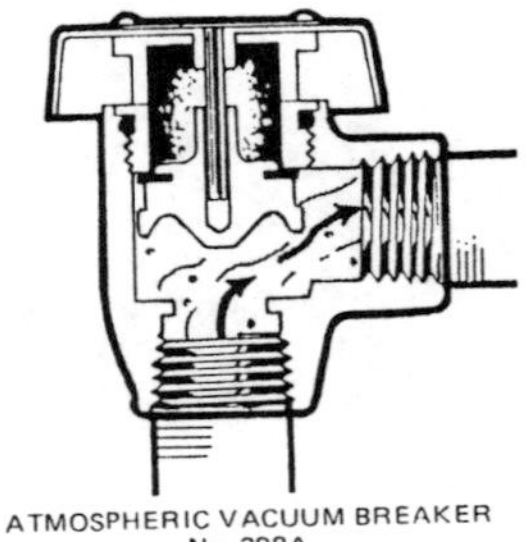

ATMOSPHERIC VACUUM BREAKER
No. 288A

3

12 Will an anti-siphon vacuum breaker protect against a backpressure backflow condition?

Absolutely not! If there is an increase in the downstream pressure over that of the supply pressure, the check valve would tend to "modulate" thus permitting the backflow of contaminated water to pass through the orifice into the potable water supply line.

13 Can an atmospheric type vacuum breaker be used on lawn-sprinkler systems?

Yes, if these are properly installed, they will protect the potable water supply. The device shall be installed 6" above the highest sprinkler head and shall have no control valves located downstream from the device.

SINGLE ZONE SYSTEM

14 Can an atmospheric type vacuum breaker be used under continuous pressure?

No! Codes do not permit this as the device could become "frozen", and not function under an emergency condition.

15 Can a pressure vacuum breaker be used on a multi-zone lawn sprinkler system?

Yes. This type of vacuum breaker can be used under continuous pressure. Therefore, if properly installed, it will protect the potable water supply. The device shall be installed 12" above the highest sprinkler head.

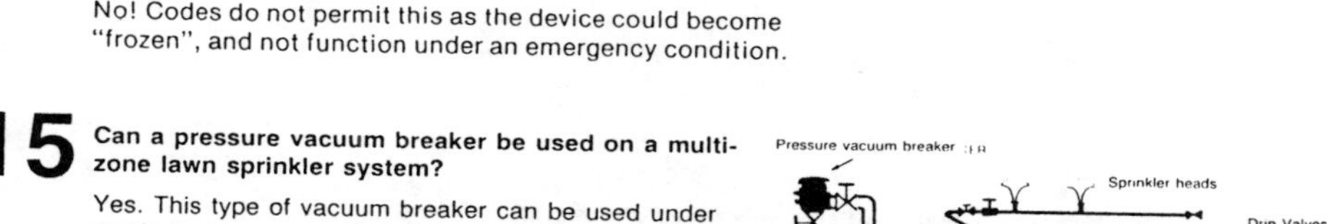

MULTI-ZONE SYSTEM

16 What is continuous pressure?

This is a term applied to an installation in which the pressure is being supplied continuously to a backflow preventive device for periods over 12 hours at a time. Laboratory faucet equipment, for example, is entirely suitable for a non-pressure, atmospheric type anti-siphon vacuum breaker because the supply is periodically being turned on and shut off. A vacuum breaker should never be subjected to continuous pressure unless it is of the continuous pressure type and clearly identified for this service.

17 Are check valves approved for use on boiler feed lines?

Most jurisdictions require backflow protection on all boiler feed lines. Some will allow a backflow preventer with intermediate vent as minimum protection for residential boilers. A reduced pressure backflow preventer is generally required on commercial and compound boilers.

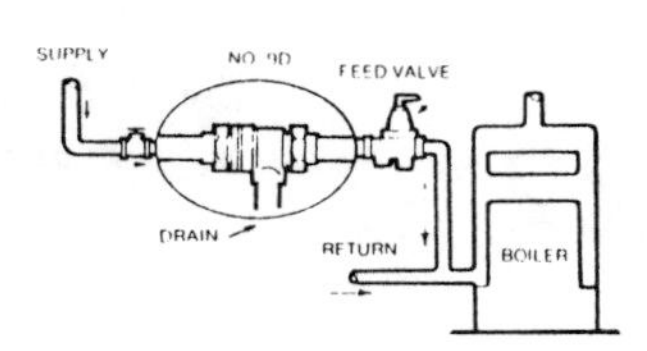

4

However, low cost, continuous pressure backflow preventers are now available which will perform with maximum protection; thus check valves are not recommended.

18 What is the difference between pollution and contamination?

Pollution of the water supply does not constitute an actual health hazard, although the quality of the water is impaired with respect to taste, odor or utility. Contamination of the water supply, however, does constitute an actual health hazard; the consumer being subjected to potentially lethal water borne disease or illness.

19 What recent case would reflect users being exposed to "pollution" of the water supply?

Pollution can sometimes be amusing. In December of 1970 in a winery in Cincinnati, Ohio the water supply valve was inadvertently left open after flushing out wine-distilling tanks. The result was that during a subsequent fermenting process, sparkling Burgundy backflowed from the vats into the city main and out of the kitchen faucets of nearby homeowners. This typical reversal of flow in water supply piping caused by the distilling tanks operating at a pressure higher than the city water supply did impair the condition of the water but did not make it dangerous. Indeed, many thought it was the best water they ever tasted.

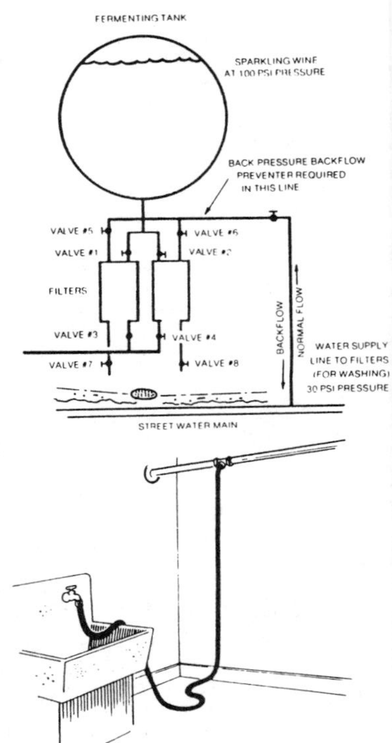

20 What recent case would reflect users being exposed to "contamination" of the water supply?

In May 1969 in a Pennsylvania college, the air conditioning system which contains dangerous chromates became blocked preventing circulation of the coolant. In an attempt to unblock the pipe line, a maintenance man inserted a hose in the pipe and attempted to dislodge the blockage by water pressure. A reversal of flow developed, allowing the chemicals in the air conditioning line to backflow through the hose and into the potable water supply. Unknowing students in other parts of the building subsequently drank what was thought to be potable water, resulting in illness to 23 persons.

21 Are there any other records of recent cases involving unprotected cross connections?

The startling fact is that cross connections are increasing at the estimated rate of 100,000 per day and there are frequently documented cases involving reverse flow.

22 What recently reported cases occurred in a plant?

In addition to the case described in "No. 19", there are additional reports but because of the possibility of litigation for these pending cases, information is difficult to obtain. However,

in 1972 in San Francisco, an industrial plant had a submerged water inlet supplying a lye vat. Immediately adjacent to this installation was the employee's shower room. Officials fortunately discovered the cross connection, but were alarmed that employees could potentially be bathing in water contaminated with lye from the vats.

23 What recent case was reported involving a school?

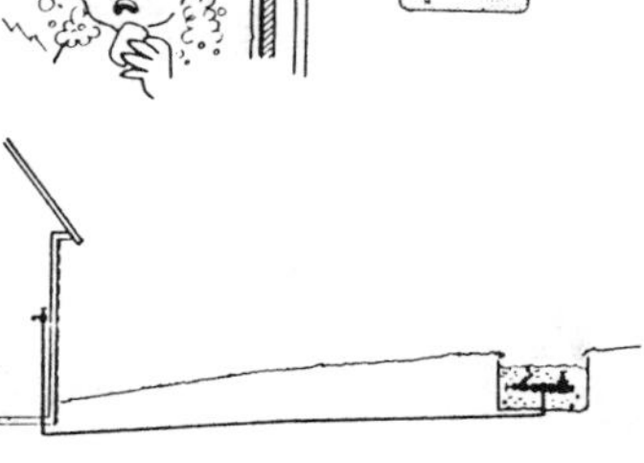

Most people are familiar with the details of the Holy Cross Football Teams' "hepatitis" incident, which was later determined to be caused by a backflow of contaminated water. It took close to nine months for officials to determine that a severe fire in nearby Worcester lowered the pressure in the football field area to the point where a back pressure backflow condition was created allowing contaminants from a sunken hose bibb pit to backflow into the field house drinking bubbler.

24 What recent case was reported involving a commercial bldg.?

Much to the surprise of the customers of a bank in Atlanta, Georgia they saw yellow water flowing from drinking fountains and green ice rolling out of cafeteria dispensing machines.

It was later reported that a pump, used for the air conditioning system, burned out; and a maintenance man, unaware of the danger, connected the system to another pump used for potable water. The result caused large doses of bichromate of soda to be forced into the potable water supply, causing the dramatic appearance of yellow water and colored ice cubes.

25 Are there any cases involving outside processing activities?

Yes, in 1972 a case occurred in a gravel pit operation in Illinois. A pump was used in the processing operation supplying 100 lbs. pressure. Contaminated water was forced back through an unprotected "prime line" overcoming the city water pressure of 45 lbs. The contaminated water entered the city main and was channeled into a nearby bottling plant. This probably would have gone undetected except that personnel in the bottling plant noticed that the water was not only dirty but was warm. City officials were immediately called which led to the discovery of the reverse flow from the gravel pit operation.

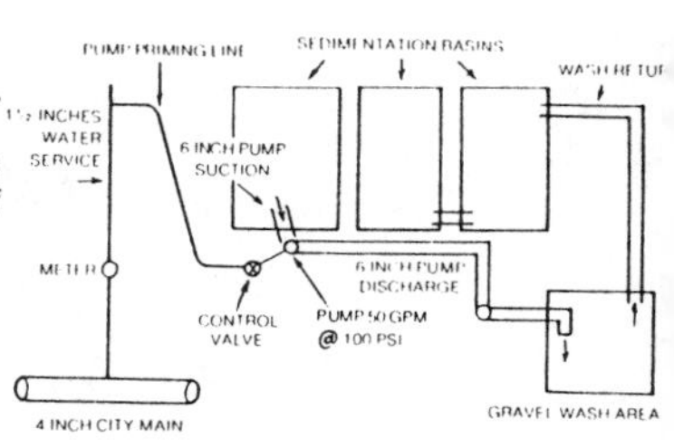

26 What other typical cases have been reported recently?

In 1972 an Automatic Car Wash injected gallons of a strong detergent solution into the city supply. The reverse flowing detergent was discovered in nearby homes more than a block away and was officially classified as being caused by an unprotected cross connection in the plumbing line.

In 1970 in Utah a Doctor reported two gold fish flowing into his bath tub. Earlier in the day he had been filling his gold fish pool with a garden hose when a back-siphonage condition developed resulting in the late emergence of the gold fish into the bath tub.

What is more significant, however, is the number of recent cases that have not been reported. With the number of unprotected cross connections in existence today, these are potential disasters which can occur any time unless adequate protective devices are installed.

27 What is meant by "Degree of Hazard"?

The degree of hazard is a commonly used phrase utilized in cross connection programs and is simply a determination on whether the substance in the non-potable system is toxic or non-toxic. Referencing No. 19 and No. 20, the winery would be a low hazard, while the air conditioning system would be a high hazard installation.

28 What is the difference between a toxic and a non-toxic substance?

Toxic substance is any liquid, solid or gas, which, when introduced into the water supply, creates, or may create, a danger to health and well-being of the consumer. An example is treated boiler water. A non-toxic substance is any substance that may create a moderate hazard, is a nuisance or is aesthetically objectionable. For example, food stuff, such as sugar, soda pop etc. Therefore, you must select the proper device according to the type of connection and degree of hazard. There are five basic devices that can be used to correct cross connection.

29 What are the five basic devices used for protection of cross connections?

The five basic devices are:

1. Air Gap
2. Atmospheric Vacuum Breakers - which also includes hose connection vacuum breakers.
3. Pressure Type Vacuum Breakers - which also includes Backflow Preventer with Intermediate Atmospheric Vent for 1/2" and 3/4" lines.
4. Double Check Valve Assembly
5. Reduced Pressure Principle Backflow Preventers.

30 What is an Air Gap?

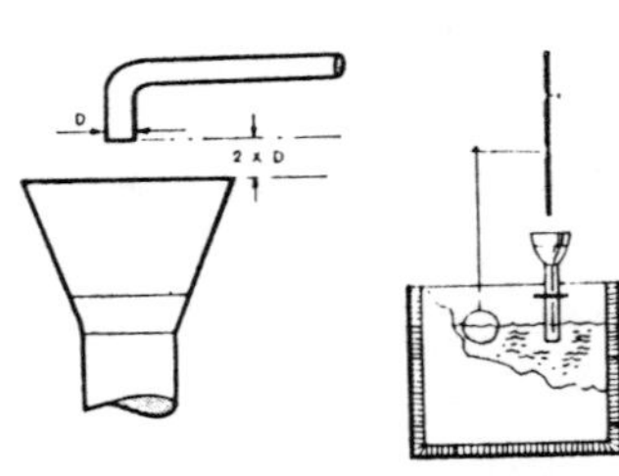

7

Air Gap is the physical separation of the potable and non-potable system by an air space. The vertical distance between the supply pipe and the flood level rim should be two times the diameter of the supply pipe, but never less than 1". The air gap can be used on a direct or inlet connection and for all toxic substances.

909AG Series

31 Where is an Atmospheric Type Vacuum Breaker used?

Atmospheric Vacuum Breakers may be used only on connections to a non-potable system where the vacuum breaker is never subjected to back-pressure and is installed on the discharge side of the last control valve. It must be installed above the usage point. It cannot be used under continuous pressure. (Also see No.11)

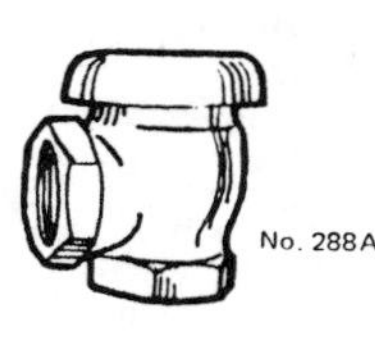
No. 288A

32 Where is a Hose Bibb Vacuum Breaker used?

Hose Bibb Vacuum Breakers are small inexpensive devices with hose connections which are simply attached to sill cocks, and threaded faucets or wherever there is a possibility of a hose being attached which could be introduced to a contaminant. However, like the Atmospheric Type Vacuum Breaker they should not be used under continuous pressure.

No. 8

33 Where is a Pressure Type Vacuum Breaker used?

Pressure Type Vacuum Breakers may be used as protection for connections to all types of non-potable systems where the vacuum breakers are not subject to back-pressure. These units may be used under continuous supply pressure. They must be installed above the usage point.

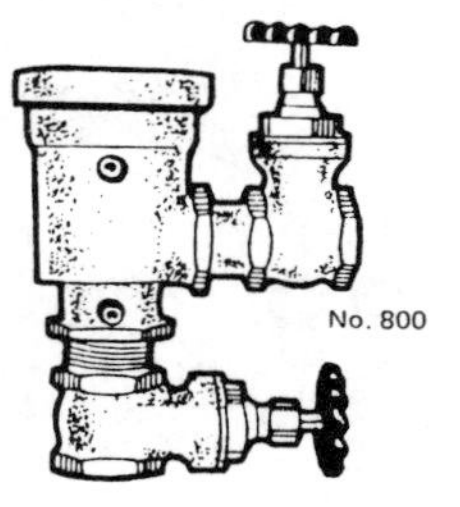
No. 800

34 Where is a Backflow Preventer with Intermediate Atmospheric Vent used?

These devices are made for 1/2" and 3/4" lines and may be used as an alternate equal for pressure type vacuum breakers. In addition, however, they provide the added advantage of providing protection against back-pressure.

No. 9D

35 Where is a Double Check Valve Assembly used?

A double check valve assembly may be used as protection of all direct connections through which foreign material might enter the potable system in concentration which would constitute a nuisance or be aesthetically objectionable, such as air, steam, food, or other material which does not constitute a health hazard.

No. 709

36 Where is a Reduced Pressure Principle Backflow Preventer used?

Reduced Pressure Zone Devices may be used on all direct connections which may be subject to back-pressure or back-siphonage, and where there is the possibility of contamination by the material that does constitute a potential health hazard.

37 What are typical applications for an Air Gap?

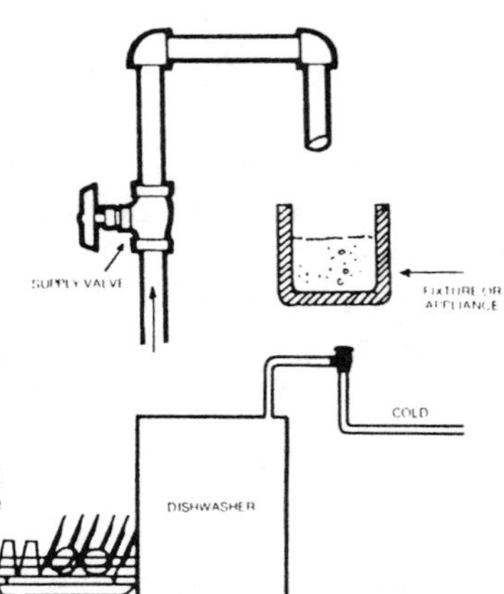

Because today's complex plumbing systems normally require continuous pressure, air gap applications are actually in the minority. It should be remembered, however, that whenever a piping terminates a suitable distance above a contaminant, this itself is actually an air gap. Air Gaps are frequently used on industrial processing application, but care should be taken that subsequent alterations are not made to the piping which would result in a direct connection.

38 What are typical applications for Atmospheric Type Vacuum Breakers?

Atmospheric Type Vacuum Breakers can be used on most inlet type water connections which are not subject to back-pressure such as low inlet feeds to receptacles containing toxic and non-toxic substances, valve outlet or fixture with hose attachments, lawn-sprinkler systems and commercial dishwashers.

39 What are typical applications for Hose Bibb Vacuum Breakers?

Hose Bibb Vacuum Breakers are popularly used on sill cocks, service sinks and any threaded pipe to which a hose may potentially be attached.

40 What are typical applications for Pressure Type Vacuum Breakers?

These applications should be similar to the Atmospheric Type Vacuum Breaker with the exception that these may be used under continuous pressure. However, they should not be subject to back-pressure.

41 What are typical applications of Backflow Preventer with Intermediate Vent?

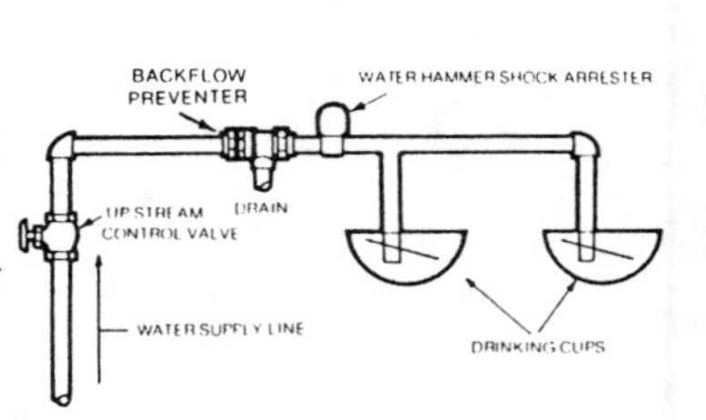

For 1/2" and 3/4" lines these devices are popularly used on boiler feed water supply lines, cattle drinking fountains, trailer park water supply connections and other similar low-flow applications. They will protect against both back-siphonage and back-pressure and can be used under continuous pressure.

42 What are typical applications for Double Check Valve Assemblies?

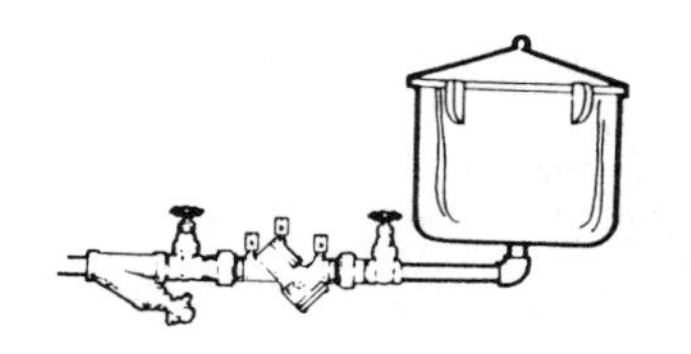

Briefly, Double Check Valve Assemblies may be used where the degree of hazard is low, meaning that the non-potable source is polluted rather than contaminated. The degree of hazard is oftentimes determined by local Inspection Departments and, therefore, such departments should be questioned in order to comply with local regulations.

43 What are typical applications for Reduced Pressure Principle Backflow Preventers?

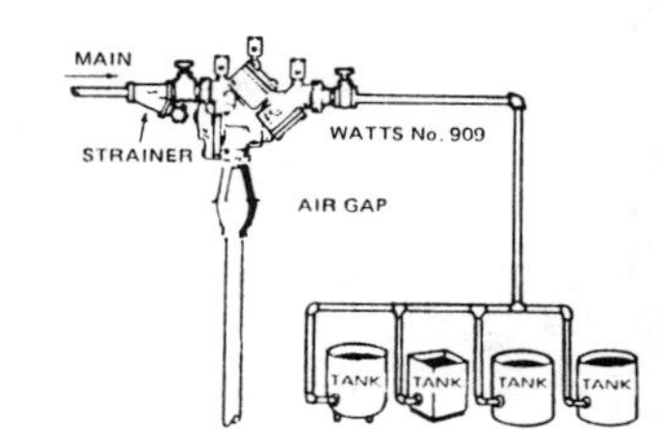

This type should be used whenever the non-potable source is more of a contaminant than a pollutant. Basically, they are applied as main line protection to protect the municipal water supply, but should also be used on branch line applications where non-potable fluid would constitute a health hazard, such as boiler feed lines, commercial garbage disposal systems, industrial boilers, etc.

44 Are there any regulations in OSHA regarding cross connections?

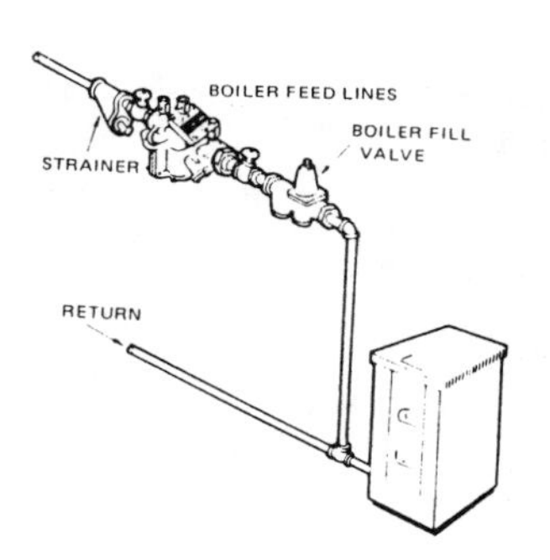

Yes, OSHA requires that no cross connection be allowed in an installation unless it is properly protected with an approved backflow prevention device. These requirements are also covered in B.O.C.A., Southern Std. Building Code, Uniform Plumbing Code and City, State and Federal Regulations.

45 What Standards are available governing the manufacture of backflow prevention devices?

Table A on Page 12 provides a summary of the various standards available relating to specific types of backflow prevention devices.

46 What is the benefit of a strainer preceding a backflow preventer?

A strainer will protect the check valves of a backflow preventer from fouling due to foreign matter and debris which may be flowing through the line. This not only protects the device but eliminates nuisance fouling and subsequent maintenance and shutdown. The use of a strainer with a water pressure reducing valve has been an accepted practice for years. The amount of pressure drop attributed to the strainer is negligible and is far outweighed by the advantages provided by the strainer.

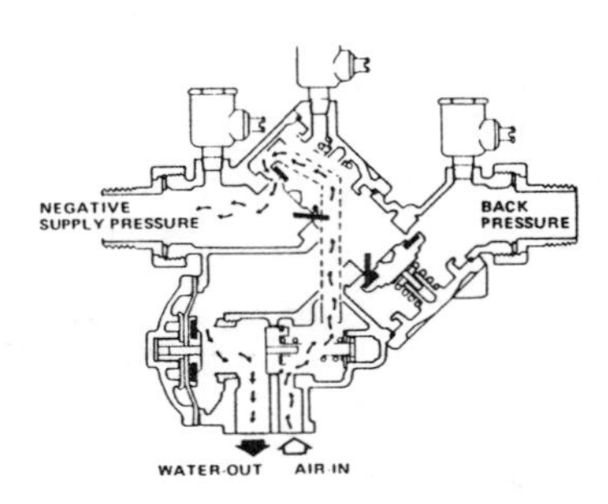

47 What would cause a reduced pressure principle backflow preventer to leak?

Leakage from a backflow preventer is normally attributed to foreign matter lodging on the seating area of either the first or second check valve. Most times this can be corrected by simply flushing the device which will dislodge any loose particles. It is, therefore, most important on new installations that the piping be thoroughly flushed before installing the unit. It should be remembered, however, that spillage does provide a "warning signal" that the device is in need of maintenance.

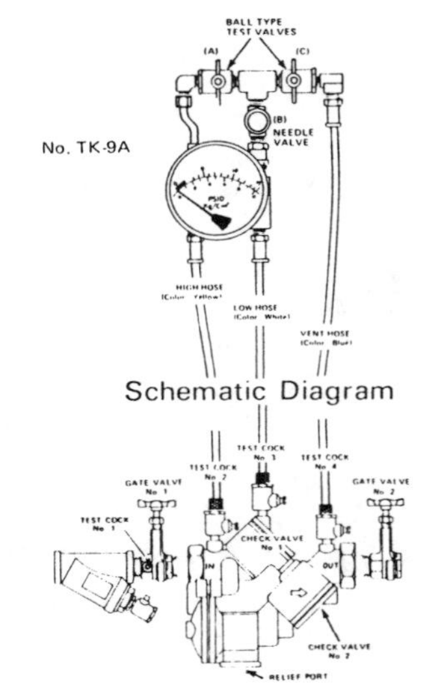

48 Is periodic testing required for reduced pressure principle backflow preventers?

Yes, and this is to ensure that the device is working properly and is a requirement of many states and cross connection control programs. Test cocks are provided on the device for this purpose and manufacturers are required to furnish field testing information.

49 Should a backflow preventer be installed in the water supply line to each residence?

Because of the growing number of serious residential backflow cases, many water purveyors are now requiring the installation of approved dual check valve backflow preventers at residential water meters. They are also educating the public concerning cross connections and the danger of backflow into the local water supply. Since water purveyors cannot possibly be responsible for or monitor the use of water within a residence, the requirements for these cross connection control programs are increasing throughout the country.

50 What is a cross connection control program?

This is a combined cooperative effort between plumbing and health officials, water works companies, property owners and certified testers to establish and administer guidelines for controlling cross connections and implementing means to ensure their enforcement so that the public potable water supply will be protected both in the city main and within buildings. The elements of a program define the type of protection required and responsibility for the administration and enforcement. Other elements ensure continuing education programs.

Watts complete line of Backflow Prevention Devices.

The Complete Concept in Cross-Connection Control and Containment

Series 8 - Backflow Preventers for Hose Bibb Installations

For ¾" H.T sill cocks and threaded faucets where a portable hose could be attached. Tested and approved under A.S.S.E. Std. 1011, and CSA Std. B64.2, ANSI A112.1.3

No. 8,8C,8A,8AC,8B,8P,S8,S8C, and NF8.
No. 8A,8AC,8,8P,S8C,NF8 and NF8C are listed by IAPMO.

No. NLF9 - Laboratory Faucet Vacuum Breaker

Double check valve with atmospheric vent. Especially made for use on laboratory faucets with gooseneck spout. Size ⅜" NPT. Male inlet.

Certified under ANSI/ASSE Std. 1035 - "Laboratory Faucet Vacuum Breakers". Listed by I.A.P.M.O.

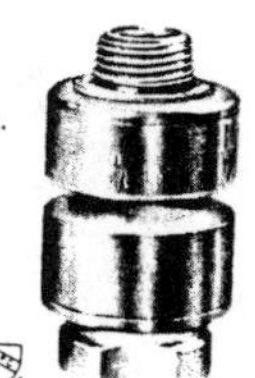

No. 9D - Backflow Preventer with intermediate atmospheric vent

Suitable for continuous pressure. Sizes ½", ¾" NPT. Meets A.S.S.E. Std. 1012 and CSA Std. B64.3.

Series 7 - Dual Check Backflow Preventer

Dual check backflow preventer for residential water supply service or individual outlets. Sizes ½"- 1¼". ANSI/A.S.S.E. Std. 1024. and CSA Std. B64.6.
No. 7C size ⅜" chrome finish. No. 7B size ¾ compact design. Series A7 angle dual check.

TABLE A
STANDARDS AND TESTING LABORATORIES FOR BACKFLOW PREVENTER PERFORMANCE TEST AND CONSTRUCTION

PRODUCT	CURRENT STANDARDS ANSI/ ASSE	CSA	FCCCHR of USC	WATTS SERIES
Atmospheric type vacuum breakers	1001	B64.1.1	Section 10	288A ,N388
Hose connection vacuum breakers	1011	B64.2		8
Backflow preventer with intermediate atmospheric vent	1012	B64.3		9D
Reduced pressure principle backflow preventer	1013	B64.4	Section 10	909, 009
Double check valve type backpressure backflow preventer	1015	B64.5	Section 10	709, 007
Vacuum breakers, pressure type	1020	B64.1.2	Section 10	800
Dual check valve backflow preventer	1024	B64.6		7
Laboratory faucet vacuum breaker	1035	B64.7		NLF9

REFERENCES:
ASSE - American Society of Sanitary Engineeering
AWWA - American Water Works Association
FCCCHR of USC - University of Southern California, Foundation for Cross Connection Control Research
NSF - National Sanitation Foundation

OTHER BACKFLOW PREVENTION DEVICES:
No. 709DCDA - double check detector assembly
No. 909RPDA - reduced pressure detector assembly
No. 288A, N388 - atmospheric vacuum breaker
No. 911 combination backflow preventer and hot water boiler fill valve.
No. N9 similar to NLF9 with female inlet.
No. 9BD backflow preventer for vending machines.
No. FHV frost-proof wall hydrants with vacuum breaker.
For Easy setter retrofit adapter with dual check backflow preventer, No. WES2-7

No. 800QT, 800MQT Pressure Type Vacuum Breaker

Anti-siphon pressure type vacuum breakers for continuous pressure piping systems. A.S.S.E. No. 1020. and B64.1.2. FCCHR of USC. Sizes ½"- 2".

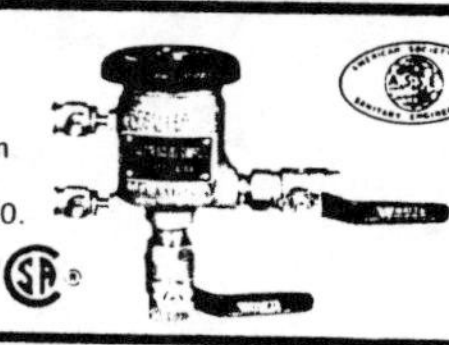

DOUBLE CHECK VALVE BACKFLOW PREVENTERS
Series 709 2½"- 10"
Series 007 ¾"- 3"

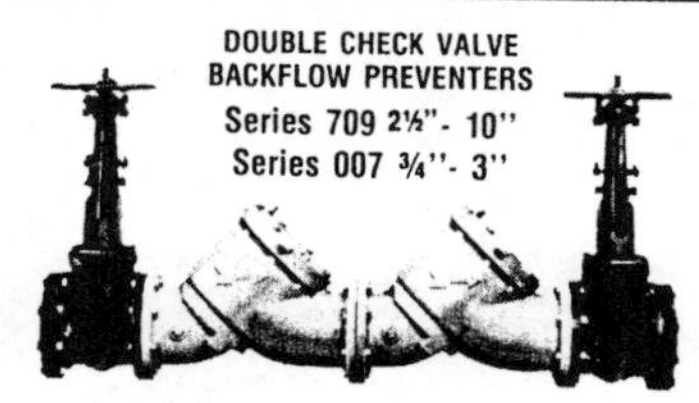

REDUCED PRESSURE ZONE BACKFLOW PREVENTERS
Series 909 ¾"- 10"
Series 009 ½"- 3"

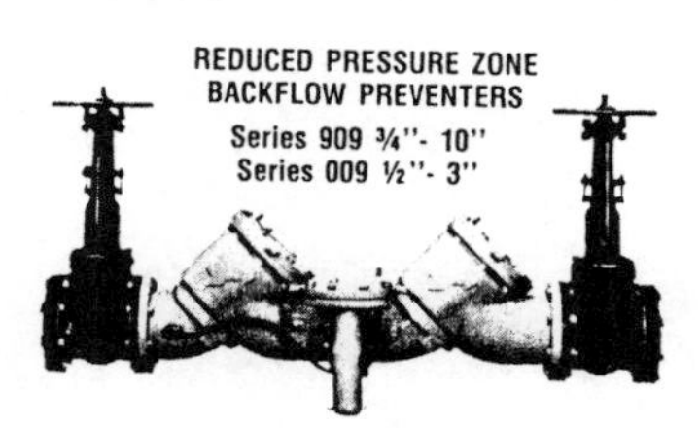

For additional information, send for C-BPD catalog.

World Class Valves

Watts Regulator Company
HDQTRS: 815 Chestnut St., N. Andover, MA 01845
MAIL: Box 628, Lawrence, MA 01842 **Telex:** 94-7460
Tel. (508) 688 1811 **Fax:** (508) 794-1848/794-1674
International Subsidiaries: Watts Regulator of Canada Ltd.
Tel (416) 851 8591 Fax (416) 851 8788
Watts Regulator (Nederland) b v Telex 844 35365

F-50-6 921
Printed in U.S.A.

Appendix D

Rates of Rainfall for Various Cities (in./hr) Informative Data Only

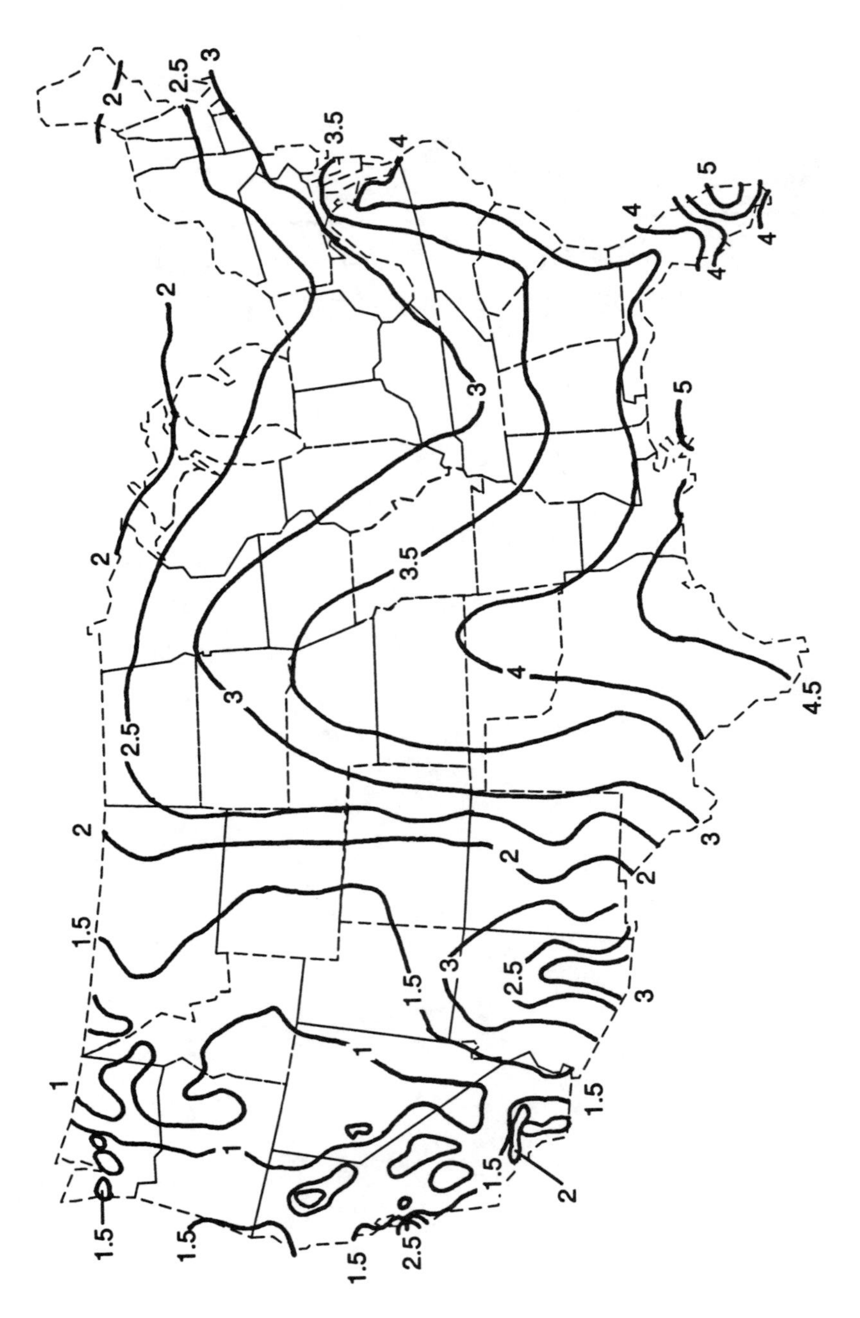

100 year 1-hour rainfall (in.)

Alabama:

Birmingham	3.8
Huntsville	3.6
Mobile	4.6
Montgomery	4.2

Alaska:

Fairbanks	1.0
Juneau	0.6

Arizona:

Flagstaff	2.4
Nogales	3.1
Phoenix	2.5
Yuma	1.6

Arkansas:

Fort Smith	3.6
Little Rock	3.7
Texarkana	3.8

California:

Barstow	1.4
Crescent City	1.5
Eureka	1.4
Fresno	1.1
Los Angeles	2.1
Needles	1.6
Placerville	1.5
San Fernando	2.3
San Francisco	1.5

Colorado:

Craig	1.5
Denver	2.4
Durango	1.8
Grand Junction	1.7
Lamar	3.0
Pueblo	2.5

Connecticut:

Hartford	2.7
New Haven	2.8
Putnam	2.6

Delaware:

Georgetown	3.0
Wilmington	3.1

District of Columbia:

Washington	3.2

Florida:

Jacksonville	4.3
Key West	4.3
Miami	4.7
Pensacola	4.6
Tampa	4.5

Georgia:

Atlanta	3.7
Dalton	3.4
Macon	3.9
Savannah	4.3
Thomasville	4.3

Hawaii:

Hilo	6.2
Honolulu	3.0
Wailuku	3.0

Idaho:

Boise	0.9
Lewiston	1.1
Pocatello	1.2

Illinois:

Cairo	3.3
Chicago	3.0
Peoria	3.3
Rockford	3.2
Springfield	3.3

Indiana:

Evansville	3.2
Fort Wayne	2.9
Indianapolis	3.1

Iowa:

Davenport	3.3
Des Moines	3.4
Dubuque	3.3
Sioux City	3.6

Kansas:

Atwood	3.3
Dodge City	3.3
Topeka	3.7
Wichita	3.7

Kentucky:

Ashland	3.0
Lexington	3.1
Louisville	3.2
Middlesboro	3.2
Paducah	3.3

Louisiana:

Alexandria	4.2
Lake Providence	4.0
New Orleans	4.8
Shreveport	3.9

Maine:

Bangor	2.2
Houlton	2.1
Portland	2.4

Maryland:

Baltimore	3.2
Hagerstown	2.8
Oakland	2.7
Salisbury	3.1

Massachusetts:

Boston	2.5
Pittsfield	2.8
Worcester	2.7

Michigan:

Alpena	2.5
Detroit	2.7
Grand Rapids	2.6
Lansing	2.8
Marquette	2.4
Sault Ste. Marie	2.2

Minnesota:

Duluth	2.8
Grand Marais	2.3
Minneapolis	3.1
Moorhead	3.2
Worthington	3.5

Mississippi:

Biloxi	4.7
Columbus	3.9
Corinth	3.6
Natchez	4.4
Vicksburg	4.1

Missouri:

Columbia	3.2
Kansas City	3.6

St. Louis 3.2
Springfield 3.4

Montana:
Ekalaka 2.5
Havre 1.6
Helena 1.5
Kalispell 1.2
Missoula 1.3

Nebraska:
North Platte 3.3
Omaha 3.8
Scottsbluff 3.1
Valentine 3.2

Nevada:
Elko 1.0
Ely 1.1
Las Vegas 1.4
Reno 1.1

New Hampshire:
Berlin 2.5
Concord 2.5
Keene 2.4

New Jersey:
Atlantic City 2.9
Newark 3.1
Trenton 3.1

New Mexico:
Albuquerque 2.0
Hobbs 3.0
Raton 2.5
Roswell 2.6
Silver City 1.9

New York:
Albany 2.5
Binghamton 2.3
Buffalo 2.3
Kingston 2.7
New York 3.0
Rochester 2.2

North Carolina:
Asheville 4.1
Charlotte 3.7
Greensboro 3.4
Wilmington 4.2

North Dakota:
Bismarck 2.8
Devils Lake 2.9
Fargo 3.1

Ohio:
Cincinnati 2.9
Cleveland 2.6
Columbus 2.8
Toledo 2.8

Oklahoma:
Altus 3.7
Boise City 3.3
Durant 3.8
Oklahoma City 3.8

Oregon:
Baker 0.9
Coos Bay 1.5
Eugene 1.3
Portland 1.2

Pennsylvania:
Erie 2.6
Harrisburg 2.8
Philadelphia 3.1
Pittsburgh 2.6
Scranton 2.7

Rhode Island:
Providence 2.6

South Carolina:
Charleston 4.3
Columbia 4.0
Greenville 4.1

South Dakota:
Buffalo 2.8
Huron 3.3
Pierre 3.1
Rapid City 2.9
Yankton 3.6

Tennessee:
Chattanooga 3.5
Knoxville 3.2
Memphis 3.7
Nashville 3.3

Texas:
Abilene 3.6
Amarillo 3.5
Brownsville 4.5
Dallas 4.0
Del Rio 4.0
El Paso 2.3
Houston 4.6
Lubbock 3.3
Odessa 3.2
Pecos 3.0
San Antonio 4.2

Utah:
Brigham City 1.2
Roosevelt 1.3
St. George 1.7
Salt Lake City 1.3

Vermont:
Barre 2.3
Brattleboro 2.7
Burlington 2.1
Rutland 2.5

Virginia:
Bristol 2.7
Charlottesville 2.8
Lynchburg 3.2
Norfolk 3.4
Richmond 3.3

Washington:
Omak 1.1
Port Angeles 1.1
Seattle 1.4
Spokane 1.0
Yakima 1.1

West Virginia:
Charleston 2.8
Morgantown 2.7

Wisconsin:
Ashland 2.5
Eau Claire 2.9
Green Bay 2.6
La Crosse 3.1
Madison 3.0
Milwaukee 3.0

Wyoming:
Cheyenne 2.2
Fort Bridger 1.3
Lander 1.5
New Castle 2.5
Sheridan 1.7
Yellowstone Park 1.4

Appendix E

Plumbing Terminology

This appendix contains a list of definitions used in plumbing codes and the plumbing industry. For general information, fire protection (FP) terminology is included. The plumbing definitions are reprinted with permission from the American Society of Plumbing Engineers *Data Book 21*, 1989.

ABS
Abbreviation for Acrylonitrile - Butadiene - Styrene.

ABSOLUTE PRESSURE
The total pressure measured from absolute vacuum. It equals the sum of gauge pressure and atmospheric pressure corresponding to the barometer (expressed in pounds per square inch).

ABSOLUTE TEMPERATURE
Temperature measured from absolute zero, a point of temperature, theoretically equal to minus 459.72°F (minus 273.18°C). The hypothetical point at which a substance would have no molecular motion and no heat.

ABSOLUTE ZERO
Zero point on the absolute temperature scale; a point of total absence of heat, equivalent to minus 459.72°F (minus 273.18°C).

ABSORPTION
The immersion in a fluid for a definite period of time, usually expressed as a percent of the weight of the dry pipe.

ACCESSIBLE
When applied to a fixture, connection, appliance, or equipment shall mean having access thereto, but which first may require the removal of an access panel, door or similar obstruction;/readily accessible shall mean direct access without the necessity of removing or moving any panel, door, or similar obstruction.

ACCESSIBLE (esp. Handicapped)
Term describing a site, building, facility, or portion there of that can be approached, entered, and/or used by physically handicapped people.

ACCESS DOOR
Hinged panel mounted in a frame with a lock, normally mounted in a wall or ceiling to provide access to concealed valves or equipment which require frequent attention.

ACCUMULATOR
A container in which fluid or gas is stored under pressure as a source of power.

ACID VENT
A pipe venting an acid waste system.

ACID WASTE
A pipe which conveys liquid waste matter containing a pH of less than 7.0.

ACME THREAD
A screw thread, the section of which is between the square and V threads, used extensively for feed screws. The included angle of space is 29° as compared to 60° of the National Coarse of U.S. Thread.

ACRYLONITRILE-BUTADIENE-STYRENE
A thermoplastic compound from which fittings, pipe, and tubing are made.

ACTIVE SLUDGE
Sewage sediment, rich in destructive bacteria, that can be used to break down fresh sewage more quickly.

ADAPTER FITTING

(1) Any of various fittings designed to mate, or fit to each other, two pipes or fittings which are different in design, when connecting the two together would otherwise not be possible. (2) A fitting that serves to connect two different tubes or pipes to each other, such as copper tube to iron pipe, etc.

ADMINISTRATIVE AUTHORITY

The individual official, board, department, or agency established and authorized by a state, county, city, or other political subdivision created by law to administer and enforce the provisions of the plumbing code.

AERATION

An artificial method in which water and air are brought into direct contact with each other. One purpose is to release certain dissolved gases which often cause water to have obnoxious odors or disagreeable tastes. Also used to furnish oxygen to water that is oxygen deficient. The process may be accomplished by spraying the liquid in the air, bubbling air through the liquid, or by agitation of the liquid to promote surface absorption of the air.

AEROBIC

Bacteria living or active only in the presence of free oxygen.

AIR BREAK

A physical separation in which a drain from a fixture, appliance, or device indirectly discharges into a fixture, receptacle, or interceptor at a point below the flood level rim of the receptacle, to prevent backflow or back-siphonage.

AIR CHAMBER

A continuation of the water piping beyond the branch to fixtures, finished with a cap designed to eliminate shock or vibration of the piping when the faucet is closed suddenly.

AIR, COMPRESSED

Air at any pressure greater than atmospheric pressure.

AIR, FREE

Air which is not contained and which is subject only to atmospheric conditions.

AIR GAP

The unobstructed vertical distance through the free atmosphere between the lowest opening from any pipe or faucet conveying water or waste to a tank, plumbing fixture receptor, or other device and the flood level rim of the receptacle. (Usually required to be twice the diameter of the inlet.)

AIR, STANDARD

Air having a temperature of 70°F, (21.1°C), a standard density of 0.0075 lb/ft (0.11 kg/m), and under pressure of 14.70 psia. (101.4 kPa). The gas industry usually considers 60°F (15.6°C) as the temperature of standard air.

AIR TEST

The test that is applied to the plumbing system upon its completion, but before the building is plastered.

ALARM (FP)
Any audible or visible signal indicating existence of a fire or emergency requiring response and emergency action on the part of the fire fighting service. Also, the alarm device or devices by which fire and emergency signals are received.

ALARM CHECK VALVE (FP)
A check valve, equipped with a signaling device, which will annunciate a remote alarm, that a sprinkler head or heads are discharging.

ALLOY
A substance composed of two or more metals or a metal and nonmetal intimately united, usually fused together and dissolving in each other when molten.

ALLOY PIPE
A steel pipe with one or more elements other than carbon which give it greater resistance to corrosion and more strength than carbon steel pipe.

AMBIENT TEMPERATURE
The prevailing temperature in the immediate vicinity, or the temperature of the medium surrounding an object.

AGA
Abbreviation for American Gas Association.

AMERICAN STANDARD PIPE THREAD
A type of screw thread commonly used on pipe and fittings.

ANSI
Abbreviation for American National Standards Institute.

ASHRAE
Abbreviation for American Society of Heating, Refrigerating and Air Conditioning Engineers.

ASME
Abbreviation for American Society of Mechanical Engineers.

ASPE
Abbreviation for American Society of Plumbing Engineers.

ASSE
Abbreviation for American Society of Sanitary Engineers or American Society of Safety Engineers.

ASTM
Abbreviation for American Society for Testing Material.

AWWA
Abbreviation for American Water Works Association.

ANAEROBIC
Bacteria living or active in the absence of free origin.

ANCHOR
A device used to fasten or secure pipes to the building or structure.

ANGLE OF BEND
In a pipe, the angle between radial lines from the beginning and end of the bend to the center.

ANGLE VALVE
A device, usually of the globe type, in which the inlet and outlet are at right angles.

APPROVED
Accepted or acceptable under an applicable specification or standard stated or cited for the proposed use under procedures and authority of the Administrative Authority.

APPROVED TESTING AGENCY
An organization established for purposes of testing to approved standards and acceptable by the Administrative Authority.

AREA DRAIN
A receptacle designed to collect surface or rain water from a determined or calculated open area.

ARTERIAL VENT
A vent serving the building drain and the public sewer.

ASPIRATOR
A fitting or device supplied with water or other fluid under positive pressure which passes through an integral orifice or "constriction" causing a vacuum.

ATMOSPHERIC VACUUM BREAKER
A mechanical device consisting of a check valve opening to the atmosphere when the pressure in the piping drops to atmospheric.

AUTHORITY HAVING JURISDICTION (FP)
The organization, office, or individual responsible for "approving" equipment, installation, or procedure.

BACKFILL
Material used to cover piping laid in an earthen trench.

BACKFLOW
The flow of water or other liquids, mixtures, or substances into the distributing pipes of a potable supply of water from any source or sources other than its intended source. (See Back-siphonage).

BACKFLOW CONNECTION
A condition in any arrangement whereby backflow can occur.

BACKFLOW PREVENTER
A device or means to prevent backflow into the potable water system.

BACKING RING
A metal strip used to prevent melted metal, from the welding process, from entering a pipe when making a butt-welded joint.

BACK-SIPHONAGE
The flowing back of used, contaminated, or polluted water from a plumbing fixture or vessel into a water supply pipe due to a negative pressure in such pipe. (See Backflow).

BACK UP
A condition where the waste water may flow back into another fixture or compartment but not backflow into the potable water system.

BACKWATER VALVE
A device which permits drainage in one direction but has a check valve that closes against back pressure. Sometimes used conjunctively with gate valves designed for sewage.

BAFFLE PLATE
A tray or partition placed in process equipment to direct or change the direction of flow.

BALL CHECK VALVES
A device used to stop the flow of media in one direction while allowing flow in an opposite direction. The closure member used is spherical or ball shaped.

BALL VALVE
A spherical shaped gate valve providing a very tight shut-off.

BARRIER FREE
(See Accessible, Handicapped).

BASE
The lowest portion or lowest point of a stack of vertical pipe.

BATTERY OF FIXTURES
Any group of two or more similar adjacent fixtures which discharge into a common horizontal waste or soil branch.

BELL
That portion of a pipe which for a short distance is sufficiently enlarged to receive the end of another pipe of the same diameter for the purpose of making a joint.

BELL AND SPIGOT JOINT
The commonly used joint in cast iron soil pipe. Each piece is made with an enlarged diameter or bell at one end into which the plain or spigot end of another piece is inserted. The joint is then made tight by cement, oakum, lead, or rubber caulked into the bell around the spigot.

BLACK PIPE
Steel pipe that has not been galvanized.

BLANK FLANGE
A soil plate flange used to seal off the flow in a pipe.

BOILER BLOW-OFF
An outlet on a boiler to permit emptying or discharge of sediment.

BOILER BLOW-OFF TANK
A vessel designed to receive the discharge from a boiler blow-off outlet to cool the discharge to a temperature which permits its safe discharge to the drainage system.

BONNET
That part of a valve which connects the valve actuator to the valve body; may also contain the stem packing in some valves.

BRANCH
Any part of the piping system other than a main, riser, or stack.

BRANCH INTERVAL
A length of soil or waste stack corresponding in general to a story height, but in no case less than 8 ft (2.4 m), within which the horizontal branches from one floor or story of a building are connected to the stack.

BRANCH TEE
A tee having one side branch.

BRANCH VENT
A branch vent is a vent connecting one or more individual vent with a vent stack or stack vent.

BRAZING ENDS
The ends of a valve or fitting which are prepared for silver brazing.

BRONZE TRIM OR BRONZE MOUNTED
An indication that certain internal parts of the valves known as trim materials (stem, disc, seat rings, etc.) are made of copper alloy.

BTU
Abbreviation for British Thermal Unit. The amount of heat required to raise the temperature of one pound (0.45 kg) of water one degree Fahrenheit (0.565°C).

BUILDING (HOUSE)
A structure built, erected, and framed of component structural parts designed for the housing, shelter, enclosure, or support of persons, animals, or property of any kind.

BUILDING (HOUSE) DRAIN
That part of the lowest piping of the drainage system which receives the discharge from soil, waste, and other drainage pipes inside the walls of the building (house) and conveys it to the building (house) sewer which begins outside the building (house) walls.

BUILDING (HOUSE) SEWER
That part of the horizontal piping of a drainage system which extends from the end of the building (house) drain and which receives the discharge of the building (house) drain and conveys it to a public sewer, private sewer, individual sewage-disposal system, or other approved point of disposal.

BUILDING (HOUSE) SEWER — COMBINED
A building (house) sewer which conveys both sewage and storm water or other drainage.

BUILDING (HOUSE) SEWER — SANITARY
A building (house) sewer which conveys sewage only.

BUILDING (HOUSE) SEWER — STORM
A building (house) sewer which conveys storm water or other drainage but no sewage.

BUILDING (HOUSE) SUBDRAIN
That portion of a drainage system which cannot drain by gravity in the building (house) sewer.

BUILDING (HOUSE) TRAP
A device, fitting, or assembly of fittings installed in the building (house) drain to prevent circulation of air between the drainage of the building (house) and the building (house) sewer. It is usually installed as a running trap.

BUBBLE TIGHT
The condition of a valve seat that, when closed, prohibits the leakage of visible bubbles.

BULL HEAD TEE
A tee the branch of which is larger than the run.

BURST PRESSURE
That pressure which can be slowly applied to the valve at room temperature for 30 seconds without causing rupture.

BUSHING
A pipe fitting for connecting a pipe with a female fitting or larger size. It is a hollow plug with internal and external threads.

BUTT-WELD JOINT
A welded pipe joint made with the ends of the two pipes butting each other, the weld being around the periphery.

BUTT-WELD PIPE
Pipe welded along a seam butted edge to edge and not scarfed or lapped.

BUTTERFLY VALVE
A device deriving its name from the wing-like action of the disk which operates at right angles to the flow. The disk impinges against the resilient liner with low operating torque.

BY-PASS
An auxiliary loop in a pipeline, intended for diverting flow around a valve or other pieces of equipment.

BY-PASS VALVE
A valve used to divert the flow to go past the part of the system through which it normally passes.

CAPACITY
The maximum or minimum flows obtainable under given conditions of media, temperature, pressure, velocity, etc. Also the volume of media which may be stored in a container or receptacle.

CAPILLARY
The action by which the surface of a liquid, where it is in contact with a solid, is elevated or depressed depending upon the relative attraction of the molecules of the liquid for each other and for those of the solid.

CATHODIC PROTECTION
(1) The control of the electrolytic corrosion of an underground or underwater metallic structure by the application of an electric current in such a way that the structure is made to act as the cathode instead of anode of an electrolytic cell. (2) The use of materials and liquid to cause electricity to flow to avoid corrosion.

CAULKING
The method of rendering a joint tight against water or gas by means of plastic substances such as lead and oakum.

CAVITATION
A localized gaseous condition that is found within a liquid stream.

CDA
Abbreviation for Copper Development Association.

CEMENT JOINT
The union of two fittings by insertion of material. Sometimes this joint is accomplished mechanically, sometimes chemically.

CESSPOOL
A lined excavation in the ground which receives the discharge of a drainage system or part thereof, so designed as to retain the organic matter and solids discharging therein, but permitting the liquids to seep through the bottom and sides.

CHAINWHEEL OPERATED VALVE
A device which is operated by a chain driven wheel which opens and closes the valve seats.

CHANNEL
That trough through in which any media may flow.

CHASE
A recess in a wall in which pipes can be run.

CHECK VALVE
A device designed to allow a fluid to pass through in one direction only.

CHEMICAL WASTE SYSTEM
Piping which conveys corrosive or harmful industrial, chemical, or processed wastes to the drainage system.

CIRCUIT
The directed route taken by a flow from one point to another.

CIRCUIT VENT
A branch vent that serves two or more traps and extends from in front of the last fixture connection of a horizontal branch to the vent stack.

CISPI
Abbreviation for Cast Iron Soil Pipe Institute.

CLAMP GATE VALVE
A gate valve whose body and bonnet are held together by a U-bolt clamp.

CLEANOUT
A plug or cover joined to an opening in a pipe, which can be removed for the purpose of cleaning or examining the interior of the pipe.

CLEAR WATER WASTE
Cooling water and condensate drainage from refrigeration and air conditioning equipment, cooled condensate from steam heating systems; cooled boiler blowdown water; waste water drainage from equipment rooms and other areas where water is used without an appreciable addition of oil, gasoline, solvent, acid, etc., and treated effluent in which impurities have been reduced below a minimum concentration considered harmful.

CLOSE NIPPLE
A nipple with a length twice the length of a standard pipe thread.

COCK
An original form of valve having a hole in a tapered plug which is rotated to provide passageway for fluid.

CODE
Those regulations, subsequent amendments thereto, or any emergency rules or regulations which the department authority having jurisdiction may lawfully adopt.

COEFFICIENT OF EXPANSION
The increase in unit length, area, or volume for 1 degree rise in temperature.

COLIFORM GROUP OF BACTERIA
All organisms considered the coli aerogenes group as set forth by the American Water Works Association.

COMBINATION FIXTURE
A fixture which combines one sink and tray or a two- or three-compartment sink or tray in one unit.

COMBINED WASTE & VENT SYSTEM
A specially designed system of waste piping, embodying the horizontal wet venting of one or more floor sinks or floor drains by means of a common waste and vent pipe, adequately sized to provide free movement of air above the flow line of the drain.

COMMON VENT
A vent which connects at the junction of two fixture drains and serves as a vent for both fixtures. Also known as a dual vent.

COMPANION FLANGE
A pipe flange to connect with another flange or with a flanged valve or fitting. It is attached to the pipe by threads, welding, or other method and differs from a flange which is an integral part of a pipe or fitting.

COMPRESSION JOINT
A multi-piece joint with cup shaped threaded nuts which, when tightened, compress tapered sleeves so that they form a tight joint in the periphery of the tubing they connect.

COMPRESSOR
A mechanical device for increasing the pressure of air or gas.

CONDENSATE
Water which has liquefied from steam.

CONDUCTOR
The piping from the roof to the building storm drain, combined sewer, or other approved means of disposal and located inside of the building.

CONDUIT
A pipe or channel for conveying media.

CONFLUENT VENT
A vent serving more than one fixture vent or stack vent.

CONTAMINATOR
A media or condition which spoils the nature or quality of another media.

CONTINUOUS VENT
A vent that is a continuation of the drain to which it connects.

CONTINUOUS WASTE
A continuous drain from two or three fixtures connected to a single trap.

CONTROL
A device used to regulate the function of a component or system.

CONTROLLER (FP)
The cabinet containing motor starter(s), circuit breaker(s), disconnect switch(es), and other control

devices for the control of electric motors and internal combustion engine driven fire pumps.

CORPORATION COCK
A stopcock screwed into the street water main to supply the house service connection.

COUPLING
A pipe fitting with female threads only used to connect two pipes in a straight line.

CRITICAL LEVEL
The point on a backflow prevention device or vacuum breaker conforming to approved standards and established by the recognized (approved) testing laboratory (usually stamped or marked CL or C/L on the device by the manufacturer) which determines the minimum elevation above the flood level rim of the fixture or receptacle served at which the device may be installed. When a backflow prevention device does not bear critical level marking, the bottom of the vacuum breaker, combination valve, or the bottom of any such approved device shall constitute the critical level.

CROSS
A pipe fitting with four branches in pairs, each pair on one axis, and the axis of right angles.

CROSS-OVER
A pipe fitting with a double offset, or shaped like letter "U" with ends turned out, used to pass the flow of one pipe past another when the pipes are in the same plane.

CROSS VALVE
A valve fitted on a transverse pipe so as to open communication between two parallel pipes.

CROSS CONNECTION
Any physical connection or arrangement between two otherwise separated piping systems, one of which contains potable water and the other water or other substance of unknown or questionable safety, whereby flow may occur from one system to the other, the direction of flow depending on the pressure differential between the two systems. (See Backflow and Back-siphonage).

CROWN
The top of the trap is termed the crown.

CROWN VENT
A vent pipe connected at the topmost point in the crown of a trap.

CS
Abbreviation for Commercial Standards.

CURB BOX
A device at the curb that contains a valve that is turned to shut off a supply line — usually of gas or water.

CVPC
Abbreviation for dichlorinated polyvinylchloride.

DAMPEN
To check or reduce; to deaden vibration.

DEAD END

A branch leading from a soil, waste, or vent pipe, building (house) drain, or building (house) sewer, which is terminated at a developed distance of 2 feet (0.6 m) or more by means of a plug or other closed fitting.

DEPARTMENT HAVING JURISDICTION

The Administrative Authority and includes any other law enforcement agency affected by any provision of this code, whether such agency is specifically named or not.

DETECTOR, SMOKE (FP)

Listed device for sensing visible or invisible products-of-combustion.

DEVELOPED LENGTH

The length along the center line of the pipe and fittings.

DEWPOINT

The temperature of a gas or liquid at which condensation or evaporation occurs.

DIAMETER

Unless specifically stated, the nominal diameter as designated commercially.

DIAPHRAGM

A flexible disk used to separate the control medium from the controlled medium and which actuates the valve stem.

DIAPHRAGM CONTROL VALVE

A control valve having a spring-diaphragm actuator.

DIELECTRIC FITTING

A fitting having insulating parts or material that prohibits flow of electric current.

DIFFERENTIAL

The variance between two target values, one of which is the high value of conditions, the other being the low value of conditions.

DIGESTION

The portion of the sewage treatment process where biochemical decomposition of organic matter takes place, resulting in the formation of simple organic and mineral substances.

DISK

That part of a valve which actually closes off the flow.

DISHWASHER

An appliance for washing dishes, glassware, flatware, and some utensils.

DISPLACEMENT

The volume or weight of a fluid, such as water, displaced by a floating body.

DISPOSER

An appliance, motor driven, for reducing food and other waste by grinding, so that it can flow through the drainage system.

DOMESTIC SEWAGE

The liquid and water borne wastes derived from the ordinary living processes, free from industrial wastes, and of such character as to permit satisfactory disposal, without special treatment, into the public sewer or by means of a private sewage disposal system.

DOSING TANK
A water tight tank in a septic system placed between the septic tank and the distribution box and equipped with a pump or automatic siphon designed to discharge sewage intermittently to a disposal field. This is done so that rest periods may be provided between discharges.

DOUBLE DISK
A two-piece disk used in the gate valve. Upon contact with the seating faces in the valve, the wedges between the disk faces force them against the body seats to shut off the flow.

DOUBLE OFFSET
Two changes of direction installed in succession or series in continuous pipe.

DOUBLE PORTED VALVE
A valve having two parts to overcome line pressure imbalance.

DOUBLE SWEEP TEE
A tee made with easy (long radius) curves between body and branch.

DOUBLE WEDGE
A device used in gate valves similar to double disk in that the last downward turn of the stem spreads the split wedges and each seals independently.

DOWN
Term referring to piping running through the floor to a lower level.

DOWNSPOUT
The rainleader from the roof to the building storm drain, combined building, sewer, or other means of disposal and located outside of the building.

DOWNSTREAM
Term referring to a location in the direction of flow after passing a reference point.

DRAIN
Any pipe which carries waste water or waterborne wastes in a building drainage system.

DRAIN FIELD
The area of a piping system arranged in troughs for the purpose of disposing unwanted liquid waste.

DRAINAGE FITTING
A type of fitting used for draining fluid from pipes. The fitting makes possible a smooth and continuous interior surface for the piping system.

DRAINAGE SYSTEM
The drainage piping within public or private premises, which conveys sewage, rain water, or other liquid wastes to an approved point of disposal, but does not include the mains of a public sewer system or a private or public sewage treatment or disposal plant.

DRIFT
The sustained deviation in a corresponding controller resulting from the predetermined relation between values and the controlled variable and positions of the final control element, also known as wander.

DROOP
The amount by which the controlled variable pressure, temperature, liquid level, or differential pressure deviates from the set value at minimum controllable flow to the rated capacity.

DROP
Term referring to piping running to a lower elevation within the same floor level.

DROP ELBOW
A small elbow having wings cast on each side, the wings having countersunk holes so that they may be fastened by wood screws to a ceiling, wall, or framing timbers.

DROP TEE
A tee having the wings of the same type as the drop elbow.

DROSS
(1) The solid scum that forms on the surface of a metal, as lead or antimony, when molten or melting, largely as a result of oxidation but sometimes because of the rising of dirt and impurities to the surface. (2) Waste or foreign matter mixed with a substance or left as a residue after that substance has been used or processed.

DRY BULB TEMPERATURE
The temperature of air as measured by an ordinary thermometer.

DRY-PIPE VALVE (FP)
A valve used with a dry-pipe sprinkler system, where water is on one side of the valve and air is on the other side. When a sprinkler head's fusible-link melts, releasing air from the system, this valve opens, allowing water to flow to the sprinkler head.

DRY WEATHER FLOW
Sewage collected during the summer which contains little or no ground water by infiltration and no storm water at the time.

DRY WELL
(See Leaching Well).

DUAL VENT
(See Common Vent).

DURHAM SYSTEM
A term used to describe soil or waste systems where all piping is of threaded pipe, tubing, or other such rigid construction, using recessed drainage fittings to correspond to the type of piping.

DURION
A high silicon alloy that is resistant to practically all corrosive wastes. The silicon content is approximately 14.5% and the acid resistance is in the entire thickness of the metal.

DWELLING
A one-family unit with or without accessory buildings.

DWV
Abbreviation for Drainage, Waste, and Vent. A name for copper or plastic tubing used for drain, waste, or venting pipe.

ECCENTRIC FITTINGS
Fittings whose openings are offset allowing liquid to flow freely.

EFFECTIVE OPENINGS
The minimum cross-sectional area at the point of water-supply discharge, measured or expressed in terms of (1) diameter of a circle, (2) if the opening is not circular, the diameter of a circle of equivalent cross-sectional area. (This is applicable to air gap.)

EFFLUENT
Sewage, treated or partially treated, flowing out of sewage treatment equipment.

ELASTIC LIMIT
The greatest stress which a material can withstand without a permanent deformation after release of the stress.

ELBOW (EL)
A fitting that makes an angle between adjacent pipes. The angle is 90 degrees, unless another angle is specified.

ELECTROLYSIS
The process of producing chemical changes by passage of an electric current through an electrolyte (as in a call), the ions present carrying the current by migrating to the electrodes where they may form new substances (as in the deposition of metals or the liberation of gases).

ELUTRIATION
A process of sludge conditioning in which certain constituents are removed by successive decontaminations with fresh water or plant effluent, thereby reducing the demand for conditioning chemicals.

END CONNECTION
A reference to the method of connecting the parts of a piping system, i.e., threaded, flanged, butt weld, socket weld, etc.

ENGINEERED PLUMBING SYSTEM
Plumbing system designed by use of scientific engineering design criteria other than design criteria normally given in plumbing codes.

EROSION
The gradual destruction of metal or other material by the abrasive action of liquids, gases, solids, or mixtures of these materials.

EVAPOTRANSPIRATION
Loss of water from the soil both by evaporation and by transpiration from the plants growing thereon.

EXISTING WORK
A plumbing system or any part thereof, which has been installed prior to the effective date of an applicable code.

EXTRA HEAVY
Description of piping material, usually cast iron, indicating piping thicker than standard pipe.

EXPANSION JOINT
A joint whose primary purpose is to absorb longitudinal thermal expansion in the pipe line due to heat.

EXPANSION LOOP
A large radius bend in a pipe line to absorb longitudinal expansion in the line due to heat.

FACE TO FACE DIMENSIONS
The dimensions from the face of the inlet port to the face of the outlet port of a valve or fitting.

FEMALE THREAD
Internal thread in pipe fittings, valves, etc., for making screwed connections.

FILTER
A device through which fluid is passed to separate contaminates from it.

FILTER ELEMENT
A porous device which performs the process of filtration.

FILTER MEDIA
The porous device which performs the process of filtration.

FIRE ALARM SYSTEM (FP)
A functionally related group of devices that when either automatically or manually activated will sound sudio or visual warning devices on or off the protected premises, signaling a fire.

FIRE DEPARTMENT CONNECTION (FP)
A piping connection for fire department use in supplementing of supply water for standpipes and sprinkler systems. (See Standpipe System).

FIRE HAZARD (FP)
Any thing or act which increases or will cause an increase of the hazard or menace of a fire to a greater degree than that customarily recognized by persons in the public service regularly engaged in preventing, suppressing, or extinguishing fire; or which will obstruct, delay, hinder, or interfere with the operations of the fire department or the egress of occupants in the event of fire.

FIRE HYDRANT VALVE (FP)
A valve that when closed, drains at an underground level to prevent freezing.

FIRE PUMPS (ALL FP)
Can Pump: A vertical shaft turbine-type pump in a can (suction vessel) for installation in a pipeline to raise water pressure.
Centrifugal Pump: A pump in which the pressure is developed principally by the action of centrifugal force.
End Suction Pump: A single suction pump having its suction nozzle on the opposite side of the casing from the stuffing box and having the face of the suction nozzle perpendicular to the longitudinal axis of the shaft.
Excess Pressure Pump: UL listed and/or FM approved, low flow high head pump for sprinkler systems not being supplied from a fire pump. Pump pressurizes sprinkler system so that loss of water supply pressure will not cause a false alarm.
Fire Pump: UL listed and/or FM approved pump with driver, controls, and accessories, used for fire protection service. Fire pumps are centrifugal or turbine type, usually with electric motor or diesel engine driver.
Horizontal Pump: A pump with the shaft normally in a horizontal position.

Horizontal Split-Case Pump: A centrifugal pump characterized by a housing which is split parallel to the shaft.

In-Line Pump: A centrifugal pump whose drive unit is supported by the pump having its suction and discharge flanges on approximately the same center line.

Pressure Maintenance (Jockey) Pump: Pump with controls and accessories used to maintain pressure in a fire protection system without the operation of the fire pump. Does not have to be a listed pump.

Vertical Shaft Turbine Pump: A centrifugal pump with one or more impellers discharging into one or more bowls and a vertical educator or column pipe used to connect the bowl(s) to the discharge head on which the pump driver is mounted.

FITTING

The connector or closure for fluid lines and passages.

FITTING COMPRESSION

A fitting designed to join pipe or tubing by means of pressure or friction.

FITTING, FLANGE

A fitting which utilizes a radially extending collar for sealing and connection.

FITTING, WELDED

A fitting attached by welding.

FIXTURE BRANCH

A pipe connecting several fixtures.

FIXTURE CARRIER

A metal unit designed to support an off-the-floor plumbing fixture.

FIXTURE CARRIER FITTINGS

Special fittings for wall mounted fixture carriers. Fittings have a sanitary drainage waterway with minimum angle of < 30° - 45° > so that there are no fouling areas.

FIXTURE DRAIN

The drain from the trap of a fixture to the junction of that drain with any other drain pipe.

FIXTURES, PLUMBING

(See Plumbing fixture).

FIXTURE SUPPLY

A water-supply pipe connecting the fixture with the fixture branch or directly to a main water supply pipe.

FIXTURE UNIT (drainage — d.f.u.)

A measure of probable discharge into the drainage system by various types of plumbing fixtures. The drainage fixture unit value for a particular fixture depends on its volume rate of drainage discharge, on the time duration of a single drainage operation, and on the average time between successive operations. Laboratory tests have shown that the rate of discharge of an ordinary lavatory with a nominal 1.25" (31.8 mm) outlet, trap and waste is about 7.5 gal/min (0.5 L/s). This figure is so near to 1 ft^3/min (0.5 L/s) that "1 ft^3/min" (0.5 L/s) has become the accepted flow rate of one fixture unit.

FIXTURE UNIT FLOW (supply — s.f.u.)
A measure of the probable hydraulic demand on the water supply by various types of plumbing fixtures. The supply fixture unit value for a particular fixture depends on its volume rate of supply, on the time duration of a single supply operation, and on the average time between successive operations.

FLANGE
In pipe work, a ring-shaped plate on the end of a pipe at right angles to the end of the pipe and provided with holes for bolts to allow fastening of the pipe to a similarly equipped adjoining pipe. The resulting joint is a flanged joint.

FLANGE BONNET
A valve bonnet having a flange through which bolts connect it to a matching flange on the valve body.

FLANGE ENDS
A valve or fitting having flanges for joining to other piping elements. Flanged ends can be plain faced, aised face, large male and female, large tongue and groove, small tongue and groove, and ring joint.

FLANGE FACES
Pipe flanges which have the entire surface of the flange faced straight across, and use either a full face or ring gasket.

FLAP VALVE
A non-return valve in the form of a hinged disk or flap, sometimes having leather or rubber faces.

FLASH POINT
The temperature at which a fluid first gives off sufficient flammable vapor to ignite when approached with a flame or spark.

FLOAT VALVE
A valve which is operated by means of a bulb or ball floating on the surface of a liquid within a tank. The rising and falling action operates a lever which opens and closes the valve.

FLOOD LEVEL RIM
The top edge of the receptacle from which water overflows.

FLOODED
A condition when the liquid therein rises to the flood-level rim of the fixture.

FLOW PRESSURE
The pressure in the water supply pipe near the water outlet while the faucet or water outlet is fully open and flowing.

FLUE
An enclosed passage, primarily vertical, for removal of gaseous products of combustion to the outer air.

FLUSH VALVE
A device located at the bottom of the tank for the purpose of flushing water closets and similar fixtures.

FLUSHING TYPE FLOOR DRAIN
A floor drain which is equipped with an integral water supply, enabling flushing of the drain receptor and trap.

FLUSHOMETER VALVE
A device which discharges a predetermined quantity of water to fixtures for flushing purposes and is

actuated by direct water pressure.

FOOT VALVE
A check valve installed at the base of a pump suction pipe. The purpose of a foot valve is to maintain pump prime by preventing pumped liquid from draining away from the pump.

FOOTING
The part of a foundation wall or column resting on the bearing soil, rock, or piling which transmits the superimposed load to the bearing material.

FRENCH DRAIN
A drain consisting of an underground passage made by filling a trench with loose stones and covering with earth. Also known as rubble drain.

FRESH-AIR INLET
A vent line connected with the building drain just inside the house trap and extending to the outer air. It provides fresh air to the lowest point of the plumbing system and with the vent stacks provides a ventilated system. A fresh air inlet is not required where a septic-tank system of sewage disposal is employed.

FROSTPROOF CLOSET
A hopper that has no water in the bowl and has the trap and the control valve for its water supply installed below the frost line.

FS
Abbreviation for Federal Specifications.

GALVANIC ACTION
When two dissimilar metals are immersed in the same electrolytic solution and connected electrically, there is an interchange of atoms carrying an electric charge between them. The anode metal with the higher electrode potential corrodes, the cathode is protected. Thus magnesium will protect iron. Iron will protect copper. (See Electrolysis).

GALVANIZING
A process where the surface of iron or steel piping or plate is covered with a layer of zinc.

GENERALLY ACCEPTED STANDARD
A document referred to in a code, covering a particular subject, and accepted by the Administrative Authority.

GRADE
The slope or fall of a line of pipe in reference to a horizontal plane. In drainage it is expressed as the fall in a fraction of an inch or percentage slope per foot length of pipe.

GREASE INTERCEPTOR
(See Interceptor).

GREASE TRAP
(See Interceptor).

GRINDER PUMP
A special class of solids-handling pumps which grind sewage solids to a fine slurry, rather than passing through entire spherical solids.

HALON 1301 (FP)
Halon 1301 (bromtrifluoromethane $CBrF_3$) is a colorless, odorless, electrically non-conductive gas that is an effective medium for extinguishing fires.

HALON SYSTEM TYPES (FP)
There are two types of systems recognized in this standard: Total Flooding Systems and Local Application Systems.
Local Application System: Consists of a supply of Halon 1301 arranged to discharge directly on the burning material.
Total Flooding System: Consists of a supply of Halon 1301 arranged to discharge into, and fill to the proper concentration, an enclosed space or enclosure about the hazard.

HANGERS
(See Supports).

HORIZONTAL BRANCH
A drain pipe extending laterally from a soil or waste stack or building drain.

HORIZONTAL PIPE
Any pipe or fitting which is installed in a horizontal position.

HUB AND SPIGOT
Piping made with an enlarged diameter or hub at one end and plain or spigot at the other end. The joint is made tight by oakum and lead or by use of a neoprene gasket or inserted in the hub around the spigot.

HUBLESS
Soil piping with plain ends. The joint is made tight with a stainless steel or cast iron clamp and neoprene gasket assembly.

INDIRECT WASTE PIPE
A pipe that does not connect directly with the drainage system but conveys liquid wastes by discharging into a plumbing fixture or receptacle which is directly connected to the drainage system.

INDIVIDUAL VENT
A pipe installed to vent a fixture trap and which connects with the vent system above the fixture served or terminates in the open air.

INDUCED SIPHONAGE
Loss of liquid from a fixture trap due to pressure differential between inlet and outlet of trap, often caused by discharge of another fixture.

INDUSTRIAL WASTE
All liquid or waterborne waste from industrial or commercial processes except domestic sewage.

INSANITARY
A condition which is contrary to sanitary principles or is injurious to health.

INTERCEPTOR
A device designed and installed so as to separate and retain deleterious, hazardous, or undesirable matter from normal wastes and permit normal sewage or liquid wastes to discharge into the disposal terminal by gravity.

INVERT
Term referring to the lowest point on the interior of a horizontal pipe.

LABELED
Equipment or materials bearing a label of a listing agency.

LATERAL SEWER
A sewer which does not receive sewage from any other common sewer except house connections.

LEACHING WELL
A pit or receptacle having porous walls which permit the contents to seep into the ground. Also known as dry well.

LEADER
The water conductor from the roof to the building (house) storm drain. Also known as downspout.

LIQUID WASTE
The discharge from any fixture, appliance, or appurtenance in connection with a plumbing system which does not receive fecal matter.

LISTED
Equipment or materials included in a list published by an organization that maintains periodic inspection on current production of listed equipment or materials and whose listing states either that the equipment or material complies with approved standards or has been tested and found suitable for use in a specified manner acceptable to the authority having jurisdiction and concerned (a listing agency).

LISTING AGENCY
An agency accepted by the Administrative Authority which lists or labels and maintains a periodic inspection program on current production of listed models. It makes available a published report of such listings in which information is included that the product has been tested and complies with generally accepted standards and found safe for use in a specified manner.

LOAD FACTOR
The percentage of the total connected fixture unit flow which is likely to occur at any point in the drainage system. Load factor represents the ratio of the probable load to the potential load and is determined by the average rates of flow of the various kinds of fixtures, the average frequency of use, the duration of flow during one use, and the number of fixtures installed.

LOOP VENT
(See Vent, loop).

MAIN
The principal artery of the system of continuous piping, to which branches may be connected.

MAIN VENT
A vent header to which vent stacks are connected.

MALLEABLE
Capable of being extended or shaped by beating with a hammer or by the pressure of the rollers. Most metals are malleable. The term "malleable iron" has also the older meaning (still universal in Great Britain) of "wrought iron," abbr. Mall.

MASTER PLUMBER
An individual who is licensed and authorized to install and assume responsibility for contractual agreements pertaining to plumbing and to secure any required permits. The journeyman plumber is allowed to install plumbing only under the responsibility of a master plumber.

MSS
Abbreviation for Manufacturers Standardization Society of the Valve and Fittings Industry, Inc.

NSF
Abbreviation for National Sanitation Foundation Testing Laboratory.

NFPA
Abbreviation for National Fire Protection Association.

OFFSET
A combination of pipe, pipes, and/or fittings which join two approximately parallel sections of the line of pipe.

OUTFALL SEWERS
Sewers receiving the sewage from the collection system and carrying it to the point of final discharge or treatment. It is usually the largest sewer of the entire system.

OXIDIZED SEWAGE
Sewage in which the organic matter has been combined with oxygen and has become stable in nature.

PB
Abbreviation for polybutylene.

PDI
Abbreviation for Plumbing and Drainage Institute.

PE
Abbreviation for polyethylene.

PERCOLATION
The flow or trickling of a liquid downward through a contact or filtering medium; the liquid may or may not fill the pores of the medium.

PITCH
The amount of slope or grade given to horizontal piping and expressed in inches of vertically projected drop per foot (mm/m) on a horizontally projected run of pipe.

PLUMBING
The practice, materials, and fixtures used in the installation, maintenance, extension, and alteration of all piping, fixtures, appliances, and appurtenances in connection with any of the following: sanitary drainage or storm drainage facilities, the venting system, and the public or private water-supply systems, within or adjacent to any building, structure, or conveyance, water supply systems, and/or the storm water liquid waste or sewage system of any premises to their connection with any point of public disposal or other acceptable terminal.

PLUMBING APPLIANCE
A plumbing fixture which is intended to perform a special plumbing function. Its operation and/or control may be dependent upon one or more energized components, such as motors, controls, heating elements, or pressure or temperature-sensing elements. Such fixtures may

operate automatically through one or more of the following actions: a time cycle, a temperature range, a pressure range, a measured volume or weight; or the fixture may be manually adjusted or controlled by the user or operator.

PLUMBING APPURTENANCES
A manufactured device, prefabricated assembly, or on-the-job assembly of component parts which is an adjunct to the basic piping system and plumbing fixtures. An appurtenance demands no additional water supply nor does it add any discharge load to a fixture or the drainage system. It is presumed that it performs some useful function in the operation, maintenance, servicing, economy, or safety of the plumbing system.

PLUMBING ENGINEERING
The application of scientific principles to the design, installation, and operation of efficient, economical, ecological, and energy-conserving systems for the transport and distribution of liquids and gases.

PLUMBING FIXTURES
Installed receptacles, devices, or appliances which are supplied with water or which receive liquid or liquid borne wastes and discharge such wastes into the drainage system to which they may be directly or indirectly connected. Industrial or commercial tanks, vats, and similar processing equipment are not plumbing fixtures, but may be connected to or be discharged into approved traps or plumbing fixtures.

PLUMBING INSPECTOR
Any person who, under the supervision of the department having jurisdiction is authorized to inspect plumbing and drainage as defined in the code for the municipality, and complying with the laws of licensing and/or registration of the state, city, or county.

PLUMBING SYSTEM
All potable water supply and distribution pipes, plumbing fixtures and traps, drainage and vent pipe, and all building (house) drains, including their respective joints and connections, devices, receptacles, and appurtenances within the property lines of the premises and shall include potable water piping, potable water treating or using equipment, fuel gas piping, water heaters, and vents for same.

POLYMER
A chemical compound or mixture of compounds formed by polymerization and consisting essentially of repeating structural units.

POOL
A water receptacle used for swimming or as a plunge or other bath, designed to accommodate more than one bather at a time.

POTABLE WATER
Water which is satisfactory for drinking, culinary and domestic purposes, and meets the requirements of the health authority having jurisdiction.

PRECIPITATION
The total measurable supply of water received directly from the clouds as snow, rain, hail, and sleet. It is expressed in inches (mm) per day, month, or year.

PRIVATE SEWAGE DISPOSAL SYSTEM
A septic tank with the effluent discharging into a subsurface disposal field, one or more seepage pits, or a combination of subsurface disposal field and seepage pit, or of such other facilities as may be permitted under the procedures set forth elsewhere in this code.

PRIVATE SEWER
A sewer which is privately owned and not directly controlled by public authority.

PRIVATE USE
Applies to plumbing fixtures in residences and apartments, private bathrooms in hotels and hospitals, rest rooms in commercial establishments containing restricted-use single fixture or groups of single fixtures and similar installations, where the fixtures are intended for the use of a family or an individual.

PUBLIC SEWER
A common sewer directly controlled by public authority.

PUBLIC USE
Applies to toilet rooms and bathrooms used by employees, occupants, visitors, or patrons, in or about any premises and locked toilet rooms or bathrooms to which several occupants or employees on the premises possess keys and have access.

PUTREFACTION
Biological decomposition of organic matter with the production of ill-smelling products and usually takes place when there is a deficiency of oxygen.

PVC
Abbreviation for polyvinyl chloride.

RAW SEWAGE
Untreated sewage.

RECEPTOR
A plumbing fixture or device of such material, shape, and capacity as to adequately receive the discharge from indirect waste pipes so constructed and located as to be readily cleaned.

REDUCED SIZE VENT
Dry vents which are smaller than those allowed by model plumbing codes.

REDUCER
(1) A pipe fitting with inside threads, larger at one end than at the other. (2) A fitting so shaped at one end that it receives a larger pipe size in the direction of flow.

REFLECTING POOL
A water receptacle used for decorative purposes.

RELIEF VENT
A vent designed to provide circulation of air between drainage and vent systems or to act as an auxiliary vent.

RESIDUAL PRESSURE (FP)
Pressure remaining in a system while water is being discharged from outlets.

RETURN OFFSET
A double offset installed to return the pipe to its original alignment.

REVENT PIPE
That part of a vent pipe line which connects directly with an individual waste or group of wastes, underneath or back of the fixture, and extends either to the main or branch vent pipe. Also known as individual vent.

RIM
An unobstructed open edge of a fixture.

RISER
A water supply pipe which extends vertically one full story or more to convey water to branches or fixtures.

RISER (FP)
A vertical pipe used to carry water for fire protection to elevations above or below grade as a standpipe riser, sprinkler riser, etc.

ROOF DRAIN
A drain installed to remove water collecting on the surface of a roof and to discharge it into the leader (downspout).

ROUGHING-IN
The installation of all parts of the plumbing system which can be completed prior to the installation of fixtures. This includes drainage, water supply and vent piping, and the necessary fixture supports.

SAND FILTER
A water treatment device for removing solid or colloidal material with sand as the filter media.

SANITARY SEWER
The conduit or pipe carrying sanitary sewage. It may include storm water and the infiltration of ground water.

SEEPAGE PIT
A lined excavation in the ground which receives the discharge of a septic tank designed to permit the effluent from the septic tank to seep through its bottom and sides.

SEPTIC TANK
A watertight receptacle which receives the discharge of a drainage system or part thereof and is designed and constructed so as to separate solids from the liquid and digest organic matter through a period of detention.

SEWAGE
Any liquid waste containing animal, vegetable, or chemical wastes in suspension or solution.

SEWAGE EJECTOR
A mechanical device or pump for lifting sewage.

SIAMESE (FP)
A hose fitting for combining the flow from two or more lines into a single stream. (See Fire Department Connection).

SIDE VENT
A vent connected to the drain pipe through a fitting at an angle not greater than 45 degrees to the vertical.

SLUDGE
The accumulated suspended solids of sewage deposited in tanks, beds, or basins, mixed with water to form a semi-liquid (sludge).

SOIL PIPE

Any pipe which conveys the discharge of water closets, urinals, or fixtures having similar functions, with or without the discharge from other fixtures, to the building (house) drain or building (house) sewer.

SPECIAL WASTES

Wastes which require some special method of handling such as the use of indirect waste piping and receptors, corrosion resistant piping, sand, oil or grease interceptors, condensers, or other pre-treatment facilities.

SPRINKLER SYSTEM (FP)

An integrated system of underground and overhead piping designed in accordance with fire protection engineering standards. The installation includes one or more automatic water supplies. The portion of the sprinkler system above ground is a network which is specially sized or hydraulically designed piping installed in a building, structure, or area, generally overhead, and to which sprinklers are attached in a systematic pattern. The valve controlling each system riser is located in the system riser or its supply piping. Each sprinkler system riser includes a device for actuating an alarm when the system is in operation. The system is activated by heat from a fire and discharges water over the fire area.

SPRINKLER SYSTEM CLASSIFICATION (Automatic sprinkler system types) (FP)

1. Wet-Pipe Systems
2. Dry-Pipe Systems
3. Pre-Action Systems
4. Deluge Systems
5. Combined Dry-Pipe and Pre-Action Systems.

Occupancy Classification: Relates to sprinkler installations and their water supplies only. They are not intended to be a general classification of occupancy hazards.

Extra Hazard Occupancies (FP): Occupancies or portions of other occupancies where quantity and combustibility of contents is very high, and flammable and combustible liquids, duct, lint, or other materials are present introducing the probability of rapidly developing fires with high rates of heat release. Extra hazard occupancies involve a wide range of variables which may produce severe fires. The following shall be used to evaluate the severity of extra hazard occupancies:

Extra Hazard (Group 1): Includes occupancies with little or no flammable or combustible liquids.

Extra Hazard (Group 2): Includes occupancies with moderate to substantial amounts of flammable or combustible liquids or where shielding of combustibles is extensive.

Light Hazard: Occupancies or portions of other occupancies where the quantity and/or combustibility of contents is low and fires with relatively low rates of heat release are expected.

Ordinary Hazard Occupancies:

Ordinary Hazard (Group 1): Occupancies or portions of other occupancies where combustibility is low, quantity of combustibles do not exceed 8 ft (2.4 m), and fires with moderate rates of heat release are expected.

Ordinary Hazard (Group 2): Occupancies or portions of other occupancies where quantity and combustibility of contents is moderate, stock piles do not exceed 12 ft (3.7 m), and fires with moderate rates of heat release are expected.

Ordinary Hazard (Group 3): Occupancies or portions of other occupancies where quantity and/or combustibility of contents is high, and fires of high rates of heat release are expected.

Sprinkler Systems — Special Types: Special purpose systems employing departures from the requirements of this standard, such as special water supplies and reduced pipe sizing, shall be installed in accordance with their listing.

SPRINKLER TYPES (FP)

Concealed Sprinklers: Recessed sprinklers with cover plates.

Corrosion-Resistant Sprinklers: Sprinklers with special coatings or platings to be used in an atmosphere which would corrode an uncoated sprinkler.

Dry Pendent Sprinklers: Sprinklers for use in a pendent position in a dry-pipe system or a wet-pipe system with the seal in a heated area.

Dry Upright Sprinklers: Sprinklers which are designed to be installed in an upright position, on a wet-pipe system, to extend into an unheated area with a seal in a heated area.

Extended Coverage Sidewall Sprinklers: Sprinklers with special extended, directional, discharge patterns.

Flush Sprinklers: Sprinklers in which all or part of the body, including the shank thread, is mounted above the lower plane of the ceiling.

Intermediate Level Sprinklers: Sprinklers equipped with integral shields to protect their operating elements from the discharge of sprinklers installed at high elevations.

Large-Drop Sprinkler: A listed sprinkler is characterized by a K factor between 11.0 and 11.5 and proven ability to meet prescribed penetration, cooling, and distribution criteria prescribed in the large-drop sprinkler examination requirements. The deflector/discharge characteristics of the large-drop sprinkler generate large drops of such size and velocity as to enable effective penetration of the high-velocity fire plume.

Nozzles: Devices for use in applications requiring special discharge patterns, directional spray, fine spray, or other unusual discharge characteristics.

Open Sprinklers: Sprinklers from which the actuating elements (fusible-links) have been removed.

Ornamental Sprinklers: Sprinklers which have been painted or plated by the manufacturer.

Pendent Sprinklers: Sprinklers designed to be installed in such a way that the water stream is directed downward against the deflector.

Quick Response Sprinklers: A type of special sprinkler.

Recessed Sprinklers: Sprinklers in which all or part of the body, other than the shank thread, is mounted within a recessed housing.

Residential Sprinklers: Sprinklers which have been specifically listed for use in residential occupancies.
Sidewall Sprinklers: Sprinklers having special deflectors which are designed to discharge most of the water away from the nearby wall in a pattern resembling one quarter of a sphere, with a small portion of the discharge directed at the wall behind the sprinkler.
Special Sprinklers: Sprinklers which have been tested and listed as prescribed in special limitations.
Upright Sprinklers: Sprinklers designed to be installed in such a way that the water spray is directed upwards against the deflector.

STACK
The vertical main of a system of soil, waste, or vent piping extending through one or more stories.

STACK GROUP
The location of fixtures in relation to the stack so that by means of proper fittings vents may be reduced to a minimum.

STACK VENT
The extension of a soil or waste stack above the highest horizontal drain connected to the stack. Also known as waste or soil vent.

STACK VENTING
A method of venting a fixture or fixtures through the soil or waste stack.

STALE SEWAGE
This contains little or no oxygen and is free from putrefaction.

STANDPIPE
A vertical pipe generally used for the storage and distribution of water for fire extinguishing.

STANDPIPE SYSTEM (FP)
An arrangement of piping, valves, hose connections, and allied equipment installed in a building or structure with the hose connections located in such a manner that water can be discharged in streams or spray patterns through attached hose and nozzles for the purpose of extinguishing a fire and so protecting a building or structure and its contents in addition to protecting the occupants. This is accomplished by connections to water supply systems or by pumps, tanks, and other equipment necessary to provide an adequate supply of water to the hose connections.

STANDPIPE SYSTEM CLASS OF SERVICE (FP)
Class I: For use by fire departments and those trained in handling heavy fire streams (2-1/2" hose).
Class II: For use primarily by the building occupants until the arrival of the fire department (1-1/2" hose).
Class III: For use either by fire departments and those trained in handling heavy hose streams (2-1/2" hose) or by the building occupants (1-1/2" hose).

STANDPIPE SYSTEM TYPE (FP)
Dry Standpipe: A system having no permanent water supply maybe so arranged through the use of approved devices as to admit water to the system automatically by opening a hose valve.
Wet Standpipe: A system having supply valve open and water pressure maintained in the system at all times.

STORM SEWER
A sewer used for conveying rain water, surface water, condensate, cooling water, or similar liquid wastes, exclusive of sewage and industrial waste.

STOP VALVE
A valve used for the control of water supply usually to a single fixture.

STRAIN
Change of shape or size of body produced by the action of stress.

STRESS
Reactions within the body resisting external forces acting on it.

SUBSOIL DRAIN
A drain which receives only subsurface or seepage water and conveys it to a approved place of disposal.

SUB-MAIN SEWER
A sewer into which the sewage from two or more lateral sewers is discharged. Also known as a branch sewer.

SUMP
A tank or pit which receives sewage or liquid waste located below the normal grade of the gravity system and which must be emptied by mechanical means.

SUMP PUMP
A mechanical device for removing liquid waste from a sump.

SUPERVISORY (TAMPER) SWITCH (FP)
A device attached to the handle of a valve, which when the valve is closed, will annunciate a trouble signal at a remote location.

SUPPORTS
Devices for supporting and securing pipe and fixtures to walls, ceilings, floors, or structural members.

SWIMMING POOL
A structure, basin, or tank containing of water for swimming, diving, or recreation.

TEMPERED WATER
Water ranging in temperature from 85°F (29°C) up to 110°F (43°C).

TRAILER PARK SEWER
That part of the horizontal piping of a drainage system which begins two feet (0.6 m) downstream from the last trailer site connection, receives the discharge of the trailer site, and conveys it to a public sewer, private sewer, individual sewage disposal system, or other approved point of disposal.

TRAP
A fitting or device so designed and constructed to provide, when properly vented, a liquid seal which will prevent the back passage or air without significantly affecting the flow of sewage or waste water through it.

TRAP PRIMER
A device or system of piping to maintain a water seal in a trap.

TRAP SEAL
The maximum vertical depth of liquid that a trap will retain, measured between the crown weir and the top if the dip of the trap.

TURBULENCE
Any deviation from parallel flow in a pipe due to rough inner wall surfaces, obstructions, or directional changes.

UNDERGROUND PIPING
Piping in contact with the earth below grade.

UPSTREAM
Term referring to a location in the direction of flow before reaching a reference point.

VACUUM
Any pressure less than that exerted by the atmosphere and may be termed a negative pressure.

VACUUM BREAKER
(See Backflow Preventer).

VACUUM RELIEF VALVE
A device to prevent excessive vacuum in a pressure vessel.

VELOCITY
Time rate of motion in a given direction and sense.

VENT, LOOP
Any vent connecting a horizontal branch or fixture drain with the stack vent of the originating waste or soil stack.

VENT STACK
A vertical vent pipe installed primarily for the purpose of providing circulation of air to and from any part of the drainage system.

VERTICAL PIPE
Any pipe or fitting which is installed in a vertical position or which makes an angle of not more than 45 degrees with the vertical.

VITRIFIED SEWER PIPE
Conduit made of fired and glazed earthenware installed to receive waste or sewage or sewerage.

WASTE
The discharge from any fixture, appliance, area, or appurtenance, which does not contain fecal matter.

WASTE PIPE
The discharge pipe from any fixture, appliance, or appurtenance in connection with the plumbing system, which does not contain fecal matter.

WATER CONDITIONING OR TREATING DEVICE
A device which conditions or treats a water supply to change its chemical content or remove suspended solids by filtration.

WATER-DISTRIBUTING PIPE
A pipe which conveys potable water from the building supply pipe to the plumbing fixtures and other water outlets in the building.

WATER HAMMER
The forces, pounding noises, and vibration which develop in a piping system when a column of non-compressible liquid flowing through a pipe line at a given pressure and velocity is stopped abruptly.

WATER HAMMER ARRESTER
A device, other than an air chamber, designed to provide protection against excessive surge pressure.

WATER MAIN
The water supply pipe for public or community use. Normally under the jurisdiction of the municipality or water company.

WATER RISER
A water supply pipe which extends vertically one full story or more to convey water to branches or fixtures.

WATER-SERVICE PIPE
The pipe from the water main or other source of water supply to the building served.

WATER SUPPLY SYSTEM
The building supply pipe, the water-distributing pipes, and the necessary connecting pipes, fittings, control valves, and all appurtenances carrying or supplying potable water in or adjacent to the building or premises.

WET VENT
A vent which also serves as a drain.

YOKE VENT
A pipe connecting upward from a soil or waste stack to a vent stack for the purpose of preventing pressure changes in the stacks.

Appendix F

Sample Plumbing Project

This appendix contains a small plumbing project prepared for two toilet rooms for a store in a mall. It includes the water and waste/vent riser diagrams for the project. The full-size drawing includes all this information plus the detail(s) plumbing fixture schedule and their characteristics, legend, and symbols. Also included is a notes section, which takes the place of the specification in a small plumbing project. This project also includes the installation of a gas line to serve the unit heater and the A/C unit.

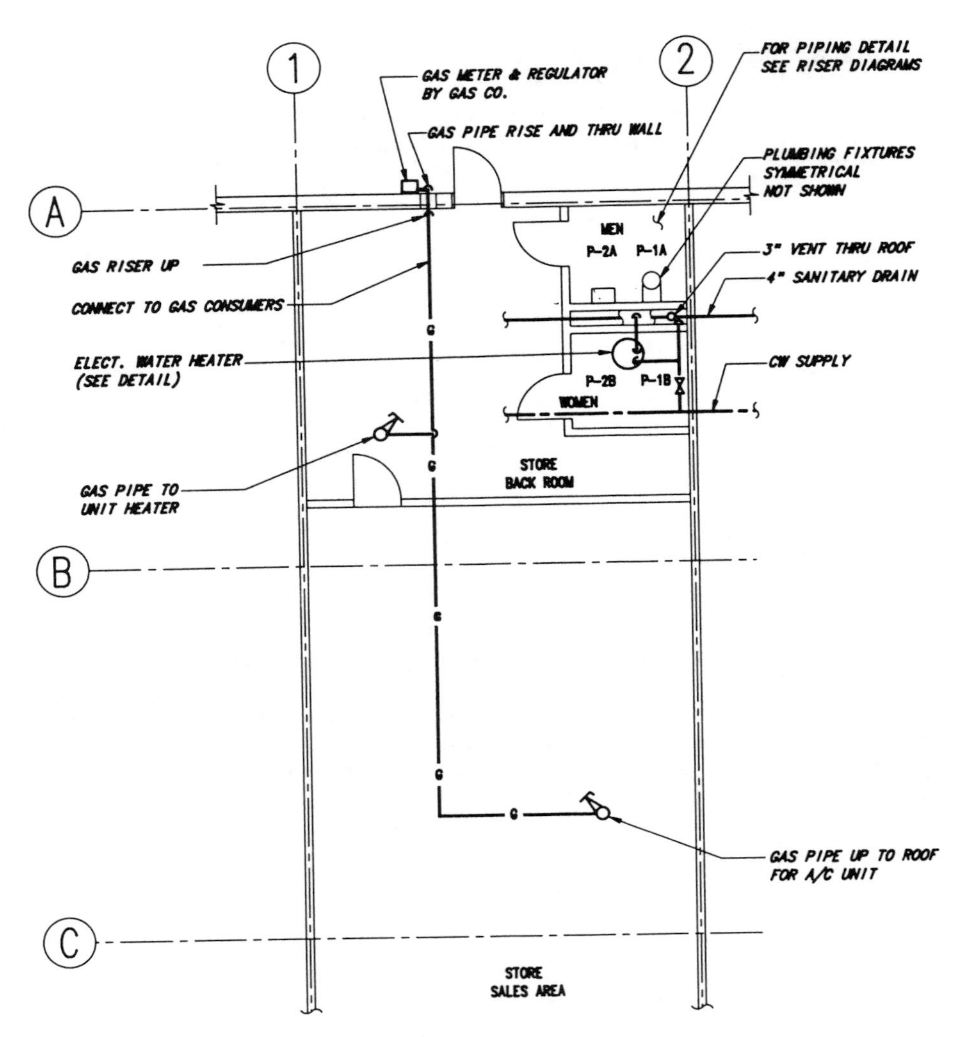

FLOOR PLUMBING PLAN FOR A RETAIL STORE

NO SCALE

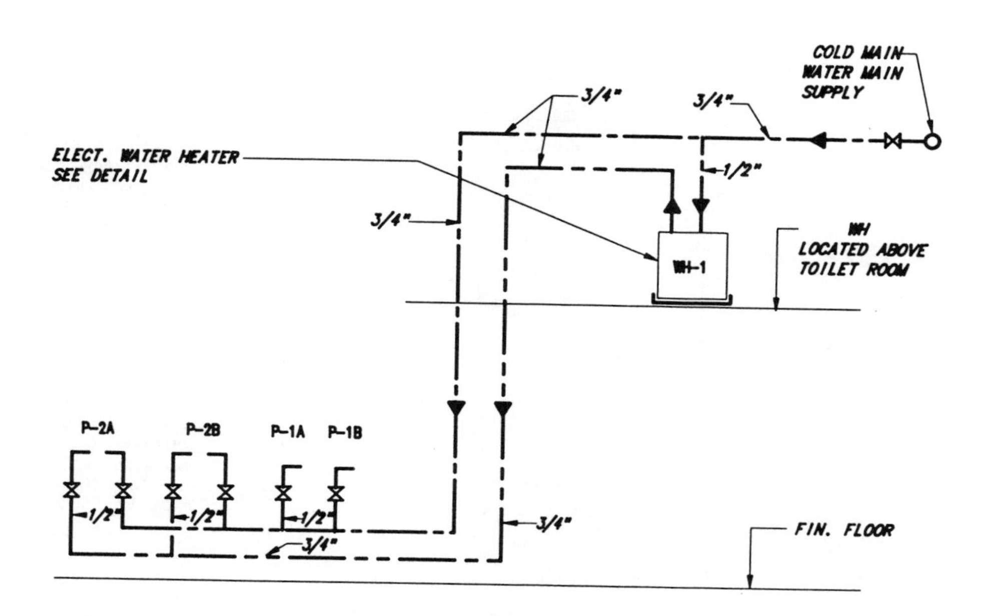

WATER RISER DIAGRAM

NO SCALE

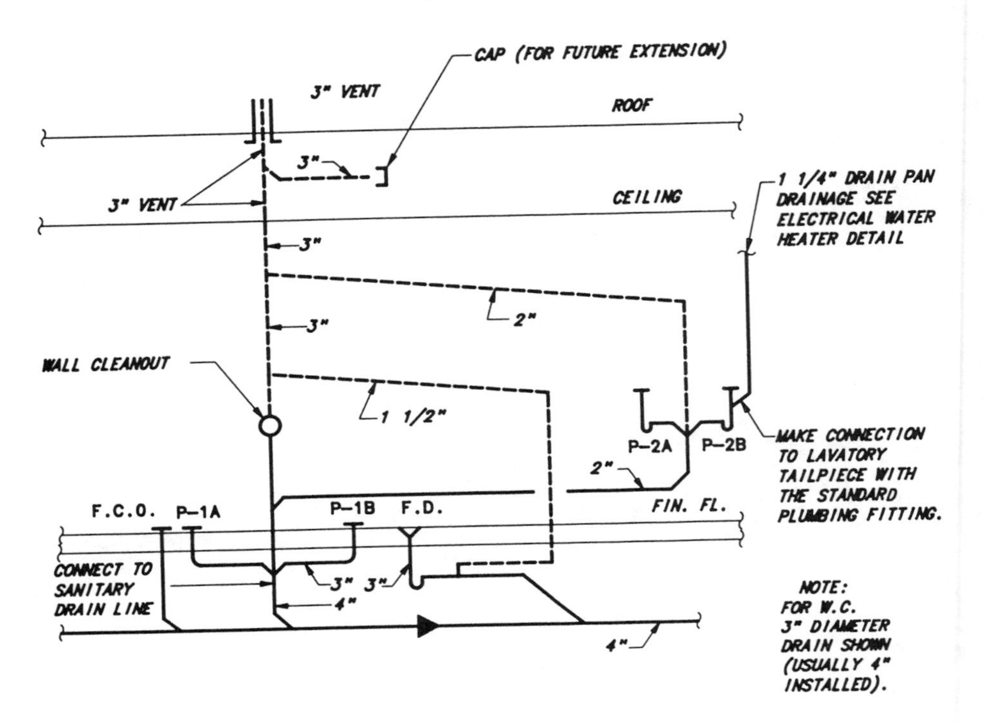

SANITARY DRAINAGE AND VENT DIAGRAM

NO SCALE

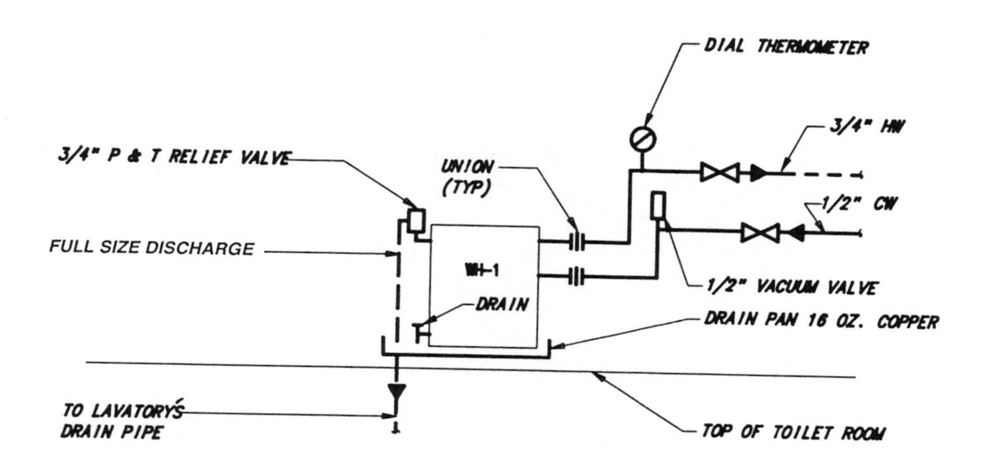

DETAIL ELECTRIC WATER HEATER

NO SCALE

PLUMBING FIXTURE SCHEDULE							
MARK	ITEM	PLUMBING CONN.				REMARKS*	SUPPLIER
		CW	HW	W	V		
P-1	WATER CLOSET	1/2"		3"	2"		
P-2	LAVATORY	1/2"	1/2"	2"	2"		
WH	WATER HEATER	1/2"	3/4"				
F.D.	FLOOR DRAIN			3"	1 1/2"		

* HANDICAP FIXTURES LOCATED IN CLOSE PROXIMITY NOT INCLUDED HERE

PLUMBING LEGEND	
——— - ——— - ———	COLD WATER (POTABLE)
——— - - ——— - - ———	HOT WATER (POTABLE)
————————	SOIL OR WASTE
- - - - - - - -	VENT
——⋈——	SHUT-OFF VALVE
——G——	GAS PIPE AND COCK ON VERTICAL PIPE
ABC	ABOVE CEILING
UF	UNDER FLOOR
P & T	PRESSURE AND TEMPERATURE
F.C.O.	FLOOR CLEAN OUT

TITLE BLOCK SAMPLE

BLOCK LOCATION: Lower Right Hand Side of Drawing

REV.	DATE	DESCRIPTION	BY	REV.	APPR.
		ISSUED FOR OR REVISION'S DESCRIPTION			

DESIGN COMPANY'S NAME

NAME OF CLIENT (COMPANY)

OTHER SPECIFICS	**CLIENT'S ADDRESS OR PROJECT'S ADDRESS**

PROJECT'S NAME

DRAWING'S TITLE **SHEET No.**		DESIGNED:
		DRAWN:
DRAWING NO.:	REV.:	REVIEWED:
SCALE:	DATE:	APPROVED:

PLUMBING FIXTURE SPECIFICATIONS

FIXTURE OR EQUIPMENT DESIGNATION	NAME	FIXTURE OR EQUIPMENT MAKE AND CATALOG NUMBER	
P-1	WATER CLOSET	CRANE PLUMBING, MODEL 3-604. SIPHON ASSISTED BLOWOUT. CLOSE COUPLED COMBINATION TOILET.	
A & B		BOWL:	3-604 EP ECONOMIZER, VITREOUS CHINA, 1/2 GALLON, QUICK ACTION FLUSH, SELF DRAINING JET, OUTLET ELONGATED RIM BOWL WITH BOLT CAPS.
		TANK:	3-610 EP ECONOMIZER VITREOUS CHINA TANK WITH FLUSHOMETER PRESSURE VESSEL WITH BUILT-IN PUSH BUTTON, PRESSURE REGULATOR AND BACKFLOW PREVENTER.
		SEAT:	CHURCH 5320.114, ELONGATED SOLID PLASTIC, CLOSED BACK/OPEN FRONT, LESS COVER, CHECK HINGE.
		SUPPLY:	POLISHED CHROME PLATED, LOOSE-KEYED ANGLE STOP, 1/2" INLET AND 3/8" O.D. X 12" LONG FLEXIBLE TUBING WITH COLLAR, WALL FLANGE AND ESCUTCHEON.
P-2	LAVATORY	CRANE PLUMBING, MODEL 1-318-V "WESTMONT" 19" X 17" VITREOUS CHINA LAVATORY, 4" CENTERS.	
A & B		FAUCET:	SYMMONS 5-60-P SCOT METERING/TEMPERATURE SELECTION, SLOW CLOSING LAVATORY FAUCET, 4" CENTERS, POP-UP DRAIN ASSEMBLY.
		SUPPLY:	POLISHED CHROME PLATED, LOOSE-KEYED ANGLE STOPS, 1/2" INLET AND 3/8" O.D. X 12" LONG FLEXIBLE TUBING OUTLET, WALL FLANGE AND ESCUTCHEON.
		P-TRAP:	CAST BRASS 1-1/4" ADJUSTABLE P-TRAP WITH CLEANOUT.
		CARRIER:	J.R. SMITH OR ZURN TO SUIT BUILDING CONDITION.
WH-1	WATER HEATER	RUDD, MODEL PE-6-1 AUTOMATIC STORAGE ELECTRIC WATER HEATER, OR EQUAL, ** GALLON CAPACITY, 120 V., ** KW.	

Appendix F

SAMPLE OF PLUMBING NOTES:

1. THE PLUMBING WORK SHALL BE PERFORMED IN STRICT ACCORDANCE WITH THE BASE BUILDING SPECIFICATION AND WITH THE LATEST EDITION OF THE PREVAILING STATE PLUMBING AND BUILDING CODES AS WELL AS ALL LOCAL REGULATIONS THAT MAY APPLY. IN CASE OF CONFLICT BETWEEN THE CONTRACT DOCUMENTS AND A GOVERNING CODE OR ORDINANCE THE MORE STRINGENT STANDARD SHALL APPLY.

2. ALL PLUMBING WORK SHALL BE COORDINATED WITH ALL OTHER TRADES BEFORE PROCEEDING WITH THE INSTALLATION.

3. INVERT ELEVATIONS AND EXACT LOCATIONS OF ALL EXISTING UTILITIES SHALL BE CHECKED BEFORE PROCEEDING WITH NEW WORK.

4. NO CHANGES ARE TO BE MADE IN PLUMBING LAYOUT WITHOUT WRITTEN PERMISSION BY THE ARCHITECT OR ENGINEER.

5. NO PIPING SHALL RUN EXPOSED IN FINISHED AREAS.

6. PLUMBING CONTRACTOR SHALL BE RESPONSIBLE FOR OBTAINING ALL NECESSARY PERMITS AND FOR PAYING RELATED FEES.

7. ROUGHING DIMENSIONS OF TOILET FIXTURES MUST BE COORDINATED WITH THE GENERAL CONTRACTOR.

8. SANITARY AND VENT PIPING SHALL BE SERVICE WEIGHT CAST IRON WITH "NO-HUB" FITTINGS.*

9. WATER SUPPLY PIPING INDOOR SHALL BE TYPE "L" COPPER.*

10. GAS PIPING SHALL BE SCHEDULE 40 BLACK STEEL WITH 150# MALLEABLE IRON FITTING. COORDINATE PIPE DIAMETER WITH HVAC CONTRACTOR.

11. INSULATE ALL WATER PIPING EXCEPT EXPOSED TO FIXTURES WITH 1/2 INCH FIBERGLASS INSULATION WITH NON-COMBUSTIBLE UL RATED VAPOR BARRIER OR CLOSED CELL FLEXIBLE INSULATION EQUAL TO ARMAFLEX II.

- 1 -

SAMPLE OF PLUMBING NOTES:

12. ALL VALVED WATER CONNECTIONS FROM MAINS TO BE ___" UNLESS OTHERWISE NOTED.

13. PROVIDE SHUT-OFF VALVES FOR WATER HEATER BRANCH. PROVIDE DIELECTRIC FITTINGS OR COUPLINGS WHEREVER DISSIMILAR METALS ARE JOINED.

14. PROVIDE SHUT-OFF VALVES AT EACH FIXTURE ON WATER SUPPLY PIPES COLD AND HOT.

15. ALL ACCESS PANELS SHALL BE BY GENERAL CONTRACTOR. PLUMBING CONTRACTOR SHALL BE RESPONSIBLE FOR THEIR LOCATION.

16. INSTALL ALL REQUIRED CLEANOUTS TO CLEAR EQUIPMENT AND FIXTURES.

17. ALL WORK SHALL BE PROPERLY TESTED, BALANCED AND CLEANED AND DISINFECTED. PROVIDE A ONE YEAR WARRANTY FROM DATE OF FINAL INSPECTION ON ALL PARTS AND LABOR.

18. ALL PLUMBING FIXTURES SPECIFIED ARE FOR INFORMATION ONLY. EQUAL EQUIPMENT MAY BE INSTALLED SECURING FIRST THE ENGINEER'S APPROVAL.

19. ALL PIPE DIAMETERS FOR INFORMATION ONLY. PLUMBING CONTRACTOR SHALL CHECK WITH THE CODE REQUIREMENTS.

20. ALL FIXTURES ARE PROVIDED WITH STANDARD TRAP AND CLEANOUT (NOT SHOWN).

* OTHER MATERIALS MAY BE SELECTED. SEE SAMPLE SPECIFICATION.

- 2 -

Appendix G

Unit and Conversion Factors

To convert from other systems of measurement to SI values, the following conversion factors are to be used. (Note: For additional conversion equivalents not shown herein, refer to ANSI Z210.1. Also issued as ASTM E380.)

Unit and Conversion Factors

a. Linear acceleration

ft/s^2 = 0.3048 m/s^2
in./s^2 = 0.0254 m/s^2

m/s^2 = 3.28 ft/s^2
m/s^2 = 9.37 in./s^2

b. Area

acre = 4046.9 m^2
ft^2 = 0.0929 m^2
in.^{2} = 0.000645 m^2 = 645.16 mm^2
mi^2 = 2,589,988 m^2 = 1.59
yd^2 = 0.836 m^2

m^2 = 0.0000247 acre
m^2 = 10.76 ft^2
m^2 = 1550.39 in.^{2}
km^2 = 0.39 mi^2
m^2 = 1.2 yd^2

c. Bending moment (torque)

pound-force·inch (lbf·in.) =
0.113 Netwon meter (N·m)
lbf·ft = 1.356 N·m

N·m = 8.85 lbf·in.
N·m = 0.74 lbf·in.

d. Bending moment (torque) per unit length

lbf·in./in. = 4.448 N·m/m
lbf·ft/in. = 53.379 N·m/m

N·m/m = 0.225 lbf·in./in.
N·m/m = 0.019 lbf·ft/in.

e. Electricity and magnetism

ampere	=	1 A
ampere-hour	=	3600 Ah
coulomb	=	1 C
farad	=	1 F
henry	=	1 H
ohm	=	1 W
volt	=	1 V

f. Energy (work)

British thermal unit (Btu) = 1055 Joule (J)
ft·lbf = 1.356 J
kWh = 3,600,000 J

J = 0.000948 Btu
J = 0.074 ft·lbf
J = 0.000000278 kWh

g. Energy per unit area per unit time

Btu/(ft^2·s) = 11,349 W/m^2

W/m^2 = 0.000088 Btu/(ft^2·s)

h. Force

ounce-force (ozf) = 0.287 N
pound-force (lbf) = 4.448 N
kilogram-force (kgf) = 9.807 N

N = 3.48 ozf
N = 0.23 lbf
N = 0.1 kgf

i. Force per unit length

lbf/in. = 175.1 N/m
lbf/ft = 14.594 N/m

N/m = 0.0057 lbf/in.
N/m = 0.069 lbf/ft

j. Heat

Btu·in./(s·ft^2·°F) = 519.2 W/(m·K)
Btu·in./(h·ft^2·°F) = 0.144 W/(m·K)
Btu/ft^2 = 11,357 J/m^2
Btu/(h·ft^2·°F) = 5.678 W/(m^2·K)
Btu/lbm = 2326 J/kg
Btu/(lbm·°F) = 4186.8 J/(kg·K)
(°F·h·ft^2)/Btu = 0.176 (K·m^2)/W

W/(m·K) = 0.002 Btu·in./(s·ft)
W/(m·K) = 6.94 Btu·in./(h·ft^2)
J/m^2 = 0.000088 Btu/ft^2
W/(m^2·K) = 0.176 Btu/(h·ft^2·°F)
J/kg = 0.00043 Btu/lbm
J/(kg·K) = 0.000239 Btu/(lbm·°F)
(K·m^2)/W = 5.68 (°F·h·ft^2)/Btu

k. Length

in. = 0.0254 m
ft = 0.3048 m
yd = 0.914 m
mi = 1609.3 m

m = 39.37 in.
m = 3.28 ft
m = 1.1 yd
m = 0.000621 mi

l. Light (illuminance)

footcandle (fc) = 10.764 1x — 1x = 0.093 fc

m. Mass

ounce-mass (ozm) = 0.028 kg — kg = 35.7 ozm
pound-mass (lbm) = 0.454 kg — kg = 2.2 lbm

n. Mass per unit area

lbm/ft^2 = 4.882 kg/m^2 — kg/m^2 = 0.205 lbm/ft^2

o. Mass per unit length

lbm/ft = 1.488 kg/m — kg/m = 0.67 lbm/ft

p. Mass per unit time (flow)

lbm/h = 0.0076 kg/s — kg/s = 131.58 lbm/h

q. Mass per unit volume (density)

lbm/ft^3 = 16.019 kg/m^3 — kg/m^3 = 0.062 lbm/ft^3
lbm/in.^{3} = 27,680 kg/m^3 — kg/m^3 = 0.000036 lbm/in.^{3}
lbm/gal = 119.8 kg/m^3 — kg/m^3 = 0.008347 lbm/gal

r. Moment of inertia

lb/ft^2 = 0.042 kg·m^2 — kg·m^2 = 23.8 lb/ft^2

s. Plane angle

degree = 17.453 mrad — mrad = 0.057 deg
minute = 290.89 μrad — μrad = 0.00344 min
second = 4.848 μrad — μrad = 0.206 s

t. Power

Btu/h = 0.293 W	W = 3.41 Btu/h
(ft·lbf)/h = 0.38 mW	mW = 2.63 (ft·lbf)/h
horsepower (hp) = 745.7 W	W = 0.00134 hp

u. Pressure (stress) force per unit area

atmosphere = 101.325 kiloPascal (kPa)	kPa = 0.009869 atm
inch of mercury (at 60°F) = 3.3769 kPa	kPa = 0.296 in. of Hg
inch of water (at 60°F) = 248.8 Pa	Pa = 0.004 in. of water
lbf/ft^2 = 47.88 Pa	Pa = 0.02 lbf/ft^2
lbf/in.^{2} = 6.8948 kPa	kPa = 0.145 lbf/in.^{2}

v. Temperature equivalent

$t_k = (t_f + 459.67)/1.8$	$t_f = 1.8\ t_k - 459.67$
$t_c = (t_f - 32)/1.8$	$t_f = 1.8\ t_c + 32$

w. Velocity (length per unit time)

ft/h = 0.085 mm/s	mm/s = 11.76 ft/h
ft/min = 5.08 mm/s	mm/s = 0.197 ft/min
ft/s = 0.3048 m/s	m/s = 3.28 ft/s
in./s = 0.0254 m/s	m/s = 39.37 in./s
mi/h = 0.447 m/s	m/s = 2.24 mi/h

x. Volume

ft^3 = 0.028 m^3 = 28.317 L	m^3 = 35.71 ft^3
in.^{3} = 16,378 mL	mL = 0.061 m^3
gal = 3.785L	L = 0.264 gal
oz = 29.574 mL	mL = 0.034 oz
pt = 473.18 mL	mL = 0.002 pt
qt = 946.35 mL	mL = 0.001 qt
acre/ft = 1233.49 m^3	m^3 = 0.00081 acre/ft

y. Volume per unit time (flow)

ft^3/min = 0.472 L/s	L/s = 2.12 ft^3/m
in.^{3}/min = 0.273 mL/s	mL/s = 3.66 in.^{3}/m
gal/min = 0.063 L/s	L/s = 15.87 gal/min

Temperature Conversion Factors

The numbers in the center column refer to the known temperature, either in °F or °C, to be converted to the other scale. If converting from °F to °C, the number in the center column represents the known temperature, in °F, and its equivalent temperature, in °C, will be found in the left column. If converting from °C to °F, the number in the center represents the known temperature, in °C, and its equivalent temperature, in °F, will be found the right column.

°C	Known temp. °F or °C	°F	°C	Known temp. °F or °C	°F
- 59	- 74	- 101	- 32.2	- 26	- 14.8
- 58	- 73	- 99	- 31.6	- 25	- 13.0
- 58	- 72	- 98	- 31.1	- 24	- 11.2
- 57	- 71	- 96	- 30.5	- 23	- 9.4
- 57	- 70	- 94	- 30.0	- 22	- 7.6
- 56	- 69	- 92	- 29.4	- 21	- 5.8
- 56	- 68	- 90	- 28.9	- 20	- 4.0
- 55	- 67	- 89	- 28.3	- 19	- 2.2
- 54	- 66	- 87	- 27.7	- 18	- 0.4
- 54	- 65	- 85	- 27.2	- 17	1.4
- 53	- 64	- 83	- 26.6	- 16	3.2
- 53	- 63	- 81	- 26.1	- 15	5.0
- 52	- 62	- 80	- 25.5	- 14	6.8
- 52	- 61	- 78	- 25.0	- 13	8.6
- 51	- 60	- 76	- 24.4	- 12	10.4
- 51	- 59	- 74	- 23.8	- 11	12.2
- 50	- 58	- 72	- 23.3	- 10	14.0
- 49	- 57	- 71	- 22.7	- 9	15.8
- 49	- 56	- 69	- 22.2	- 8	17.6
- 48	- 55	- 67	- 21.6	- 7	19.4
- 48	- 54	- 65	- 21.1	- 6	21.2
- 47	- 53	- 63	- 20.5	- 5	23.0
- 47	- 52	- 62	- 20.0	- 4	24.8
- 46	- 51	- 60	- 19.4	- 3	26.6
- 45.6	- 50	- 58.0	- 18.8	- 2	28.4
- 45.0	- 49	- 56.2	- 18.3	- 1	30.2
- 44.4	- 48	- 54.4	- 17.8	0	32.0
- 43.9	- 47	- 52.6	- 17.2	1	33.8
- 43.3	- 46	- 50.8	- 16.7	2	35.6
- 42.8	- 45	- 49.0	- 16.1	3	37.4
- 42.2	- 44	- 47.2	- 15.6	4	39.2
- 41.7	- 43	- 45.4	- 15.0	5	41.0
- 41.1	- 42	- 43.6	- 14.4	6	42.8
- 40.6	- 41	- 41.8	- 13.9	7	44.6
- 40.0	- 40	- 40.0	- 13.3	8	46.4
- 39.4	- 39	- 38.2	- 12.8	9	48.2
- 38.9	- 38	- 36.4	- 12.2	10	50.0
- 38.3	- 37	- 34.6	- 11.7	11	51.8
- 37.8	- 36	- 32.8	- 11.1	12	53.6
- 37.2	- 35	- 31.0	- 10.6	13	55.4
- 36.7	- 34	- 29.2	- 10.0	14	57.2
- 36.1	- 33	- 27.4	- 9.4	15	59.0
- 35.5	- 32	- 25.6	- 8.9	16	60.8
- 35.0	- 31	- 23.8	- 8.3	17	62.6
- 34.4	- 30	- 22.0	- 7.8	18	64.4
- 33.9	- 29	- 20.2	- 7.2	19	66.2
- 33.3	- 28	- 18.4	- 6.7	20	68.0
- 32.8	- 27	- 16.6	- 6.1	21	69.8

°C	Known temp. °F or °C	°F	°C	Known temp. °F or °C	°F
- 5.6	22	71.6	25.6	78	172.4
- 5.0	23	73.4	26.1	79	174.2
- 4.4	24	75.2	26.7	80	176.0
- 3.9	25	77.0	27.2	81	177.8
- 3.3	26	78.8	27.8	82	179.6
- 2.8	27	80.6	28.3	83	181.4
- 2.2	28	82.4	28.9	84	183.2
- 1.7	29	84.2	29.4	85	185.0
- 1.1	30	86.0	30.0	86	186.8
- 0.6	31	87.8	30.6	87	188.6
0	32	89.6	31.1	88	190.4
0.6	33	91.4	31.7	89	192.2
1.1	34	93.2	32.2	90	194.0
1.7	35	95.0	32.8	91	195.8
2.2	36	96.8	33.3	92	197.6
2.8	37	98.6	33.9	93	199.4
3.3	38	100.4	34.4	94	201.2
3.9	39	102.2	35.0	95	203.0
4.4	40	104.0	35.6	96	204.8
5.0	41	105.8	36.1	97	206.6
5.6	42	107.6	36.7	98	208.4
6.1	43	109.4	37.2	99	210.2
6.7	44	111.2	37.8	100	212.0
7.2	45	113.0	38	100	212
7.8	46	114.8	43	110	230
8.3	47	116.6	49	120	248
8.9	48	118.4	54	130	266
9.4	49	120.2	60	140	284
10.0	50	122.0	66	150	302
10.6	51	123.8	71	160	320
11.1	52	125.6	77	170	338
11.7	53	127.4	82	180	356
12.2	54	129.2	88	190	374
12.8	55	131.0	93	200	392
13.3	56	132.8	99	210	410
13.9	57	134.6	100	212	414
14.4	58	136.4	104	220	428
15.0	59	138.2	110	230	446
15.6	60	140.0	116	240	464
16.1	61	141.8	121	250	482
16.7	62	143.6	127	260	500
17.2	63	145.4	132	270	518
17.8	64	147.2	138	280	536
18.3	65	149.0	143	290	554
18.9	66	150.8	149	300	572
19.4	67	152.6	154	310	590
20.0	68	154.4	160	320	608
20.6	69	156.2	166	330	626
21.1	70	158.0	171	340	644
21.7	71	159.8	177	350	662
22.2	72	161.6	182	360	680
22.8	73	163.4	188	370	698
23.3	74	165.2	193	380	716
23.9	75	167.0	199	390	734
24.4	76	168.8	204	400	752
25.0	77	170.6	210	410	770

Appendix H

Sample Specification

This appendix consists of a sample master specification. New specifications are not usually developed for each project. Instead, design companies, government agencies, etc., use the same specification for different projects. When a new project is started, additional items or articles are added to the existing specification to clarify the project, and information that is not pertinent to the project is eliminated.

SECTION 15400

PLUMBING - TABLE OF CONTENTS

2 MATERIALS (OR PRODUCTS)

3 CONSTRUCTION METHODS

SECTION 15400

PLUMBING

1 GENERAL

1.1 DESCRIPTION

1.1.1 Provide all plant facilities, labor, materials, tools, equipment, appliances, transportation, supervision, and related work necessary or incidental to complete the work specified in this Section and as shown on the Drawings.

1.1.2 The Drawings indicate the extent and general arrangements of the Plumbing Systems. If any departures from the Drawings are deemed necessary by the Plumbing Contractor, details of such departures and the reasons therefore shall be submitted to the Owner and Engineer for approval. No such departures shall be made without prior written approval of the Owner and Engineer. Equipment and piping arrangements shall provide adequate and acceptable clearances for entry, servicing, and maintenance.

1.1.3 Plumbing Contractor shall be responsible for consulting with the General Contractor and indicating locations of openings for his work.

1.2 WORK INCLUDED

1.2.1 Sanitary drainage and vent piping systems.

1.2.2 Roof storm drainage systems.

1.2.3 Hot, cold, and recirculating water supply system.

1.2.4 Gas piping system.

1.2.5 Compressed air system.

1.2.6 Acid waste piping system.

1.2.7 Exterior water service to 5 feet outside of building.

1.2.8 Plumbing fixtures and trim.

1.2.9 Floor drains, area drains, and roof drains.

1.2.10 Hot water heaters.

1.2.11 Sleeves, inserts, and hangers.

1.2.12 Required valved water piping for heating and air conditioning equipment.

1.2.13 Water make-up for cooling tower.

1.2.14 Sprinkler test drain receptors.

1.2.15 Emergency generator exhaust, cooling water, and waste piping and diesel fuel oil piping.

1.2.16 Insulation.

1.2.17 Pipe identification, valve tags, and charts.

1.2.18 Copper drip pans.

1.2.19 Core drilling.

1.2.20 Operating and maintenance manuals.

1.2.21 Lubricants.

1.2.22 Special tools.

1.2.23 Equipment drive guards.

1.2.24 All supplementary steel for piping and equipment supports.

1.2.25 Final connections to all existing storm, sanitary, water, and gas lines.

1.2.26 Plumbing connections to existing systems.

1.2.27 Hot water booster system.

1.2.28 Apply for, obtain, and pay for all permits, certificates, inspections, and approvals required in connection with work under this Section.

1.2.29 Roughing and final connections for all kitchen and food service equipment.

1.2.30 Roughing, installation, and final connections of shampoo sinks.

1.2.31 Disinfection of interior domestic water piping system.

1.2.32 Automotive garage drainage system including oil separator, waste oil tank, and piping.

1.2.33 LP gas piping system including final connection to boiler ignition.

1.2.34 Removal and/or relocation of existing plumbing fixtures, piping, and equipment.

1.2.35 Single and double post automotive lift piping system.

1.2.36 Sump pump.

1.2.37 Exterior concrete grease trap.

1.3 WORK NOT INCLUDED

a. The following items of work are to be done by others and shall not be included in the work of this Section. However, it shall be the responsibility of this Contractor to supply the Subcontractors with the necessary information, Drawings, and supervision so that they can properly complete their phase of the installation.

1.3.1 Excavating and backfilling for waste oil tank.

1.3.2 Concrete structures (manholes, access covers, sandtraps, catch basins, pits, etc.), for equipment and/or integral parts of plumbing systems including excavating and backfilling.

1.3.3 Access doors — Plumbing Contractor is to be responsible for locations.

1.3.4 Toilet accessories, including soap dispensers.

1.3.5 Cutting and patching.

1.3.6 Electrical wiring and mounting of starting and control equipment for electrically operated plumbing equipment.

1.3.7 Roof drain flashing.

1.3.8 Finish painting of all exposed piping and hangers.

1.3.9 Temporary water and extensions by trades requiring same.

1.3.10 Foundation drains and sand traps.

1.3.11 Electric heat tracing.

1.3.12 Concrete equipment bases.

1.3.13 All excavating and backfilling for plumbing lines both inside and outside of building. Plumbing Contractor is responsible for supervision of proper bedding and backfilling to 12" above crown of pipe.

1.3.14 Equipment flues (gas flues).

1.3.15 Flashing for pipes penetrating roof.

1.3.16 Refrigerant piping.

1.3.17 Water service from main to within 5'-0" of building line.

1.3.18 Storm and sanitary laterals from site drains to approximately within 5'-0" from building line.

1.3.19 Excavating and backfilling with selected gravel material for exterior foundation wall drainage system.

1.3.20 Condensate piping from air conditioning equipment to floor drain.

1.4 VISITING THE PREMISES

1.4.1 The Plumbing Contractor, before submitting a bid on the work, must visit the site and familiarize himself with all visible existing conditions.

1.4.2 The submission of a bid will be considered an acknowledgment on the part of the bidder of his visitation to the site. The Plumbing Contractor shall be responsible for the installation of the work as shown on the Drawings and specified herein.

1.5 DEFINITIONS

1.5.1 *Furnish and Install* or *Provide* means to supply, erect, install, and connect up to complete for readiness for regular operation, the particular work referred to.

1.5.2 *Plumbing Contractor* means the Plumbing Contractor and any of his subcontractors, vendors, suppliers, or fabricators.

1.5.3 *Piping* includes pipe, all fittings, valve hangers, and other accessories relative to such piping.

1.5.4 *Concealed* means hidden from sight in chases, furred spaces, shafts, hung ceilings, embedded in construction, or in crawl spaces.

1.5.5 *Exposed* means not installed underground or *concealed* as defined above.

1.5.6 *Invert Elevation* means the elevation of the inside bottom of pipe.

1.6 QUALITY ASSURANCE

1.6.1 The work shall be executed in strict accordance with the latest edition of the prevailing State Plumbing and Building Codes (or BOCA as applicable) and all local regulations that may apply. In case of conflict between the Contract Documents and a governing code or ordinance, the more stringent requirements shall apply.

1.6.2 Unless otherwise specified or indicated, materials and workmanship shall conform with the latest edition of the following codes and standards:

a. American National Standards Institute (ANSI).

b. Underwriters' Laboratories, Inc. (UL).

c. American Society for Testing and Materials (ASTM).

d. National Fire Protection Association (NFPA).

e. American Gas Association (AGA).

f. National Electric Code.

1.6.3 If any work is performed and subsequent changes are necessary to conform to the ordinances, the changes shall be made at the Plumbing Contractor's expense.

1.6.4 All new plumbing equipment shall be designed to conform to applicable state and local energy codes. Pipe insulation, water heaters, mixing valves, flow control fittings, operating costs of equipment, shall be selected with operating efficiencies and design conditions to meet applicable energy codes.

1.6.5 Availability of a *Certificate of Approval* from the Board of Examiners of Plumbers and Gas Regulatory Board or a similar governing authority shall be a prerequisite to scheduling a final inspection to this Contract. Non-availability of these certificates may be grounds for cancellation and postponement of the scheduled inspection. A copy of the certificate shall be submitted to the Engineer.

1.7 WORKMANSHIP AND MATERIALS

1.7.1 Workmanship shall be of the best quality and none but competent mechanics skilled in their trades shall be employed. The Plumbing Contractor shall furnish the services of an experienced superintendent, who will be constantly in charge of the erection of the work, until completed and accepted.

1.7.2 Unless otherwise hereinafter specified, all materials and equipment under this Section of the Specifications shall be new, of best grade, and as listed in printed catalogs of the manufacturer. Each article of its kind shall be the standard product of a single manufacturer.

1.7.3 Whenever the words *or equal*, *equivalent equipment*, *acceptable*, or other words of similar intent or meaning are used, implying that judgment is to be exercised.

1.7.4 The Engineer shall have the right to accept or reject material, equipment, and/or workmanship and determine when the Plumbing Contractor has complied with the requirements herein specified.

1.7.5 All manufactured materials shall be delivered and stored in their original containers. Equipment shall be clearly marked or stamped with the manufacturer's name and rating.

1.7.6 Reference to standards are intended to be the latest revision of the standard specified.

1.8 MANUFACTURERS' RECOMMENDATIONS

1.8.1 Equipment installed under this Section of the Specifications shall be installed according to manufacturers' recommendations, unless otherwise shown on the Drawings or herein specified.

1.8.2 The plumbing system shall be capable of supplying the required quantity of domestic (hot and cold) water for the various utilities, provide immediate disposal of necessary quantities of sanitary waste and storm water, and shall include any required water treatment where specified.

1.9 FIXTURE & EQUIPMENT SUBMITTALS AND SAMPLES

1.9.1 The Plumbing Contractor shall prepare a complete submittal of all plumbing fixtures, piping, and equipment included under this Section.

1.9.2 The submittal shall show, in detail, the size and arrangements of all parts including piping and method of support, the relation of the work of other trades, elevations, and all other details required for the proper installation of the work.

1.9.3 Any of the brand names specified shall be used and furnished by the Plumbing Contractor; if the Plumbing Contractor desires to use products other than those specified he shall so state in his bid, giving details of the proposed substitution and the adjustment to be made in his bid if the substitution is approved. If substitution does not reflect a price advantage, the specified product shall be used.

1.9.4 Whenever deemed necessary, Engineer may request samples of equipment and materials.

1.9.5 After receiving approval of equipment manufacturers, and prior to delivery of any material to job site, and sufficiently in advance to allow Engineer ample time for checking, six copies of complete fixtures and equipment submittals or catalog cuts, showing manufacturer's name, model number, construction, size , arrangement, operating clearances, performing characteristics, and capacity of material and equipment shall be forwarded to the Architect. Each item of equipment proposed shall be a standard catalog product of an established manufacturer and of equal quality, finish, and durability to that specified.

1.9.6 Submittals, drawings, specifications and catalogs submitted for approval shall be properly labeled indicating specific service for which material or equipment is to be used, section and article number of specifications governing, Plumbing Contractor's name, and name of job and owner. Catalog items shall be clearly marked in ink. Data of general nature will not be accepted.

1.9.7 Plumbing Contractor shall furnish all necessary templates, patterns, etc., for installing work and for the purpose of making adjoining work conform and shall furnish setting plans and shop details to other trades as required.

1.9.8 Submittals

a. Shop Drawings: Shop drawings shall include drawings, schedules, performance charts, instructions, brochures, diagrams, and other information to illustrate the requirements and operation of the system. Shop drawings shall be provided for the complete plumbing system including piping layout and location of connections; schematic (elementary) diagrams and wiring diagrams or connection and interconnection diagrams.

b. Equipment catalog cuts or shop drawings of the following fixtures and equipment must be submitted for approval:

1) Floor cleanouts.

2) Wall cleanouts.

3) Floor drains.

4) All toilet room fixtures (including trim, supports, and carriers).

5) Electric water coolers.

6) Electric water heaters.

7) Hot water storage tank and supports.

8) Circulators.

9) Service sinks.

10) All lavatories and sinks located outside toilet rooms.

11) Pipe and equipment insulation.

12) Gate valves, globe valves, hose bibbs, balancing valves, check valves, backwater valves, pressure and temperature relief valves, backflow preventers, vacuum breakers, pressure-reducing valves, strainers, gauges, and thermometers.

13) Exterior wall hydrants.

14) Roof, marquee drains, and gutter drains.

15) Oil separator.

16) Waste oil tank.

17) Acid neutralizing tank.

18) Grease trap.

19) Water booster system.

20) Gas-fired water heaters.

c. Operating and Maintenance Manual: The Contractor shall furnish operating and maintenance instructions for review, for each item of equipment outlining the step-by-step procedures required for system periodic maintenance. After review, furnish six copies of each manual bound in hardback binders or an approved equivalent.

1.10 SUBSTITUTION OF MATERIAL OR EQUIPMENT

1.10.1 If apparatus or materials substituted for those specified necessitate changes or additional connections, piping supports, or construction; same shall be provided and the Plumbing Contractor shall assume the cost and the entire responsibility thereof.

1.10.2 The Architect's permission to make such substitutions shall not relieve the Plumbing Contractor from full responsibility for the work.

1.11 INTERRUPTION OF SERVICES

1.11.1 While work is in progress, except for designated short intervals during which connections are to be made, continuity of service shall be maintained to all existing systems. Interruptions shall be coordinated with the Owners as to time and duration. The Contractor shall be responsible for any interruptions to service and shall repair any damages to existing systems caused by his operations.

1.12 BASES AND SUPPORTS

1.12.1 All concrete bases and supports will be provided by the General Contractor.

1.12.2 The Plumbing Contractor shall furnish to the General Contractor all required foundation sizes, bolts, washers, sleeves, plates, and templates, for the plumbing equipment.

1.12.3 The size of the foundation bolts shall be as recommended by the manufacturer.

1.12.4 The foundation bolts shall be set in pipe sleeves, which are held in place by a template. Sleeves shall be 2-1/2" diameter larger than the bolt to allow movement for final positioning of the bolts.

1.12.5 All equipment shall be set on the foundations and shimmed level with steel shims and grouted up under base for uniform bearing.

1.12.6 All plumbing equipment shall be so installed and shall so operate that no noise or vibration shall be transmitted to any part of the building beyond the room or rooms in which equipment is located.

1.12.7 All metal supports shall be furnished and installed by the Plumbing Contractor.

1.12.8 The anchoring of all equipment to the structure shall comply with all applicable requirements of the codes pertaining to earthquake proofing.

1.13 GIVING INFORMATION

1.13.1 The Plumbing Contractor shall keep himself fully informed as to the shape, size, and position of all openings and foundations required for his apparatus and shall give full information to the General Contractor sufficiently in advance of the work, so that all such openings and foundation may be built in advance. He shall also furnish all sleeves and supports herein specified or required, so the General Contractor may build same in place.

1.13.2 Plan all work so that it proceeds with a minimum of interference with other trades. Inform the General Contractor of all openings required in the building construction including roof openings for the installation of his work. Provisions shall be made for all special frames, openings, and pipe sleeves as required. The cutting, patching, and core drilling made necessary by his failure to properly direct and supervise such work at the correct time shall be done and paid for by the Plumbing Contractor.

1.14 OBTAINING INFORMATION

1.14.1 The Plumbing Contractor shall obtain detailed information from the manufacturers of apparatus, which he is to provide, for the proper methods of installation. He shall also obtain all information from the General Contractor and other Subcontractors to ensure full comprehension of the work to be done and to ensure coordination between work under this Section and all other work under this Contract.

1.15 CLEANING & PAINTING EQUIPMENT AND MATERIALS

1.15.1 The Plumbing Contractor shall provide for the safety and good condition of all materials and equipment until final acceptance by the Owner; protect all materials and equipment from damage; provide adequate and proper storage facilities during the progress of the work; provide protection for bearings, open connections, pipe coils, pumps, compressors, and similar equipment.

1.15.2 All fixtures, piping, finished surfaces, and equipment shall have all grease, adhesive labels, and foreign materials removed.

1.15.3 All pumps, motors, tanks, and all other factory manufactured and assembled apparatus shall be factory coated with one coat of primer and one coat of machinery enamel except where specified finished is specified herein.

1.15.4 All piping shall be drained and flushed to remove grease and foreign matter. Pressure regulating assemblies, traps, flush valves, and similar items shall be thoroughly cleaned. Remove and thoroughly clean and reinstall all liquid strainer screens after the system has been in operation ten (10) days.

1.15.5 Air, oil, and gas piping shall be blown out with clean compressed air or inert gas.

1.15.6 When connections are made to existing systems, the Plumbing Contractor shall do all cleaning and purging of the existing systems required to restore them to the condition existing prior to the start of work.

1.16 PIPE MARKERS

1.16.1 All piping shall be color-coded and labeled with W.H. Brady pipe markers. Use Brady marker style 1 on all pipes 3" diameter and larger. Apply markers on the two lower quarters of the pipe and where view is not obstructed. Apply pipe marker and arrow marker at each valve to show contents and direction of flow. When a pipe passes through a wall, pipe marker and arrow marker shall be applied on the pipe on both sides of the wall. arrow markers must point away from pipe markers and in direction of flow. Apply pipe markers and arrow markers every 50 feet along continuous lines. Apply pipe marker and arrow marker on each riser and T-joint.

1.17 VALVE TAGS AND PIPE MARKER

1.17.1 On all valves and controls, furnish identifying numbered metal tags fastened to stem or handle by heavy brass S-hooks. Tags shall be 2" diameter, 14 gauge aluminum with stamped number filled in with black paint.

1.17.2 Four (4) sets of separate charts shall be furnished, one in each service manual, and one under glass showing each pipe system in diagrammatic form with all valves and controls numbered to correspond to numbered metal tags. Chart shall also include size of valve, type, function, and manufacturer.

1.18 OPERATING AND MAINTENANCE MANUALS

1.18.1 Operating Instructions: Provide operating instructions to the Owner with respect to operation functions and maintenance procedures for all equipment and systems installed. The cost of such instruction shall be included in the contract price. Three days shall be a minimum time for instruction.

1.18.2 Maintenance Manuals: At the completion of the project, four complete manuals containing the following shall be turned over to the Architect:

a) Complete shop drawings of all equipment.

b) Operation description of all systems.

c) Names, addresses, and telephone numbers of all suppliers of the systems and service agents.

d) Preventive maintenance instructions for all systems.

e) Spare parts list of all system components.

1.18.3 All information shall be in one book.

1.18.4 These maintenance manuals will be reviewed by the Architect.

1.19 AS-BUILT RECORD DRAWINGS

1.19.1 The Plumbing Contractor shall keep an accurate record of all deviations in the work as actually installed from work as shown on design drawings, paying particular attention to dimensioning all underground utility lines, utility structures, their offsets, valves, and grades. All inside and outside buried lines shall be dimensioned from column lines and elevation of lines noted.

1.19.2 As work progresses, the Plumbing Contractor shall keep a record of all deviations and on Monday of each week, these shall be neatly and correctly entered in colored crayon on paper print of each design drawing affected and print kept available at site for inspection.

1.19.3 At substantial completion of the work, the Engineer shall provide the Plumbing Contractor with reproducible transparencies of all Drawings at which time the Plumbing Contractor shall transfer all changes made during construction onto such transparencies. These transparencies shall thereupon constitute As-Built Drawings and shall be delivered to the Engineer prior to final payment.

1.20 GUARANTEE

1.20.1 All materials, equipment, fixtures, piping, and devices shall be guaranteed to be free from mechanical defects or faulty workmanship for a period of 1 year from the date of written acceptance by the Engineer for the Owner.

1.20.2 The Engineer is aware and the bidders are hereby made aware that certain manufacturer's equipment guarantees are valid only for a period of 1 year from the date of shipment of installation and will therefore not be valid until the date of guarantee set forth herein. It shall therefore be noted that this Plumbing Contractor shall be fully responsible for all material, equipment, and labor for the full guarantee period as set forth herein (i.e., 1 year from date of acceptance). If the manufacturer's guarantee on his equipment is more than 1 year, the Plumbing Contractor shall be responsible for the full length of the guarantee period.

1.20.3 Labor and material required to fulfill the requirements of this guarantee shall be furnished to the Owner by this Contractor at no additional cost.

1.20.4 The Plumbing Contractor shall be responsible for servicing the system for a period of 12 months after acceptance. Guarantee shall include servicing of the equipment during this time and answering any necessary trouble calls.

1.20.5 Before acceptance, Owner may call for any specific test necessary to determine if the system is functioning as intended. That is to see if any excess noise or vibration is apparent or other objectionable items are apparent.

2 MATERIALS

a. Materials shall be new, unused, best of their respective kinds, and free from defects.

b. Reference to Specifications of recognized authorities, to establish bases of quality: Latest edition in force at date of bidding.

2.1 PIPE

2.1.1 Pipe and fittings shall conform to the latest ANSI, ASTM, ASME, and Commercial Standards (CS).

2.1.2 Each length of pipe, each pipe fitting, trap, material, and device used in the plumbing system shall have a cast stamped or indelibly marked on it, the marker's name or mark, weight and quality of the product when such marking is required by the approved standard that applies.

2.1.3 Soil waste, vent, and storm (above ground) use one of the following:

a. Service weight cast iron and fittings, hub and spigot pattern (coated or plain) with caulked joints. ASTM-A74-75.

b. Hubless cast iron pipe and fittings with approved clamps. ASTM C564-70 and CISPI 301-78.

c. ABS (Acrylonitrile-Butadiene-Styrene) Schedule 40 pipe and fittings. ASTM D2661-78.

d. PVC (Polyvinyl Chloride) Schedule 40 pipe and fittings. ASTM D2665-78.

Use for sump pump and sewage ejector discharge smaller than 3". 3" and larger shall be flanged cast iron.

e. Schedule 40 ASTM A53 grade A galvanized steel pipe with cast iron plain or galvanized drainage pattern fittings for waste, cast iron, or galvanized malleable iron fittings for vents conforming to ANSI B16.3 Class 150 threaded or grooved end fittings for storm water only. ASTM A120-78 & B16.12-77.

f. Hard drawn Type L copper tubing with cast brass drainage fittings. ASTM B88-78.

g. Schedule 40 chrome plated red brass I.P.S. (Exposed piping at fixtures.)

h. Storm and sanitary piping above ground shall be ductile iron with mechanical joints in accordance with AWWA C151, Class 51. Joints shall be in accordance with ANSI/AWWA C111/A21.11.

2.1.4 Soil, waste, vent, and storm (below ground):

a. Extra heavy weight cast iron hub and spigot pattern with caulked joints or resilient gaskets. Pipe shall be coated with two coats of hot tar or asphaltum. ASTM A74-75 & C563-76.

b. ABS (Acrylonitrile-Butadiene-Styrene) Schedule 40 pipe and fittings. ASTM D266-78.

c. PVC (Polyvinyl Chloride) Schedule 40 pipe and fittings. D2665-78.

d. Hard drawn Type K copper tubing with cast bronze drainage fittings. ASTM B88-78 & ANSI B16.18-77.

e. Ductile iron, bituminous coated with push on mechanical joints in accordance with AWWA C151, Class 51.

2.1.5 Indirect waste piping:

a. Hard drawn Type L copper tubing with cast brass drainage fittings. ASTM B88-78 & ANSI B16.18-78.

2.1.6 Water piping (above ground):

a. Type L hard-drawn copper tubing with wrought copper fittings. ASTM B88-78.

2.1.7 Water piping (below ground):

a. Type K hard-drawn copper tubing with brazed joints. ASTM B88-78.

2.1.8 Water piping (exterior water service 2" and smaller):

a. Type K soft tempered copper tubing with brazed joints. ASTM B88-78.

2.1.9 Water piping (exterior water service 3" and larger):

a. Class 250 cast iron cement lined pipe and fittings — mechanical gasket joints with bituminous seal coat. ASTM A377-77.

2.1.10 Acid waste piping (above and below ground):

a. GSR Fuseal polypropylene pipe and fittings as manufactured by R & G Sloane Manufacturing Co.

b. Acid resisting cast iron pipe and fittings (Durion).

2.11.1 Gas piping (2-1/2" and larger):

a. Schedule 40 black steel pipe with welding fittings and joints. ASTM A120-78.

2.1.12 Gas piping (2" and smaller):

a. Schedule 40 black steel pipe with 150# malleable iron fittings. ASTM A120-78 & ANSI B16.3-77.

2.1.13 Gas piping (on roof):

a. Schedule 40 black steel pipe with welding fittings and joints. ASTM A120-78.

2.1.14 Emergency generator exhaust piping:

a. Schedule 80 black steel pipe with welding fittings and joints. ASTM 120-78.

2.1.15 Oil piping (hoists) (above and below ground):

a. Schedule 40 galvanized Yoloy pipe with 300# galvanized malleable iron fittings.

2.1.16 Compressed air piping (above ground):

a. Schedule 40 galvanized steel pipe with 300# galvanized malleable iron and fittings.

2.1.17 Compressed air piping (below ground):

a. Schedule 40 galvanized steel pipe with 300# galvanized malleable iron and fittings.

2.2 JOINTS AND CONNECTIONS

2.2.1 Caulked (Cast iron soil pipe):

a. Caulked joint for cast iron bell and spigot soil pipe shall be firmly packed with oakum or hemp and filled with molten lead not less than 1" deep and not to extend more than 1/8" below the rim of the hub. No paint, varnish, or other coatings shall be permitted on the jointing material until after the joint has been tested and approved. Lead shall be run in one pouring and shall be caulked tight.

b. Joints in acid-resistant cast iron pipe shall be made with a wax-free asbestos rope packing and molten lead as specified for cast iron soil pipe.

2.2.2 Gasketed (Cast soil pipe):

a. A positive seal, one piece elastomeric compression type gasket may be used for joining hub and spigot cast iron soil pipe as an alternate for lead or oakum joints. The joint is formed by inserting an approved gasket in the hub. The inside of the gasket is lubricated and the spigot end of the pipe is pushed into the gasket until seated, this effecting a positive seal.

b. A positive-seal, one piece elastomeric compression type gasket for joining hub and spigot cast iron soil pipe may be used for a drainage and waste system above and below ground.

c. Compression gaskets for cast iron soil pipe shall be neoprene, marked as such, with ASTM C564 and the CI symbol of Cast Iron Soil Pipe Institute to indicate the gasket meets the standard.

2.2.3 No-Hub (Cast iron soil pipe):

a. Joints for hubless cast iron soil pipe shall be made with an approved neoprene gasket and stainless steel or cast iron retaining sleeve.

b. No-Hub gaskets shall be marked with the manufacturer's name, ASTM C564-76, the work "No-Hub", nominal diameter, and the CI symbol of the Cast Iron Soil Pipe Institute indicating it meets the standard. Stainless steel couplings for No-Hub shall be marked "All Stainless", name and manufacturer, words "No-Hub", nominal diameter, and the CI symbol indicating it conforms to CISPI Standard 301-78.

c. Installation of the hubless cast iron soil pipe system shall in accordance with CISPI Pamphlet 100 — *Installation Suggestions for CI No-Hub Pipe and Fittings.*

2.2.4 Soldered or sweat:

a. Soldered or sweat joints for tubing shall be made with approved fittings. Surfaces to be soldered or sweated shall be properly cleaned and reamed. The joints shall be properly fluxed and made with approved solder. Joints in copper water tubing shall be made by appropriate use of approved brass or wrought copper water fittings in accordance with ANSI B16.22, properly sweated or soldered together.

Flanges shall be cast bronze in accordance with B16.24. The joining between tube and fittings shall be made with brazing filler metals complying with ANSI/ASW A58. Fluxes and brazing filler metal shall be lead-free. Threaded joints at valves shall be made with lead-free polyetraflu-orethylene sealant.

2.2.5 Threaded:

a. Threaded joints shall conform to American National Taper Pipe Thread ANSI B2.1-68. All burrs shall be removed. Pipe ends shall be reamed or filed out to size of bore, and all chips shall be removed. Screwed joints may be made up with Teflon tape. Pipe joint cement and paint shall be used only on male threads.

b. All close and shoulder nipples shall be extra heavy.

2.2.6 Slip:

a. Slip joints shall be made using approved packing or gasket material, or approved joint brass compression rings. Ground joint brass connections which allow adjustment of tubing but provide a rigid joint when made up shall not be considered as slip joints. Slip joints may be used on the house side of the trap only.

2.2.7 Plastic (ABS & PVC pipe):

a. Joints in plastic piping shall be made with approved fittings by solvent welded connections. Solvent welded connections shall be made only with a solvent cement manufactured specifically for the materials to be joined.

2.2.8 Plastic (Polypropylene pipe):

a. Joints shall be made by the heat fusion method in accordance with the manufacturer's recommendations.

2.2.9 Grooved:

a. Joints using grooved end pipe and grooved type couplings shall be installed in accordance with the manufacturer's instructions.

2.2.10 Unions:

a. Unions in the water supply system shall be metal-to-metal with ground seats.

b. Unions on drainage systems may be used only in the trap seal or on the inlet side of the trap. Unions shall have metal-to-metal ground seats.

c. Dielectric unions/couplings: Insulated union/couplings shall be provided for connecting dissimilar materials. Union shall have a water impervious insulation barrier capable of limiting galvanic current to one percent of the short circuit current in a corresponding bimetallic joint. When dry, insulation barrier shall be able to withstand a 600-volt breakdown test.

2.2.11 Brazed:

a. Brazed joints shall be made in accordance with the provisions of Section 6 of the Code for Pressure Piping. ANSI B31.1-1955.

2.2.12 Welded:

a. Joints to be welded shall be cleaned free from rust and scale and welded by certified welders who qualify according to Section 6 of the Code for Pressure Piping. ANSI B31.1-1955.

2.2.13 Joints between different piping materials:

a. Cast iron to copper tube: Every joint between cast iron and copper tube shall be made by using a brass caulking ferrule and properly soldering the copper tube to the ferrule or other approved methods.

b. Cast iron to vitrified clay: Every joint between cast iron piping and vitrified clay piping shall be made either of hot poured bitumastic compound or by a performed elastomeric ring. This ring shall, after ramming, completely fill the annular space between the cast iron spigot and the vitrified clay hub.

c. Copper tube to threaded pipe joints: Every joint from copper tube to threaded pipes shall be made by the use of brass converter fittings or dielectric fittings. The joint between the copper pipe and the fitting shall be properly soldered, and the connection between the threaded pipe and fitting shall be made with a standard pipe size screw joint.

d. Lead to cast iron or steel: Every joint between lead and cast iron or steel pipe shall be made by means of wiped joints to a caulking ferrule, soldering nipple, bushing, or by means of a mechanical adapter.

e. Threaded pipe to cast iron: Every joint between wrought iron, steel, or brass, and cast iron pipe shall be either caulked or threaded or shall be made with approved adapter fittings.

f. Special joints for drainage piping: Different types of drainage piping materials shall be joined either by adapter fittings or by means of an acceptable prefabricated sealing ring or sleeve as specifically approved by the Plumbing Code.

g. ABS or PVC DWV to other materials: ABS or PVC DWV joints to other materials shall be subject to the following requirements:

   (1) Threaded joints: ABS or PVC DWV joints, when threaded, shall use the proper male or female threaded adapter. Only approved thread tape or lubricant seal, or other approved material as recommended by the manufacturer, shall be used. Threaded joints shall not be over-tightened. After hand tightening the joint, one-half to one full turn with a strap wrench will be sufficient.

   (2) Cast iron hub joints: Joints may be made by caulking with lead and oakum or by use of a compression gasket that is compressed with the plastic pipe is inserted into the cast iron hub end of the pipe. Adapters are not required for this connection.

   (3) Cast iron spigot ends: Schedule 40 steel pipe or copper DWV tube: Joints between these materials and plastic shall be joined with an approved adapter fitting.

h. Connections between earthenware of any fixture and flanges in soil and waste piping shall be made absolutely gas and watertight with the closet setting compounds and gaskets which must be absolutely gas and fireproof, watertight, stain proof, containing neither oil nor asphaltum, and which will not rot, harden or dry under any extreme climatic change, and must adhere on wet surfaces.

i. Any fitting or connection on a drainage system which has an enlargement chamber, or recess with a ledge, shoulder, or reduction of pipe area, that offers an obstruction to flow through the drain, is prohibited.

2.3 VALVES

2.3.1 Valves shall be furnished and installed in all branches serving more than two pieces of equipment or group of plumbing fixtures, on both sides of equipment such as tanks, pumps, meters, etc., for isolating of branch mains eliminating the necessity of interrupting service to the entire building structure for maintenance purposes and where indicated on the Drawings. Valves shall be installed with the best workmanship and appearance and shall be grouped so that all parts are easily accessible from a minimum number of access doors. Manufacturer's figure numbers are specified to indicate type and quality and construction, but products of listed approved manufacturers may be substituted for those specific numbers shown.

2.3.2 All valves of similar service to be of the same manufacturer and shall have manufacturer's name or trade mark and the working pressure stamped or cast on the body.

2.3.3 Water:

a. Gate valves 2-1/2" and smaller; 125 psi, solid wedge, non-rising stem, bronze body, seats, and disk.

(1) Solder end — Jenkins 1240 or equal, Fairbanks, or Lunkenheimer.

(2) Threaded end — Jenkins 47 or equal, Fairbanks, or Lunkenheimer.

b. Gate valves 3" and larger; iron body with bronze trim, OS&Y, flanged ends conforming to MSS SP-85 class 150.

(1) Jenkins 651-A or equal, Nibco-Scott, Fairbanks, or Lunkenheimer.

c. Check valves 2-1/2" and smaller; 125 psi, bronze body disk, swing type.

(1) Solder end — Jenkins 1222 or equal, Fairbanks, or Lunkenheimer.

(2) Threaded ends — Jenkins 92-A or equal, Fairbanks, or Lunkenheimer.

d. Check valves 3" and larger; 125 psi, iron body with bronze trim, flanged ends.

(1) Jenkins 624 or equal, Fairbanks, or Lunkenheimer.

e. Globe valves 2-1/2" and smaller; 150 psi, bronze body, renewable composition disk.

(1) Solder ends — Jenkins 1200 with No. 112A disk or equal, Nibco-Scott, Fairbanks, or Lunkenheimer.

(2) Threaded end — Jenkins 106A with No. 119A disk or equal, Nibco-Scott, Fairbanks, or Lunkenheimer.

f. Globe valves 3" and larger; 125 psi, iron body with bronze trim, renewable composition disk, flanged ends.

(1) Jenkins 142 or equal, Nibco-Scott, Fairbanks, or Lunkenheimer.

g. Balancing valves — Jenkins Bros. No. 106A with No. 119A disk and No. 344 throttling but. Nibco-Scott or Fairbanks.

h. Drain valves — 3/4" all bronze end gate valve, Jenkins No. 372 with No. 658 cap and chain or equal, Kennedy Valve Mfg. Co. or Walworth Co.

i. Approved manufacturers: Jenkins, Nibco-Scott, Fairbanks, Lunkenheimer, or Hammond.

2.3.4 Compressed air:

a. Globe valve:

(1) Screwed — Jenkins Bros. No. 106A with No. 294-S disk or equal, Crane Co. or Fairbanks Co.

(2) Solder end — Jenkins Bros. No. 1200 with No. 294-S disk or equal, Crane Co. or Fairbanks Co.

2.3.5 Diaphragm-actuated check valve:

a. Diaphragm-actuated check valve shall be globe type valve with a diaphragm actuator hydraulically operated by line pressure. The valve shall be 200 lb WOG, iron body, bronze mounted with screwed ends. Body material shall conform to ASTM Standard Specification for Gray Iron Castings, Designation A48-76. The main valve and pilot control trim shall conform to ASTM Standard Specification for Steam or Valve Bronze Castings, Designation B61-76.

2.3.6 Solenoid Valves:

a. Solenoid valves shall be bronze body, screwed-end, single-integral-seat, full pipe-area, globe-type valves, with renewable composition disk seats. Valves shall be suitable for operation on 120-volt, single-phase, 60-Hertz current, and designed to close when energized. The solenoid valves shall be manufactured by Automatic Valve Co., Inc., Indianapolis, IN; J.D. Gould Co., Indianapolis, IN; Automatic Switch Co., Florham Park, NJ; Magnatrol Valve Corp., Hawthorne, NJ; or be an acceptable equivalent product.

2.3.7 Gas valve:

a. Valves on gas lines shall be brass or bronze plug-cocks with screwed ends for sizes up to and including 1-1/2" and shall be lubricated-type plug valves with iron bodies and flanged ends for sizes over 1-1/2" Lubricated-type plug valves shall have short bodies and tapered plugs. A box of stick lubricant shall be furnished for each six lubricated plug valves of each size of lubricant fittings.

b. Gas cocks for shutoff at each gas-connected fixture shall be Boston key cocks, or acceptable equivalent product approved by the local utility company. Provide Owner with wrenches to fit each size.

2.3.8 Pressure reducing valves:

a. Pressure-reducing valves 2-1/2" and smaller shall be self-contained, bronze body, single-port valves with a spring-loaded diaphragm. The valves shall be manufactured by Fisher Governor Co., Marshalltown, Iowa; A.W. Cash Valve Mfg. Corp., Decatur, IL; Watts Regulator Co., Lawrence, MA; or be an acceptable equivalent product.

2.4 SPECIALTIES

2.4.1 Non Freeze Wall Hydrants:

a. Josam Series 71150-80 cast brass, heavy duty, box type wall hydrant with polished brass face — 3/4" npt straight hose outlet — T-handle key.

2.4.2 Hose bibbs:

a. For concealed piping to be Nibco Figure #63 or #662.

b. For exposed piping to be Nibco Figure #72 or #73.

c. For toilet rooms — Chicago Faucet Co. No. 952 or approved equal, chrome plated sill faucet with vacuum breaker, 3/4" hose thread outlet with lockshield cap, No. 293-4 removable T-handle and 3/4" flanged female threaded inlet.

2.4.3 Shock absorbers:

a. Josam Series 1485, Zurn Z-1700, J.R. Smith Series 5000 or approved equal. Each shock absorber shall be located in accordance with the manufacturer's recommendations.

b. Sizing of shock absorbers shall be in accordance with Plumbing and Drainage Institute "Standard PDI WH201."

2.4.4 Air vents:

a. Bell and Gossett #7 or approved equal. Install at all high points at hot and cold pipe runs and risers which may become air bound.

2.4.5 Vacuum breakers:

a. Watts Regulator Co. Series 288A or approved equal (sinks and hose bibbs).

b. Watts Regulator Co. Series 9D or approved equal (boiler feed lines).

2.4.6 Vacuum relief valves:

a. Watts Regulator Co. Series 36A or approved equal (hot water heater supply).

2.4.7 Pressure & temperature relief valve:

a. Watts Regulator Co. 100 X L or equal, ASME rated ANSI (hot water heaters).

2.4.8 Pressure reducing valves:

a. Watts Regulator Co. U5BLP or approved equal (dishwashers).

b. Watts Regulator Co. U135B or approved equal (1-1/2" to 2").

2.4.9 Backflow preventers (reduced pressure type)

Backflow preventers shall be of the reduced pressure principle type conforming to the applicable requirements of AWWA C506. Furnish a certificate of Full Approval or a current certificate of Approval for each design, size and make of backflow preventer being provided for the project. The certificate shall be from the Foundation for Cross-Connection Control and Hydraulic Research, University of Southern California, and shall attest that this design, size, and make of backflow preventer has satisfactorily passed the complete sequence of performance testing and evaluation for the respective level of approval. A certificate of Provisional Approval is not acceptable in lieu of the above.

a. Watts Regulator Series 909 or approved equal. ASSE 1013-71.

2.4.10 Backflow preventers (double check valve type):

a. Watts Regulator Series 700 or approved equal. ASSE 1015-72 (3/4" to 2").

2.4.11 Water meters:

a. Water meters shall be positive displacement, disk- type meters with bronze body and parts. They shall have straight reading dials. Meters shall be made by Neptune Meter Co., Long Island City, NY; Hersey-Sparling Meter Co., Dedham, MA; Rockwell Mfg. Co., Pittsburgh, PA; or an acceptable equivalent or of a type required by the Owner.

2.4.12 Strainers:

a. Strainers shall be placed ahead of each control valve and elsewhere as specified or indicated on the Drawings. Strainers shall be screwed or flanged as specified for valves. Bodies shall be of the T, S, or Y type designed for not less than 125-lb working pressure. Screens shall be bronze, Monel, or stainless steel with perforations as follows:

Strainer Size	Perforation Size
3/4" to 2" inclusive	1/32"
Over 2-1/2"	1/16"

2.4.13 Flow indicators:

a. Flow indicators shall be provided for installation in the sealing water supply pipe to pumps where indicated on the Drawings. Each indicator shall consist of a bronze body with threaded ends and a sight glass with nylon rotor or a tapered glass tube with brass float capable of indicating less than 1 gpm of flow. Indicators shall be manufactured by Jacoby-Tarbox Corp., Yonkers, NY; Eugene Ernst Products Co., Farmingdale, NJ; Schuttle and Koerting Co., Cornwell Heights, PA; or be an acceptable equivalent product.

2.4.14 Shock absorbers:

Shock absorbers shall be series 1485 shock absorbers made by Josam Mfg. Co.; Shockstop made by Wade, Inc.; Shocktrol made by Zurn Industries, Inc.; or an acceptable equivalent product and shall be provided on both hot and cold water supplies to each fixture.

2.4.15 Thermometers and pressure gauges:

a. Thermometers:

(1) Furnish and install where indicated on the Drawings and where specified herein separable well type dial thermometers.

(2) All thermometers shall be installed in such a manner as to cause a minimum of restriction to flow in the pipes and so that they can be easily read from the floor.

(3) Thermometers shall be 4" (minimum in diameter) hermetically sealed, bi-metal dial type with raised jet black figures, stainless steel stems, and brass separable sockets.

(4) The accuracy of all thermometers shall be within one percent of the scale range.

(5) Thermometers shall be installed where indicated on the Drawings and in the following systems locations:

(a) Each hot water system circulating pump discharge piping.

(b) Outlet piping of each systems hot water heaters.

(6) The scale range for thermometers shall be as follows:

Service	Temperature Range
Hot water supply	50 to 200°F.
Hot water return	50 to 200°F.

b. Pressure gauges:

(1) Furnish and install where indicated on the Drawings and where specified herein, bourdon spring type pressure gauges.

(2) All pressure gauges shall be installed so as to be easily readable from the floor in a standing position without parallax.

(3) Each gauge shall have dull black steel or aluminum casings with chrome plated bezels or rims. The gauges shall have white faces with black filled engraved numerals. The face diameter of dials on gauges installed within 8' from the finish flow shall be not less than 4-1/2" all other gauges installed above 8' from the finished floor shall have a 6" diameter dial face.

(4) The accuracy of all gauges shall be within one percent of the scale range.

(5) All gauges on water lines shall be fitted with pressure snubbers.

(6) A bronze needle valve shall be installed on the system side of each gauge. Provide test tee between valve and gauge.

(7) A pressure gauge shall be installed in the suction and discharge of each water booster pump, hot water circulating pumps. Additional pressure gauges shall be installed where indicated on the Drawings.

(8) The scale range of pressure gauges shall be as follows:

Service	Pressure Range
City pressure water (cold, hot, and return systems)	0 to 100 psig
Boosted pressure water (cold, hot, and return systems)	0 to 200 psig

2.4.16 Acid neutralizing tank:

a. Furnish and install GSR Polyprophylene #NT-55 neutralizing tank. See Detail on Drawings. Furnish and install 600 lb limestone chips as called for on Drawings.

2.4.17 Oil interceptor:

a. Furnish and install an oil interceptor as detailed on the Drawings. Installation shall be in accordance with all applicable codes and local ordinances.

b. Interceptor to be Josam GA40 having a flow rate of 25 gpm and shall be modified to have a 4" inlet and outlet connection. See Drawings for piping Details. (Size listed is for example only.)

2.4.18 Flush tank (Junk battery box):

a. Eljer 171-0802 vitreous china wall hung flush tank complete with automatic flush valve and 3/8" I.P.S. supply valve.

2.4.19 Drip pan:

a. Provide water tight drip pans with 6" deep sides, 16 gauge sheet copper reinforced and properly supported under water and drainage piping running over or near electrical apparatus, such as switchboards, control board, or in electrical equipment rooms, etc., provide pans with 1-1/2" drain outlet and piped to spill over floor drains, service sink, or as directed.

2.5 SLEEVES & ESCUTCHEONS

2.5.1 Each and every pipe and conduit, regardless of material, which passes through a concrete slab, foundation wall, masonry wall, roof, or other portion of the building structure and shall pass through a sleeve.

2.5.2 Except as hereinafter specified, all sleeves shall be constructed from either electric metallic tubing or light weight steel pipe, and shall be installed flush on both sides of the surface penetrated. Above grade sleeves shall be constructed from 22 gauge galvanized steel and shall be flush on both sides of the surface penetrated. The sleeves shall be sized to allow free passage of the pipe to be inserted, and where the pipe is to be insulated, the sleeves shall be large enough to pass the insulation.

2.5.3 Sleeves passing through walls or floors on or below grade or in moist areas shall be constructed of Schedule 40 black steel pipe and shall be designed with 150 lb black steel welding flange in the center to form a waterproof passage. Paint sleeves with 1 coat of bitumastic inside and outside. After the pipes have been installed in the sleeves, the void space around the pipe shall be caulked with jute twine and fitted with an asphalt base compound to insure a waterproof penetration. Sleeves provided for piping between floors and through fire walls or smoke partitions shall be installed with approved packing between sleeves and piping to provide fire stops.

2.5.4 Provide pipes passing through membrane waterproofed floors with sleeves as described above. Sleeves shall be located in 6" high concrete curb and shall be flush with top if finished curb. All other sleeves shall be extended 1" above finished floor.

2.5.5 Plumbing Contractor shall coordinate the location of all sleeves with the General Contractor. Sleeves passing through floor under a partition wall must be located by figured dimensions in the field.

2.5.6 Provide a separate pipe sleeve for each pipe passing through floor, wall, partition, etc. Do not run 2 or more pipes through one common pipe sleeve.

2.5.7 Unless otherwise noted, provide exposed pipe, both bare and covered with chrome plated Beaton & Caldwell 3-A or approved equal cast brass escutcheons where they pass through walls, partitions, floors, or ceilings; on bare pipes, held in place by set screws and on covered pipes by internal spring tension.

2.5.8 Where sleeves, hubs, or fittings project slightly from wall, partitions, floor or ceilings, provide special deep type escutcheons to cover each case.

2.6 HANGERS AND SUPPORTS

2.6.1 All piping shall be supported from the building structure by means of approved hangers and supports. Piping shall be supported to maintain required grading and pitching of lines, to prevent vibration and to secure piping in place, and shall be so arranged as to provide for expansion and contraction. Chain, perforated strap, bar, or wire hangers are not permitted.

a. No piping shall be hung from the piping of other trades.

b. Branches shall have separate supports and no branch 5'-0" or longer shall be without support.

Do not support piping from ductwork, conduit, or other trades.

2.6.2 Do not hang pipe hangers from bottom chord of roof joists. Hanger must be installed at or near panel point of roof joists. 6" and larger pipes running parallel to roof joints must be supported by two roof joists. Plumbing Contractor shall furnish and install angle iron bridging between joists of adequate size to securely support pipe hanger. Bridging for pipe hangers must also be supported by the top chord of the roof joist and must be installed at or near panel point. Use approved type brackets to support piping racked along walls. Support piping running just above floor on pipe saddle supports. Pipe hangers shall not penetrate waterproof floor membrane or roof deck. Copper tubing shall be supported by copper plated hangers. Where overhead construction does not permit fastening of hanger rods in required locations, provide additional steel framing as required and approved.

2.6.3 Maximum spacing of hangers on runs of pipe having no concentrations of weight shall be as follows:

a. Schedule — hanger spacing in feet/pipe material

Pipe Size	Steel or Alloy	Copper or Brass
1/2"	7	5
3/4"	7	5
1"	7	5
1-1/4"	10	6
1-1/2"	10	8
2"	12	8
2-1/2"	12	10
3"	12	10
3-1/2"	12	10
4"	14	12
5"	14	12
6"	14	12
8"	20	16
10"	20	—
12"	20	—

2.6.4 Plastic piping 1-1/2" or less shall be supported at 3' intervals; 2" and over at 4' intervals.

2.6.5 Maximum spacing of hangers on soil pipe shall be 5' and hangers shall be provided at all changes in direction. For pipes exceeding 5'-0" length, they shall be placed at intervals equal to the pipe length but not exceeding 10'-0". Hangers for No-Hub piping shall be provided at least every other joint except when the developed length between hangers exceeds 4'-0" they shall be provided at each joint.

2.6.6 Where codes having jurisdiction require closer spacing, the hanger spacing shall be as required by code in lieu of the distances specified herein.

2.6.7 Provide hangers at a maximum distance of 2 feet from all changes in direction (horizontal and vertical) on both sides of concentrated loads independent of the piping.

2.6.8 Friction clamps shall be installed at the base of all plumbing risers and at each floor. Friction clamps <u>shall not</u> be supported from or rest on floor sleeves. 1-1/4" and smaller vertical shall be supported at 8'-0" maximum.

2.6.9 Hangers in general for all horizontal piping shall be Clevis type hangers. These hangers shall be sized to provide for insulation protectors as hereinafter specified.

2.6.10 Hangers for uncovered (non-insulated) copper and brass piping shall be factory applied copper plated steel Clevis hangers. Rods and nuts used with these hangers shall also be factory applied copper plated.

2.6.11 Where three or more pipes are running parallel to each other factory-fabricated gang type hangers with pipe saddle clips or rollers may be used in lieu of the herein specified Clevis hangers. These hangers shall be sized to provide for insulation protectors as hereinafter specified. Pipe saddle clips shall be not less than 16 gauge metal and shall be copper when installed with non-insulated copper piping.

a. Trapeze type hangers shall be made up of angles bolted back to back or channels for supporting parallel lines of piping. Trapeze type hangers shall be supported with suspension rods having double nuts, and securely attached to construction with inserts, beam clamps, steel fish plates, cantilever brackets, lag screws, or other approved piping attached along walls.

2.6.12 Field painting or spraying of hangers, rods, and nuts in lieu of copper plating will not be accepted.

2.6.13 All suspended horizontal piping shall be supported from the building by mild steel rod connecting the pipe hanger to inserts, beam clamps, angle brackets, and lag screws as required by the building construction in accordance with the following:

Pipe size in inches:	3/4 - 2"	2-1/2 - 3-1/2"	4 - 5"	6"	8 - 12"
Rod diameter in inches:	3/8"	1/2"	5/8"	3/4"	7/8"

2.6.14 All hangers on insulated lines shall be sized to fit the outside diameter of the pipe insulation. Provide pipe covering protection saddles at all hangers on the insulated lines. Provide 16 gauge sheet metal shield, 12" long and covering 180 degrees of arc on the covering at all hangers on insulated lines.

2.6.15 Remove rust from all ferrous hanger equipment, (hangers, rods, and bolts) and dip hangers and supports in zinc chromate primer before installation.

2.6.16 Piping at all equipment and control valves shall be supported to prevent strains or distortion in the connected equipment and control valves. Piping at equipment shall be supported to allow for removal of equipment, valves, and accessories with a minimum of dismantling and without requiring additional support after these items are removed.

2.6.17 All piping installed under this Section of the Specifications shall be independently supported from building structure and not from the piping, ductwork, or conduit of other trades.

2.6.18 All supplementary steel including factory fabricated channels and associated accessories throughout the project for this Section of the Specifications both suspended and floor mounted shall be furnished and installed by the Plumbing Contractor and shall be subject to the approval of the Architect.

2.6.19 Safety retaining clips shall be installed with all beam clamps.

2.6.20 Lay exterior underground piping on only solid undisturbed ground, except where crossing another trench or excavation adjacent to building wall or foundation, and there, or on unsuitable ground, support piping on approved foundations of concrete or brick piers or cradles as directed. Bottoms of trenches shall be tamped hard, graded to secure required pitch, and shaped to give substantial uniform support to lower third of full length of pipe, with minimum recesses excavated for bells and joints.

a. Support and protect underground piping so that it remains in place without settling and without damage during and from backfilling. Replace any piping damaged.

2.6.21 Interior underground piping in fill or where firm bearing is not available shall receive added support by the use of approved Clevis type hangers suspended from structural slab fish-plate and secured by nuts and washers above and below fish-plate. Fish-plate shall be imbedded in concrete slab a minimum of 3" from bottom of slab. Hangers shall be placed at each hub and a maximum of 5'-0" on centers on the pipe runs. With the exception of fish-plates, the entire hanger assembly including nuts, shall be galvanized steel. All of assembly with exception of fish-plate shall be bitumastic coated. Rod diameters and fish-plate sizes shall be as listed below:

Pipe Size	Rod Diameter	Fish Plate Size
2"	3/8"	4" x 4" x 1/4"
2-1/2"	1/2"	4" x 4" x 1/4"
3"	1/2"	4" x 4" x 1/4"
4"	5/8"	4" x 4" x 1/4"
5"	5/8"	6" x 6" x 1/4"
6"	3/4"	6" x 6" x 1/4"
8"	7/8"	6" x 6" x 1/4"
10"	7/8"	6" x 6" x 1/4"
12"	7/8"	6" x 6" x 1/4"

2.6.22 Pipe supports shall be of the following type and figure number manufactured by F&S, Central Buldex Div., Brooklyn, NY, FEF & Mason, Carpenter & Peterson, or Grinnel, or as hereinafter indicated:

Pipe Hanger Schedule

	F&S	F&M	GRINNELL
Beam clamp	55	282	—
Multi-J hook plate	9293	239	—
Clevis hanger	86	239	260
120° shield	980	80	—
Pipe saddle	900 Series	170 & 1700 Series	180 Series
Rigid trapeze	170	—	Std. 45
U-Bolt	37	176	137
Adj. steel pipe stanchion	421	291	259
Welded steel bracket	800 or 801	151 or 155	195 or 199
Single bolt riser clamp	91 or 93	241	261

Double bolt clamp riser	92	—	Std. 40
Structural pipe stanchion	720	—	—
Double bolt pipe clamp	89	261	295
Welded beam attachment	988	—	66
Welded beam attachment w/ B&N	966A	251	66
Insert	180A 180B	—	280
Continuous slotted insert	150	190	—

2.7 CLEANOUTS

2.1.7 Cleanouts shall be provided in soil, waste, and storm drainage piping at change in directions, at foot of stacks, or other points so that all portions of the lines will be readily accessible for cleaning or rodding out.

2.7.2 The maximum horizontal distance between cleanouts, in piping 4" in diameter and smaller, shall be no more than 50' apart; in piping 5" in diameter and larger they shall be no more than 100' apart.

2.7.3 Cleanouts shall be of the same size as the pipe installed up to 4" in diameter and not less than 4" in diameter for piping larger than 4" in diameter.

2.7.4 Traps not included with fixtures and in accessible locations shall be provided with a brass trap screw protected by the water seal, and will be regarded as a cleanout.

2.7.5 Bodies of cleanout ferrules in bell and spigot piping shall be standard pipe sizes conforming in thickness to that required for pipe and fittings, and shall extend not less than 3/4" above the hub of the pipe. The cleanout plug shall be of cast brass and shall be provided with a raised nut 3/4" high. Cleanouts in copper waste piping shall be soldered brass cleanout fittings with extra heavy brass screw plugs of the same size at the line. Clearance in threaded waste piping shall be cast iron drainage T pattern 90° branch fitting with extra heavy brass screw plugs of the same size of the pipe.

2.7.6 Test tees with brass cleanout plugs shall be installed at the foot of all vertical coil, waste, acid waste, and roof conductor lines and at each floor. Wherever cleanouts on vertical lines occur concealed behind finished walls, they shall be extended to back of finish wall and a wall plate shall be provided.

2.7.7 Cleanouts shall be installed to clear all mechanical equipment and Owner's fixtures.

2.7.8 Cleanouts shall be as manufactured by Josam, Zurn, Smith, or Wade. Plumbing Contractor shall furnish Owner with one T-handle for recessed plugs for every five wall and floor cleanouts.

2.7.9 Walls and partitions cleanouts: Josam Series No. 58710-3-15-21-22 iron caulking ferrule with brass raised plug covered with chrome plated bronze wall plates.

2.7.10 Finished floor cleanouts (not waterproof): Josam Series 56000-21-22. Provide carpet marker in carpeted areas.

2.7.11 Machinery stock and unfinished rooms cleanouts (waterproofed): Josam Series 56000-2-21-22-41. Install 24" x 24" 16 oz copper membrane.

2.7.12 Machinery and unfinished rooms cleanouts (not waterproofed): Josam Series 56000-2-21-22.

2.7.13 Above ceiling cleanouts: Josam Series 58500-20 cast iron caulking ferrule with raised head brass plug.

2.7.14 Automotive garage: Josam Series 56040-5-15 with ductile iron cover secured to plug by countersunk bronze screw.

2.7.15 Approved manufacturers: Josam, Smith, Wade, Zurn.

2.8 FLOOR DRAINS & AREA DRAINS

2.8.1 General:

a. Installation of floor drains in areas and locations as called for on Drawings. Floor drains shall be equipped with trap primer connections where shown on Drawings.

b. Floor drains in toilet rooms or located adjacent to partitions must be coordinated with General Contractor. Dimensions locating floor drains in food service areas must be verified in the field.

c. Type of floor drains listed below are as manufactured by Josam and Smith. Equal drains as manufactured by Zurn, Smith, and Wade are acceptable.

d. *NOTE*: All floor drains except those which are installed in first floor slab shall be equipped with flashing clamp device and 24" x 24" 16 oz copper membrane.

2.8.2 Type A:

a. Josam 3003-6A-80 with sediment bucket — caulk bottom outlet, cast iron body with Nikaloy strainer set 1/4" below finished floor.

2.8.3 Type B:

a. Josam 32220 in sizes indicated on Drawings with sediment bucket — caulk bottom outlet, cast iron body with polished bronze top in unfinished areas, Nikaloy tops in finished areas.

2.8.4 Type C:

a. Josam Series 30003 cast iron bottom outlet with E5 cast iron 6" round Hub, complete, with 5" diameter brass strainer.

2.8.5 Type D:

a. Josam 30003 cast iron body — caulked bottom outlet with E1 7" diameter bronze top and strainer, set rim flush with finished floor.

2.8.6 Type E (Automotive garage area depending on the battery type)

a. Smith 2450-M cast iron floor drain with 4" bottom outlet, solid bottom suspended sediment bucket, ductile iron gate.

2.8.7 Type F (Battery room floor drain):

a. Durion plate #5501 AB.

2.8.8 Trap primer:

a. Install trap primer equal to Josam Series 1465 where shown on Drawings or where required by Plumbing Code.

2.9 ROOF DRAIN (Controlled Flow)

2.9.1 Zurn Z-105-10EC or approved equal for sloped roof construction Dura-coated cast iron body with extension, roof sump receiver, and underdeck clamp control flow weir shall be linear functioning with integral membrane flashing clamp/gravel guard and aluminum dome. Sizes and locations shall be as indicated on Drawings.

2.10 ROOF DRAIN

2.10.1 Zurn Z-100-C-E or approved equal Dura-coated cast iron body with combination membrane flashing clamp/gravel guard and low silhouette Ploy-Dome. Sizes and locations shall be as indicated on Drawings.

2.11 ROOF DRAIN (Gutters)

2.11.1 Zurn ZRB-180 or approved equal Dura-coated cast iron body, plain bronze dome strainer, and membrane flashing clamp. Size and locations shall be as indicated on the Drawings.

2.12 MEMBRANES

2.12.1 Provide 16 oz copper clashing extending at least 10" around all drains and cleanout deck plates above first floor and water proof sleeves in membrane waterproofed floors and flashing sleeves; and securely held by clamping device equal to Josam Series 26420.

2.13 INSULATION

2.13.1 General:

All pipe covering specified herein for piping systems shall be furnished and installed by a competent Pipe Covering Contractor responsible to the Plumbing Contractor. Before covering is applied, all pressure tests shall have been performed and approved, with all surfaces to be covered shall have been cleaned. Pipe covering and auxiliaries shall be kept dry during storage and application. Adhesives, cements, and coatings shall not be applied when the ambient temperature is below 40°F.

a. The jacket shall have a pressure sealing lap adhesive to eliminate the use of staples, adhesives, or bands, and installation shall be in accordance with manufacturers' recommendations.

b. All insulation shall have composite (insulation, jacket, or facing, and adhesive used to adhere the facing or jacket to the insulation) fire and smoke hazard ratings of NFPA 90A as determined by Underwriters' Laboratories procedure, ASTM E-84-50T, NFPA 255, and UL 723 not exceeding: Flamespread — 25; Smoke developed — 50.

   (1) Accessories, such as adhesives, mastics, cements, and tapes for fittings shall have the same component rating as listed above.

c. Pipe covering shall be continuous and shall be carefully fitted with side and end joints butted tightly and staggered. Valves, fittings, flanges, and accessories shall have the same thickness of pipe covering as the adjacent pipe. Pipe covering for these items shall be factory molded type, field fabricated or asbestos cement.

d. Valve bodies and bonnets shall have covering applied up to the packing gland.

f. Pipe covering for flanges shall overlap the adjoining pipe by a minimum of 3" on each side.

g. The end joints of each section of the installed pipe covering shall be tightly butted. Where pipe covering ends at equipment of fixtures, end caps on the covering shall be installed.

h. Where vapor barrier mastic is installed on flanges, valves, fittings, roof drain bodies, and strainers, the thickness shall be equal to the covering on the adjoining piping. The cement shall be neatly troweled on. Over the cement, furnish and install an 8 oz canvas jacket, which shall be embedded in a fire retardant vapor proof adhesive with a 3" minimum overlap at all seams. The canvas jacket shall be smoothed out during the application to avoid wrinkles and gaps.

i. All piping with a factory and/or field applied canvas jacket over the pipe covering shall be finished with a fire-resistant coating have a vapor transmission of not more than 1.0 perm.

j. Furnish and install at each pipe hanger and or pipe support on covered lines a section of foam glass, cork, or calcium silicate in lieu of the fiberglass covering specified herein. The inserted section shall be of the same thickness as the adjacent pipe covering and finished as specified herein for fittings.

k. Provide non-combustible insulation on all domestic cold water lines, hot water supply, and circulating pipes horizontal roof leaders and emergency generator exhaust pipe and drains subject to condensation.

2.13.2 Cold & hot water & recirculating lines:

a. Insulate cold and hot water piping with fiberglass pipe insulation with Fire Retardant Vapor Barrier Jacket (FRJ). Pipe insulation to be sealed with a fire resistive adhesive.

b. Valves and fittings shall be insulated with Zeston or approved equal, Hi-Lo Temp insulation of thickness equal to adjacent piping and covered with Zeston one piece PVC insulated fitting covers bound with Zeston Z-tape of a width recommended by manufacturer. Valves larger than 3" shall be insulated with Zeston Hi-Lo Temp Insulation of thickness equal to adjacent piping and covered with Zeston performed insulated foam valve covers sealed with Zeston Vapor Barrier Mastic Adhesive and Z-tape on exposed edges.

c. Thickness of insulation shall be 1" for all cold water piping and hot water piping 2" and under. Thickness of insulation of hot water piping 2-1/2" and larger shall be 1-1/2". (Check the water temperature.)

2.13.3 Horizontal roof leaders:

a. Insulate the horizontal roof drain piping with 1" thick fiberglass duo-temp insulation same as water piping. Vertical at roof drain, roof drain body, and elbow at the end of the horizontal run are to be insulated.

2.13.4 Electric water cooler drains, kitchen equipment drains, equipment drains, and all piping subject to condensation shall be insulated with 1" thick fiber-glass duo-temp insulation.

2.13.5 Emergency generator exhaust:

a. Insulate the emergency generator exhaust piping and muffler with 2" thick calcium silicate high temperature (1200°F) insulation, similar to Owens-Corning Kaylo-10. Each layer to be securely wired on with all joints staggered. Fittings and flanges to be insulated with molded or mitered segments of pipe insulation securely wired on. Piping exposed to view to be finished with a 12 oz asbestos cloth or glass cloth with lagging adhesive insulated non-combustible material.

2.13.6 Cold water meter:

a. Shall be insulated with Fiberglass Aerocor with foil facings 1-1/2" sealed and tied with jute twine or wire, a smooth coat of asbestos cement with an open weave glass cloth jacket applied with BF 30-35 adhesive.

2.13.7 All pipe insulation exposed to view in mechanical rooms shall be covered with 8 oz canvas jacket.

2.13.8 Hot water storage tank:

a. Shall be insulated with 9 lb 1-1/2" fiberglass board. Bring edges of insulation boards on tanks into firm contact and cut or score where necessary to fit the shape and contour of the vessel. Fill all voids in the insulation with insulating cement. Hold insulation in place with 1/2", 25 gauge galvanized steel bands not more than 12" centers. When in-sulation has been installed apply a flooding brush coat of Foster 30-35 or as approved to the entire surface. Into the wet coating embed one layer of open weave glass cloth smoothed out in wallpapering manner to avoid wrinkles and holidays, and overlap at all fabric seams to a minimum of 2". Apply a final finish coat of Sealfas or as approved.

2.14 PLUMBING FIXTURES & EQUIPMENT

2.14.1 Fixtures shall be best quality regular selection genuine white vitreous china, acid-resisting enameled cast iron, or stainless steel as specified; free from cracks, dents, crazes, chips, twists, discoloration, and other defects. Surfaces of enameled iron fixtures not required to be enameled; factory coat of white paint. Fixtures shall have manufacturer's guarantee label or trademark indicating first quality. Acid resisting (AR) enameled ware shall bear manufacturer's symbol signifying acid-resisting material. Fixtures shall be Crane, American Standard, Kohler. Flush valves shall be Sloan "Royal." Toilet seats shall be Church, Olsonite, or Beneke. All fixtures shall be of the same manufacturer.

2.14.2 Exposed pipe, fittings, traps, escutcheons, valves, valve handles, and accessories, both above and below fixtures shall be CP brass (covering tubes not permitted except as noted). Brass tubing shall not be lighter than No. 17 gauge. Water supplies and drainage nipples to wall shall have cast brass escutcheons with set screw. Exposed fixture traps shall be equipped with cleanout plugs. Supply fixtures shall be complete with renewable seats, composition washers, all-metal indexed handles, and integral or separate screw driver or lock-shield stops. All lavatory faucets, except self-closing type, shall be equipped with flow controls to limit the flow to no more than 3 gpm.

2.14.3 All materials specified to be chromium plated shall be thoroughly cleaned and polished before plating and plate shall be heavily, thoroughly, and evenly applied; guaranteed not to strip or peel.

2.14.4 All fixtures and equipment shall be supported and fastened in satisfactory manner. Where wall hung fixtures are secured to interior masonry walls or partitions they shall be fastened with 1/4" through bolts provided with nuts and washers at back. Bolt heads and nuts shall be hexagon chromium plated brass.

2.14.5 Where secured to concrete or exterior brick walls, they shall be fastened with brass bolts or machine screws in lead sleeve type expansion shields and shall extend at least 3" into solid concrete or brick work, except fixtures specified to be supported on chair carriers.

2.14.6 Thoroughly clean fixtures and fittings when directed. Fixtures shall be in perfect condition at completion of job and any fixtures not in perfect condition at completion of job and any fixtures not in perfect condition at the time, due to damage during construction or any other cause, shall be replaced by Contractor at no additional charge. Replace all toilet seats which are temporarily used during construction.

2.14.7 The Plumbing Contractor shall be responsible for providing those portions of the fixtures fittings or trimming which are not provided with the fixtures but which are required for the complete installation. All fixtures shall be carefully checked to determine which portions must be provided to complete the installation.

a. Where escutcheons are not furnished with plumbing fixtures, the Plumbing Contractor shall supply them.

2.14.8 Mounting height for all plumbing fixtures shall be established by the Architect. Consult with Architect for the determination of same before installing the plumbing fixtures.

2.14.9 Refer to architectural and plumbing drawings for the quantities of fixtures to be furnished under this Section of the Specifications, which shall be deemed to include all plumbing fixtures shown of the types described hereinafter. Final location shall be determined from architectural drawings.

2.14.10 The following is a description of the fixtures to be furnished. The Plumbing Contractor will be responsible for quantities required for the job. All fixtures are to be complete with all fittings and accessories required for a complete job. All fixtures to be white unless otherwise noted.

2.14.11 Fixture Schedule:

a. Fixtures and trim listed below are American Standard except as noted.

(1) Approved "or equal" manufacturers: Kohler, Crane, Eljer.

P-1 Water closet:

"Afwall" 2477.016 vitreous china, wall hung, siphon jet bowl, with Church 295C hung black plastic seat.

Flush valve to be Sloan Royal 110FYO-3 with water saving feature. Mount handle 36" above finished floor. J.R. Smith, Josam, Zurn combination carrier and drainage fitting to suit condition.

P-2 Water closet:

"Water Saver Madera" 2221.018.

Seat: Church 9500NSSC — black plastic seat.

Flush Valve: Sloan Royal 110 FY 0-3 with water saving feature. (Mount handle 36" above finished floor.)

Bolt Caps: 481310.016.

P-3 Water closet:

"Elongated Water-Saver Cadet" 2109.395.

Seat: Church 295C — black plastic seat.

Supply pipe: 3405.016.

Bolt caps: 481310.016.

P-4 Urinal:

"Washbrook" 6500.011.

Flush valve: Sloan Royal 186.11.

Carrier: J.R. Smith, Josam, Zurn to suit building conditions.

P-5 Urinal:

"Lynbrook" 6530.018.

Flush valve: Sloan Royal 180-YVO.

Carrier: J.R. Smith, Josam, Zurn to suit conditions.

P-6 Urinal:

"Jetbrook" 6570.014.

Flush valve: Sloan Royal 180-YVO.

Carrier: J.R. Smith, Josam, Zurn to suit conditions.

P-7 Urinal:

"Water-Saver Trimbrook" 6560.015.

Flush valve: Sloan Royal 186 with water saving feature.

Carrier: J.R. Smith, Josam, Zurn to suit conditions.

P-8 Lavatory:

"Lucerne" 0350.132. 20" x 18".

Supply & drain fittings: 2379.018.

Trap: 4420.030. 1-1/4" x 1-1/2".

Carrier: J.R. Smith, Josam, Zurn to suit conditions.

P-9 Lavatory:

"Lucerne" 0350.132. 20" x 18".

Supply & drain fittings: Speakman S-4151.

Wall supplies: 2302.149.

Trap: 4420.030. 1-1/4" x 1-1/2".

Carrier: J.R. Smith, Josam, Zurn to suit building conditions.

P-10 Lavatory:

"Scotian" 0360.057. 20" x 18".

Supply & drain fittings: Vandalproof combination 2248.714.

Wall supplies: 2302.081.

Trap: 4420.030. 1-1/4" x 1-1/2".

P-11 Lavatory (Wheel chair):

"Wheel Chair" 9140.013. 20" x 27"

2238.129 — Gooseneck faucet — 4" handles aerator.

7536.014 — Extended supply pipes with loose key stops.

2340.016 — Tailpieces.

7723.018 — Cast grid drain assembly.

4401.061 — 1-1/4" x 1-1/2" adjustable P-Trap.

Carrier: J.R. Smith, Josam, Zurn to suit building conditions.

P-12 Service sink:

"Akron" 7696.016. 24" x 20" cast iron, acid resisting enameled inside with wall hanger 7798.176 trap standard, 8379.026 rim guard, and Speakman S-5811-RCP service sink faucet (rough chrome plated).

P-13 Service sink in battery room (Check with electrical engineer.):

"Lakewell" 7692.049. 22" x 18" acid resisting enameled cast iron service sink with rim guard. P-trap to be Polypropylene (GSR #7203A) (no substitution).

P-14 Mop service sink:

Basin to be Fiat Model MSB 3624 or approved equal, molded stone with vinyl bumper guards and 3" cast brass drain fitting with flat strainer.

Supply fitting: U/R84114 rough plated with pail hook, hose end, vacuum breaker, integral stops, and wall brace.

P-15 Wash Fountain:

Bradley Type CFB 54" semi-circular wash fountain with precast stoneware bowl — "Granito" color. Pedestal to be bonderized steel finished with high temperature baked, acid resisting enamel color to harmonize with bowl.

P-16 Dispensary sink:

Sink to be equal to "Elkay" LR-1918 — 18 gauge stainless steel with sound deadening coating. Faucet to be Elka LK-2200. Strainer and tailpiece to be LK-35.

4401.010 P-Trap.

2303.154 angle stop valves and supplies.

P-17 Electric water cooler:

White-Westinghouse Model WELC14 or approved equal, wall mounted cooler. Cabinet shall conceal all plumbing and shall be metallic bronze vinyl finish.

Cold water standard rated capacity to be 14.5 gph.

Cooler to be equipped with remote connection to serve wheel chair fountains.

P-18 Electric water cooler:

White-Westinghouse Model WEWCO7-OP or approved equal, all stainless steel. Mounting height shall be as indicated by the Owner.

P-19 Electric water cooler:

White-Westinghouse Model WEEC10 or approved equal.

Cold water standard rating capacity to be 10.0 gph.

P-20 Sink:

Elkay LR 2522 or approved equal, single compartment 18 gauge.

Type 304 — self rimming.

LK 221 faucet with retractable spray.

LK 35 drain outlet with tailpiece.

4401.013 P-Trap.

8205.023 angle stop valves and supplies.

2.14.12 Fixtures, fittings, accessories, & supplies:

a. Water closet, shall be an ultra low usage water saver type in conformance with the proposed ANSI Standard for 1.5 gpm per flush. Performance shall be certified by an independent testing laboratory. Manufacturers shall submit adequate documentation that the water closet will provide satisfactory sanitation and a substantial life with minimum maintenance. Flushing shall be accomplished with a pressure assisted device and a quiet action chrome flush valve with nickel silver handle, screw driver angle stop with protecting cap and vacuum breaker is preferred. Closet shall be blow out type elongated bowl, wall hung, in white vitreous china with a top spud. Seat shall be solid black plastic, open front with concealed stainless steel check hinge. Adjustable closet fitting and supports shall be cast iron with foot supports, corrosion resistant coupling, adjustable face plate, vent connection, extended fitting to meet modular spacing of water closet centers.

b. Handicap water closets, shall be equal to the regular one above but for handicapped use rim height shall be set at 19" above finished floor.

c. Urinals shall be 18-1/2" wide, 27-1/4" in height and 12" deep and manufactured of white vitreous china, washout type 1 gpm urinal with extended shields, integral flush spreader, 3/4" top spud, outlet connection threaded two inch inside with bolts, washers and hangers. Connect with special exposed, one gpm, quiet action, chrome water conserving flush valve, complete with screwdriver angle stop with protecting cap, vacuum breaker, non hold open nickel silver handle, 3/4" outlet spud coupling with flange and cast brass wall escutcheon with set screw. Supports shall be coated cast iron with rectangular steel uprights and adjustable support plates. Support system shall have cast iron feet for floor bolting and mounting fasteners. Bottom plate shall have bearing jacks.

d. Handicap urinals shall be the same as the regular one above but mounting height shall be 15" above the finished floor to the lip of the urinal. Fixtures shall be in accordance with ANSI A117.1.

e. Lavatories may be, as one example, 20" wide by 18" deep, white vitreous china, front overflow lavatory, less soap depressions, and modified for concealed carrier arms. Provide drain plug with integral perforated grid and 1-1/4" tailpiece, pair of threaded offset tailpieces, pair of 3/8" threaded supplies with loose key angle stops, reducers and cast brass escutcheons with set screws, 1-1/4" by 1-1/2" cast brass P-trap with cleanout, slip inlet, and threaded outlet, 1-1/2" by 6" strap nipple, and cast brass escutcheon with set screw. Single lever handle chrome faucet with 1/2" threaded inlets and coupling nuts on 4" centers fitted with a vandal proof 3/4 gpm flow regulated spout aerator. Carrier shall be coated cast iron type with foot supports, rectangular steel uprights with concealed arms, and sleeves mounted on adjustable headers.

f. Handicap lavatories shall be 20" wide by 27" deep, white vitreous china, front overflow wheelchair lavatory for concealed carrier arms. Provide drain plug assembly with integral perforated grid, 1/4" threaded offset tailpieces, pair of 3/8" threaded extended supplies with loose key stops, reducers and cast brass escutcheons with set screws, 1-1/4" by 1-1/2" cast brass P-trap with cleanout, slip inlet and threaded outlet, 1-1/2" by six inch trap nipple, and cast brass escutcheon with set screw. Provide chrome finish gooseneck with chrome faucets with four inch handles, 3/4 gpm flow regulated spout aerator and 1/2" threaded inlets and coupling nuts on four inch centers. Set bottom of rim 29" above finished floor. Carrier shall be equal to that specified for lavatory above. Lavatories shall be punched to receive soap dispensers. Fixtures shall be in accordance with ANSI A117.1.

g. First aid lavatory may be, as one example, 20" wide by 18" deep, white vitreous china, less soap depression with faucets, gooseneck spout, four inch wrist blade handles, grid strainer, 1-1/4" by 1-1/2" cast brass P trap with cleanout plug, 3/8" threaded supplies with loose key angle stops, reducers, and cast brass escutcheons with set screws. Carrier shall be equal to that specified for lavatory above.

h. Service sink, shall be 24" wide by 20-1/2" deep enameled cast iron. Sink shall be drilled for mounting faucet through the back, have drain channels in the bottom and have stainless steel rim guard. Sink shall be complete with 3" floor mounted cast iron P-trap with cleanout plug, cast iron wall plate hanger with through the wall fasteners, and 3/4" brass valve set with stops in shanks and with threaded hose spout and bucket hook and vacuum breaker. Brace spout to the wall and provide outlet strainer.

i. Wall mounted electric water coolers, shall be surface mounted, rated to produce 14 gph of 50°F water with inlet water at 80°F and ambient temperature at 90°F. Cooler top shall be stainless steel with one- piece chrome plated shielded bubbler. Finish shall be selected to conform with the finished in its location. Water coolers are to be in accordance with ARI 1010-84.

j. Electric water cooler for handicap shall be surface mounted, wheel chair level model (rim set 30" above finished floor), stainless steel cabinet and top, having a cooling capacity of 7 gph from 80°F inlet water to 50°F drinking water at room temperature of 90°F. Coolers are to be in accordance with ARI 1010-84.

k. Emergency eye wash shall have a stainless steel receptor mounted on 1-1/4" galvanized iron pipe pedestal with 9" floor flange. Waste shall be 1-1/4" and water supply shall be 1/2" I.P.S. eye/face wash shall have soft Buna N covered plastic twin heads and circular chrome plated spray ring to bathe face. Unit shall be activated by push-to-operate stay open ball valve.

l. Emergency eye wash with shower, shall be free standing, floor mounted design with 1-1/4" galvanized pipe support with 9" floor mounted flange and 1-1/4" water supply to unit.

(1) Shower head shall be ABS green plastic, 10" diameter, with instant action, stay-open ball valve and PULL rod.

(2) Eye/face wash fountain shall have a stainless steel bowl with soft Buna N covered plastic with twin heads and circular chrome plated spray ring to bathe face. Unit shall be activated by PUSH-to-OPERATE stay open ball valve.

2.14.13 Emergency Shower:

a. Emergency shower unit shall be a manual deluge-type safety shower complete with self-closing valve, chain, and ring. The shower head shall be installed on a drop pipe through the ceiling and set approximately 7 ft above the floor. The shower head, valve, chain, and ring shall have a rough nickel plate finish. The exposed portion of the drop pipe shall have a chrome finish or be provided with an acceptable chrome-plated sleeve. The unit shall be Model SE-206CP made by Speakman Co., Wilmington, DE; Model 824-VCP made by Western Drinking Fountains Inc., San Leandro, CA, Model 8123VPCP made by Haws Drinking Faucet Co., Berkeley, CA; or be an acceptable equivalent product.

2.14.14 Emergency Eyewash Fountain:

a. The emergency eyewash fountain shall be a free standing aerated eye/face wash with floor mounting flange. Each unit shall be complete with quick-opening, stay-open, full-flow ball valve activated by a push handle; stainless steel bowl; heavy chrome plating on all exposed parts; multiple outlets which supply converging streams to form a curtain of aerated water. Unit shall be Model 420 made by Speakman Co., Wilmington, DE; Model 761 made by Western Drinking Fountains Inc., San Leandro, CA; Model 7761 made by Haws Drinking Faucet Co., Berkeley, CA; or an acceptable equivalent product.

2.14.15 Emergency Shower and Eyewash:

a. The emergency shower and eyewash unit shall be a free-standing type complete with deluge shower head, self-closing valve, chain and strap, aerated eyewash with stainless-steel bowl, hand and foot operated, stanchion, floor flange and all interconnected fittings. The unit shall be Model SE-610 made by Speakman Co., Wilmington, DE, Model 9300-SS made by Western Drinking Fountains Inc., San Leandro, CA, Model 8300 MOD made by Haws Drinking Faucet Co., Berkeley, CA; or an acceptable equivalent product.

2.14.16 Following is a Schedule of Plumbing Fixtures which may be used. Cut and details shall be presented to the Owner for approval. Fixture model numbers are for reference only.

SCHEDULE OF PLUMBING FIXTURES

	Manufacturer and Catalog No.			
Fixture & Accessories	American Standard	Kohler	Crane	Remarks
Water closet	250.011	K-4430-ET	3-444	Wall hung siphon jet vitreous china; chair carrier support
Water closet	2222.016	K-4282-ET	3-297	Floor mounted siphon jet vitreous china; bolt caps
Seat	5334.024	K-4671-C	Olsonite 10-CC	Black, open front without cover
Flush valve	Sloan (Royal) 110-FYO	Delany 402-AVE	Watrous W100YV	
Service sink	7695.018	K-6716	7-564	24" by 20" enameled C.I.; w/wall hangers
Supply fitting	8341.075	K-8905	8-3756	

Manufacturer and Catalog No. (Continued)

Fixture & Accessories	American Standard	Kohler	Crane	Remarks
Trap standard	7798.075	K-6673	7-620	
Rim guard	8379.176	K-8936	8-7508	
Lavatory	4869.020	K-2861	1-506-V	20" by 18" enameled C.I.; chair carrier support
Supply and drain fitting	2103.224	K-7408-TL	8-2065 w/ 8-5222	
Supply pipes	2303.105	K-7606	8-5003	Loose key
Trap	4420.030	K-9004	8-5270	1-1/4" by 1-1/2"

2.14.17 Toilet Accessories (required to be installed by others):

The toilet accessories shall be installed as herein specified and in locations acceptable to the Engineer.

a. Paper holders

Paper holders shall be provided in each water closet stall and shall be surface mounted, die cast zinc, copper-nickel-polished chrome-plated, single roll toilet tissue holder, Model 505 manufactured by Bradley Washfountain Co., Menomonee Falls, WI, Meriden, CT; Model 964 by Scott Paper Co., Philadelphia, PA; or an acceptable equivalent product.

b. Clothes hooks

Clothes hooks shall be provided on each water closet stall door and shall be surface mounted, forged or cast brass, nickel plated, polished chrome finish, clothes hook with bumper. Hook shall be Model 915 manufactured by Bradley Washfountain Co., Menomonee Falls, WI; Model G manufactured by Charles Parker Co., Meriden, CT; No. B-212 manufactured by Bobrick Washroom Equipment, Inc., Ballston Bake, NY; or an acceptable equivalent product.

c. Mirror-shelf combination

Mirror-shelf units 18" by 24" high, shall be provided over each service sink and lavatory and shall be surface mounted. The mirror shall be No. 1 quality, 1/4" polished plate glass, electrolytically copper plated, edges and back fully padded, guaranteed for 10 years against silver spoilage. Back plate shall be of 20-gauge galvanized steel. The frame shall be 3/4" by 3/4" angle of 18-gauge satin finish stainless steel with 18 gauge reinforcing brackets at each end. The corners of the frame and shelf shall be fully welded and polished. The unit shall snap over concealed mounting plate and be fastened with theft proof concealed setscrews. The unit shall be Model 705 manufactured by Bradley Washfountain Co., Menomonee Falls, WI; Model 53020-SS manufactured by Watrous Inc., Bensenville, IL; or an acceptable equivalent product.

d. Paper towel dispenser

Paper towel dispensers shall be provided at each service sink and shall be made of 22-gauge satin-buffed stainless steel and shall be key locked. The unit shall be Model B-262-C made by Bobrick Washrooms Equipment, Inc., Ballston Bake, NY; No. W1103-C made by Watrous Inc., Bensenville, IL; No. 6945M made by Charles Parker Co., Meriden, CT; or an acceptable equivalent product.

e. Waste receptacle

Waste receptacle shall be provided at each service sink and lavatory and shall be a round, white-enameled steel waste basket approximately 14-1/2" high having a 26-qt capacity.

f. Soap dispenser

Liquid soap dispenser shall be provided over each service sink and lavatory and shall be a wall-mounted unit having a suspended, removable reservoir of at least 1-qt capacity and built-in plunger-type valve. It shall have a tamperproof mounting. All exposed metal parts shall be chromium plated. It shall be a Model 656 dispenser made by No. 20 made by Charles Parker Co., Meriden, CT; Model W-832 by Watrous Inc., Bensenville, IL; or an acceptable equivalent product.

2.14.18 Equipment Schedule:

a. Electric Water Heater

Heater to be equal to A.O. Smith DEN-52 with a 52 gallon glass lined storage tank. Heater be equipped with (1) 9000W element 480/3/60, P & T relief valve, thermostat, drain valve, and shall be UL approved.

Heater recovery shall be 37 gph at 100°F rise.

Please note that sizes are for example only.

Approved manufacturers: Jackson, A.O. Smith, Ruud, Hubbell.

b. Electric water heater (EWH)

Heater to be equal to Ruud RP 6P4-1 with a 6 gallon glass lined storage tank. Heater to be equipped with (1) 6000W element 480/3/60, P & T relief valve, thermostat, drain valve, and shall be UL approved.

Heater recovery shall be 25 gph at 100°F rise.

Plumbing Contractor shall provide steel mounting shelf.

Sizes shown are for example only.

Approved manufacturers: Jackson, A.O. Smith, Ruud, Hubbell.

c. Gas fired water heater:

(1) Water heater shall be Model BT-199/GC-167 as manufactured by the A.O. Smith Corporation, Kankakee, IL or approved equal. Water heater shall be of glass lined design, and gas fired, equipped to burn natural gas and design certified by the American Gas Association under Volume III tests for commercial heaters for delivery of 110°F water and shall be approved by the National Sanitation Foundation. Heaters shall have an input rating of 199.000 Btu/h and a recovery capacity of 167 gph at a temperature rise of 100°F. with a storage capacity of 86 gallons. Heater shall be equipped with 1-1/2" water inlet and outlet openings, (1) 2-3/4" x 3-3/4" boiler-type handhole cleanout, and shall have a working pressure of 150 psi.

(2) Water heater shall be equipped with a self-generating integrated control system consisting of two immersion-type thermostats, the lower of which shall be adjustable to 180°F. The upper thermostats shall operate as an automatic recycling control with atop setting of 195°F. Heater shall be provided with an automatic gas shut-off device and 100% safety shut-off in event pilot flame is extinguished; a gas pressure regulator set for the type of gas supplied; coated steel burners; an approved draft diverter and extruded magnesium anode rods rigidly supported for cathodic protection.

(3) A pressure and temperature relief valve shall be furnished and installed by the Plumbing Contractor.

d. All water heaters shall conform to the ASHRAE 90A-1980 Standard for energy efficiencies and all applicable state or local Energy Conservation Codes.

e. All electric water heaters with multiple heating coils shall be wired for simultaneous operation.

f. All electric water heaters should be started up by an authorized service representative of the manufacturer.

g. Hot water storage tank:

(1) Glass lined storage tank shall be A.O. Smith Glass lined Storage Tanks or equal. Tanks shall be 30" diameter by 72" and have a nominal capacity of 200 gallons. Tanks shall have threaded openings as shown on Drawings. Exterior of the tank shall be thoroughly cleaned and painted. Interior of the tank shall be glass lined with an alkaline borosilicate composition which has been fused to the steel base by firing at a temperature range of 1400°F to 1600°F. Glass coating shall be continuous over the entire inner surface of the tank, handhole, and manhole covers. Cathodic protection shall be provided. Tank shall be constructed in accordance with ASME Boiler Construction Code Section VIII Unfired Pressure Vessels and shall have a working pressure of 125 psi. Tanks shall be guaranteed against corrosion for a period of 5 years.

h. Hot water circulating pump:

(1) Bell & Gossett or approved equal, all-bronze, in line, centrifugal pump with starter. Pump shall be sized to pump 7 gpm at 12 ft head, 1/6 hp, 60 cycle, 120 volt, 1 phase, 1750 rpm.

Note sizes shown are for example only.

(2) Furnish and install in the hot water system recirculating line where indicated on the Drawings, single stage immersion type aquastat for control of the pump. Aquastat switch shall have an immersion bulb screwed into a separable hot water container brass well in the recirculation line and shall be arranged for conduit connection. Electric control of the aquastat shall have an adjustable range from 100°F to 240°F and shall be set at the most remote fixture served by the hot water distribution system.

(3) Wiring for pumps and aquastats will be furnished and installed by Electrical Contractor.

i. Water Booster System:

(1) Paco Miniflo Model 817 with characteristics to be given by the engineer. Pump is PACO Type L close coupled, bronze fitted end suction with mechanical shaft seal. Pump shall be fitted with a stainless steel shaft.

(2) System will include motor starter or contactor in Nema I enclosure, HOA, thermal switch, pressure switch, transformer where required, pressure tank, suction check valve, pressure relief valve, system mounting frame, and interconnecting wiring.

(3) Operation: Upon demand, water flows from the pressure tank until the system pressure drops below the field adjustable pressure switch setting. The pump operates during demand and automatically shuts off several minutes after demand ends when temperature of water in pump case rises to set point of the temperature switch. This control system allows the PACO Miniflo to run continuously during demand periods, eliminating pressure surges and fluctuations.

(4) Maximum operating pressure is 100 psi and ambient operating temperatures of 32° to 100°F.

2.14.19 Sump pumps

Sump pumps shall be suitable for installation in sumps of the dimensions indicated on the Drawings.

a. Suitable centrifugal type

The sump pump shall be a submersible, vertical type, close-coupled centrifugal pump and a motor in a hermetically sealed, waterproof assembly. The pump shall have a casing of cast iron or bronze, a bronze impeller, a bronze strainer with openings small enough to protect the pump from injury but with net area greater than that of the pump inlet, and an integrally mounted electrode or diaphragm switch-control. The controls and cables shall be waterproof. If electrode control is used, the electrodes shall be of brass or stainless steel.

b. Controls

Pump shall be automatically controlled by a NEMA I pedestal mounted float switch complete with float, float rod, and float buttons to activate high water alarm.

c. Basin and Cover:

(1) Furnish fiberglass basin characteristics and location of basin inlet shall be as detailed on the Drawings.

(2) Furnish a steel circular cover for a basin to accommodate pump controls and a suitable inspection opening.

(3) Provide 2 NEMA I alarm panels (wall mounted type) with bell light and silencer, one of which will be mounted in a remote location.

d. Submersible recessed impeller

The sump pump shall be submersible recessed impeller vertical, close-coupled, hermetically sealed, waterproof unit and shall be not smaller than 3" size. The pump casing shall be of cast iron with integrally cast feet to support the pump, and with a vertical flanged discharge connection. The impeller shall be of cast iron and of the open, recessed torque-flow type.

Special diameter impellers shall be provided as required to meet any intermediate ratings which are not in accordance with standard pump impeller ratings. The impellers shall be capable of passing a sphere not less than a 2" diameter.

Pump shall be the Wemco pump manufactured by Arthur G. McKee & Co., San Francisco, CA or Morris Pumps Inc., Baldwinsville, NY.

Each pump shall be controlled by mercury float switches consisting of two normally open mercury switches, encapsulated in epoxy resin and having a polypropylene float casing. The switches shall be provided with acceptable adjustable corrosion resistant supports in the sump to provide operating levels desired for starting and stopping the pumps and sounding a high water alarm as indicated on the Drawings.

Where duplex units are specified, the controls shall be arranged to manually alternate the pumps by means of a sequence selector switch and to automatically start the second pump if the first cannot handle the load.

The controls shall be furnished by the manufacturer of the pump and shall include level switches, and sufficient submersible power and control cables with compression fittings for connection to a terminal box. The pump controls shall include thermal protection and motor moisture detection systems and sensing panel with waterproof cable and compression fittings from the motor to the panel. The motor detection alarm and high-water alarm wiring to a remote alarm shall be provided under the appropriate electrical sections and as indicated on the Drawings.

The starters, NEMA Type 4 terminal boxes, spring return HAND-OFF-AUTO switches, connecting conduit, and wiring from the terminal box or enclosure to the starters will be furnished and installed under the appropriate electrical sections.

The motors shall be hermetically sealed and waterproof, and shall be suitable for operation on 3-phase, 60-Hertz current.

The motors shall have sufficient capacity to operate the pumps at any point on the head-capacity curve without exceeding its nameplate rating.

Motors shall also be capable of continuously operating in air, having a maximum ambient temperature of 100°F, at the indicated pump performance ratings.

e. Submersible non-clog type

The sump pump shall be a submersible vertical type, close-coupled, hermetically sealed, waterproof unit. The pump casing shall be of cast iron or bronze with integrally cast feet to support the pump with a vertical flanged discharge connection. The impeller shall be of bronze and of the non-clog type requiring no inlet strainer.

Special diameter impellers shall be provided as required to meet any immediate ratings which are not in accordance with standard pump impeller ratings. The impeller shall be capable of handling spheres having no less than 2" diameter.

Pump shall be manufactured by Weil Pump Co., Chicago, IL; Deming Division Crane Co., Salene OH; Clow Corp., Florence, KY.; or an acceptable equivalent product.

Each pump shall be controlled by mercury float switches consisting of two normally open mercury switches, encapsulated in epoxy resin and having a polypropylene float casing. The switches shall be provided with acceptable adjustable corrosion resistant supports in the sump to provide operating levels desired for starting and stopping the pumps and sounding a high water alarm as indicated on the Drawings.

Where duplex pumps are specified, the controls shall be arranged to manually alternate the pumps by means of a sequence selecter switch and to start the second pump if the first cannot handle the load.

The controls shall be furnished by the manufacturer of the pump and shall include level switches and sufficient submersible power and control cables with compression fittings for connection to a terminal box. The pump controls furnished shall conform to the requirements specified under the appropriate electrical sections and as indicated on the Drawings.

The starters, NEMA Type 4 terminal boxes, spring return HAND-OFF-AUTO switches, connecting conduit, and wiring from the terminal box or enclosure to the starters will be furnished and installed under the appropriate electrical sections.

The motors shall be hermetically sealed and waterproof and shall be suitable for operation on 3- phase, 60-Hertz current. The cables shall be of the 4-wire type having the fourth wire grounded.

The motors shall have sufficient capacity to operate the pumps at any point on the head-capacity curve without exceeding its nameplate rating.

Motors shall also be capable of continuously operating in air having a maximum ambient temperature of 100°F at the indicating pump performance ratings.

f. Wet-pit type

The sump pump shall be a submerged, vertical, wet-pit type sump pump of polyvinyl chloride or polypropylene construction resistant to the corrosive effects of the materials being pumped.

The unit shall consist of a vertical pump, motor, extension shaft, float switch, polypropylene angle frame, and coverplate suitable for the sump pit. The pump shall be of the capacity listed on the Drawings. The pump impeller and suction screen shall be of all plastic construction and the stainless-steel shaft shall be enclosed by a polypropylene sleeve. The pump motor shall be of the totally enclosed, fan cooled, vertical type with mounting flange and non-overloading at any pump operating point. Pump shall be provided with pure ceramic and carbon sleeve bearings lubricated by the liquid being pumped. The pump controls shall be furnished by the pump manufacturer and shall include a pedestal-mounted float switch assembly with float, rod guides and adjustable stops to control the liquid level, all of a material resistant to the corrosive effects of the material being pumped. All switches shall be mounted in NEMA Type 4 enclosures. The controls furnished shall conform to the requirements specified under the appropriate electrical sections. The pump shall be manufactured by Vanton Pump and Equipment Corp., Hillside, NJ; Sethco Mfg. Co., Freeport, NY; Serfilco, Chicago, IL.; or an acceptable equivalent product.

2.14.20 Sump pumps and sewage ejectors

The pumps shall be submersible vertical type, close-coupled, hermetically sealed, waterproof units. The pump casing shall be of cast iron or bronze with integrally cast free to support the pumps with vertical flanged discharge connections. The impellers shall be of bronze and of the non-clog type requiring no inlet strainers. Pumps shall be manufactured by Weil Pump Co., Chicago, IL; Deming Division Crane Co., Salem, OH; Clow Corp., Florena, KY.

The motors shall be hermetically sealed and waterproof and shall be suitable for operation on 3-phase, 60-Hertz current. The cables shall be of the 4-wire type having the fourth wire grounded. Pump capacities shall be as scheduled on the Drawings.

Each pump shall be controlled by mercury float switches consisting of two normally open mercury switches, encapsulated in epoxy resin and having a polypropylene float casing. The switches shall be provided with acceptable adjustable corrosion resistant supports in the sump to provide operating levels desired for starting and stopping the pumps and sounding a high water alarm as indicated on the Drawings.

The controls shall be furnished by the manufacturer of the pumps and shall include level switches and sufficient submersible power and control cables with compression fittings for connection to the control cabinet.

The control equipment shall include a duplex control panel in a NEMA Type 4, weatherproof enclosure with hanged gasketed door. The panel shall include the following:

a. Fused disconnect switches with lockout handles through cover.

b. Magnetic starters with overload and low-voltage protection.

c. TEST-ON-AUTO selecter switches.

d. Alternator.

e. Control circuit transformers.

f. Green pump running lights.

g. Overload reset buttons.

h. Alarm bell and alarm light mounted on panel door.

i. Alarm silencing switch.

j. Numbered and wired terminal strip.

The controls furnished shall conform to the requirements specified under the appropriate electrical sections.

The pump manufacturer shall furnish with each set of duplex pumps a duplex cast-iron basin cover and flush curb frame suitable for installation on a 4 ft by 4 ft concrete sump pit. The cover shall be complete with openings for the pump floor plates, control cable plate, manhole, and 4" vent connection. The pump floor plates shall cover the openings through which the pumps may be installed and withdrawn. The power cables to the pump motors shall go through the pump plates. The control cables to the diaphragm switches shall go through the cover plate.

Pump floor plates shall be provided with openings for pump discharge pipes.

2.14.21 Pneumatic sewage ejectors

The ejectors shall be of the duplex package type consisting of two ejector receivers with two motor-driven air compressors with automatic controlling devices arranged to alternate in operation so that one receiver can store incoming sewage while the other receiver is discharging. All components shall be assembled by the ejector manufacturer before shipment.

Each ejector shall have a capacity as indicated on the Drawings which shall be the continuous rated flow into the ejector station.

Each receiver shall be cast iron with dished top and bottom heads and shall be designed for 50 psi minimum working pressure. Each receiver shall be provided with inlet and discharge openings of sizes indicated on the Drawings. The inlet and discharge openings shall be at the bottom, on opposite sides of each receiver. A 16" minimum diameter manhole shall also be provided in the top of each receiver. The unit shall be provided with a center inlet and discharge cross connection to provide a common inlet and common discharge.

Each inlet and discharge connection shall terminate with flanges for attaching the inlet and discharge valves. There shall be no moving parts in the ejector receivers. The interior of each receiver shall be smooth and free from projections. The interior and exterior of each receiver shall have two coats of inert epoxy resin having a total minimum thickness of 8 mils.

Each inlet and discharge shall be provided with special iron body, bronze-mounted, flanged, swing check valve with access for cleaning, and an iron body, bronze-mounted, double-disk, flanged gate valve of the non-rising stem type.

Each ejector control shall consist of two vertical insulated stainless steel 3/8" minimum diameter electrodes, one induction relay, with HAND-OFF-AUTO selector switch, one air-activated 3-way diaphragm valve, regulating valve and strainer, automatic alternator and adjustment timer to be energized at the start of the ejector cycle to determine the length of the cycle from 0 to 30 seconds.

The air control valves shall be electrically actuated, of the quick acting type and shall be automatically positioned to vent each receiver to the atmosphere when the compressors are not operating. The high level electrode in the receiver shall sense the sewage level and start the compressor motor, and the air valve shall be positioned to automatically allow air into the receiver.

The automatic alternators shall alternate the lead ejector with the switch in AUTO position, and fail safe operation shall provide for the automatic starting of the lag ejector if the lead ejector fails. Each ejector shall be provided with a combination starter with circuit breaker and control transformer in a NEMA Type 4 enclosure, complete with connecting conduit and wiring from the terminal box to the starters. An interlock shall also be provided to prevent both units from operating simultaneously.

Each ejector shall be provided with an air-cooled, single-stage, two-cylinder, reciprocating type air compressor mounted on a cast-iron base with supports no less than 3 feet high. The compressor shall be complete with mechanical shaft seals, flexible coupling and automatic lubricating system. Accessories to be supplied with each compressor shall include discharge safety valve, relief valve, check valve, and inlet air filter and silencer. Each compressor shall be connected to a totally enclosed, fan-cooled motor by an adjustable V-belt drive. The motor shall be non-overloading to the design operating conditions.

The units shall be manufactured by Smith & Loveless, Inc., Kansas City, MO; Clow Corp., Waste Treatment Div., Florence, KY; Can Tex Industries, Mineral Wells, TX; or be an acceptable equivalent product.

2.14.22 Seal-water pumping unit

The sealing water unit shall be a self-contained, shop-assembled duplex unit consisting of a suction tank, two pumps, and a tank-mounted float valve. A continuously running pump shall draw water from an open tank. The tank shall be filled through a two-position float-controlled valve. The valve shall be open wide at a low level and close tight at a high level. The pump shall be equipped with a recirculation line to prevent overheating when operating during periods of no demand.

The tank shall be a vertical, galvanized, approximately 50-gallon capacity, open top, cylindrical steel tank of all-welded construction with suction, drain, overflow and bypass connections, all of which shall be threaded. Mounting bases for the pumps shall be welded to the side of the tank near the bottom. The pump suction and recirculation piping shall be shop assembled.

The float valve shall have a discharge which terminates about 2" above the top rim of the tank. The float valve shall be a bronze valve, hydraulically operated by a float-actuated pilot valve. The float shall be made of copper and shall be not less than 5" diameter. The valve shall be sized to discharge the full rated capacity of both pumps at not more than 10 psi pressure differential.

The pumps shall be close-coupled, horizontal or vertical, end suction, single-stage turbine pumps. The main frame shall be integral with the end bell of the motor, and the motor bearings shall also be the pump bearings. The pump shaft shall be an extension of the motor shaft and shall be protected from wear at the stuffing box and from contact with liquid by a removable bronze sleeve.

Pump capacities shall be as scheduled on the Drawings. Motors shall be suitable for operation on the electric service indicated on the Drawings.

The pump casing shall be made of close-grained cast iron and shall be tested under a hydrostatic pressure of at least 150% of the rated shutoff head. An air vent shall be fitted on the high point of the casing and the stuffing box shall be fitted with a mechanical seal.

The impeller shall be of bronze and of the turbine type.

The unit shall be manufactured by Interstate Machine Co. (IMCO), St. Louis, MO; Pacific Pumping Co., Oakland, CA; Chicago Pump, Chicago, IL; Aurora Pump, Aurora, IL; or an acceptable equivalent product.

A pressure controller switch shall be furnished for mounting on the discharge of the pumps at each seal-water pumping unit to stop the pumps and sound an alarm when the pump discharge pressure drops below the normal operating pressure. The switches shall be of the highly sensitive external knob and dial adjustment type with all wetted parts of Type 347 stainless steel and with Teflon diaphragms. They shall be single switches with a range of 0 to 100 psig, an on-off differential of 0.25 to 0.75 psi, and a proof pressure of 225 psig.

The switch shall be a UL listed single-pole double-throw snap-action switch, having a rating of 15 amp, 125/250 volts, alternating current resistive, with three terminal screws. The enclosure shall be a NEMA Type 4, gasketed enclosure, of die cast aluminum construction.

2.14.23 Cleaning unit

The cleaning unit shall be a fully automatic self-contained, hydraulic pressure combination cleaner; Model 250-GES made by Malsbary Mfg. Co., Subsidiary of Carlisle Corp., North Uniontown, PA; Model 2100-C GESR made by Jenny Div., Homestead Industries, Coraopolis, PA; or an acceptable equivalent product.

2.15.1 Acid neutralizing sumps

Acid neutralizing sumps shall be of the size indicated on the Drawings. The sumps shall be formed of a dense acid proof chemical stoneware vitrified in one piece complete with bell connections and an initial fail of lump limestone 1" to 2" in size.

2.15.2 Air gap

Air gap for glassware washer shall be manufactured by Frost Co., Kenosha, WI; Eastman Central Co., Division of United Brass Corp., Los Angeles, CA; or be an acceptable equivalent product.

2.15.3 Shock absorbers

Shock absorbers shall be a series 1485 Shock Absorber made by Josam Mfg. Co.; Shockstop made by Wade, Inc.; Shocktrol made by Zurn Industries, Inc.; or an acceptable equivalent product.

2.15.4 Fresh air intakes

Fresh air intakes shall be cast brass with threaded connection and extra heavy polished brass convex grille, and shall be Series 1860 made by Josam Mfg., Co.; Service W-3880G made by Wade, Inc.; Series Z-1388 made by Zurn Industries, Inc.; or an acceptable equivalent project.

3 CONSTRUCTION METHODS

3.1 SANITARY DRAINAGE SYSTEMS

3.1.1 Scope:

a. Complete sanitary drainage systems as shown on Drawings, including all soil and waste mains, branch lines, risers, cleanouts, traps, vent piping system, hanger, supports, and final connections to fixtures and equipment as specified hereafter.

b. All below floor slab sanitary soil, waste, and vent piping unless otherwise noted shall be extra heavy weight cast iron pipe. Above floor waste and vent pipe shall be service weight cast iron or cast iron no-hub or galvanized standard weight steel pipe with galvanized threaded cast iron drainage fittings and galvanized threaded malleable iron or galvanized cast iron vent fittings for vent piped 2" and smaller.

c. Exposed drainage piping fixtures and equipment including sinks shall be chromium plated in toilets and finished areas.

3.1.2 General:

a. Horizontal sanitary soil, storm, and drainage piping shall be installed at a uniform grade as required by the Plumbing Code. Vent piping shall be graded to provide venting of the plumbing fixture and freedom from condensation. Changes in direction in drainage piping shall be made by the appropriate use of 45° wyes, half wyes, long sweep quarter, sixth, eighth, or sixteenth bends. Sanitary tees or short quarter bends may be used on vertical stacks of drainage lines where the change in direction of flow is from the horizontal or the vertical, except that long turn tee shall have a common drain. Straight tees, elbows, and crosses may be used on vent lines. No change in direction of flow greater than 90° shall be made. Reduction of the size of drainage piping in the direction of flow is prohibited. Connections between copper tubing and cast iron hub, shall be made by means of cast bronze sweat joint spigots caulked into hubs. Connections between threaded pipe and cast iron hub shall be by means of a tapped cast iron spigots caulked into hubs. All plumbing vents shall be increased before passing through the roof. Cleanouts shall be installed in sewer, waste, and drain lines where shown on the contract Drawings and as required by the Code and field conditions. Cleanouts shall be provided at the case of each soil and waste stack, and downspouts; at intervals of 50' maximum distance in straight runs, at each change in direction, and at the end of all branches. Cleanouts shall not be located in direct traffic areas. Cleanouts shall be of the same size as the pipe up to 4", no less than 4" for pipe up to 6", and no less than 6" for pipe over 6". All cleanouts shall be so located as to provide easy rodding. All cleanouts shall be accessible. Extend cleanouts to grade, floor, or wall.

b. Each fixture and piece of equipment requiring connections to the soil and waste systems shall be equipped with a code approved trap. The trap shall be placed as near to the fixture as possible and no fixture shall be double trapped. All shower, floor, and drains connected to sanitary sewers shall be trapped. Position drains so that they are readily accessible and easy to maintain.

c. Provide flashing for each plumbing vent through the roof. The flashing shall extend in and around the vent pipe and turn down into the vent at least 2" and shall extend over the roof deck at least 1 foot in each direction from the vent. Coordinate with floor waterproofing work and roofing as necessary to interface drains with waterproofing. Install flashing collar, flange or seepage part so that no leakage occurs between drain and adjoining materials. Maintain integrity of waterproof membranes where penetrated.

d. Protect pipe, openings, valves, and fixtures from dirt, foreign objects, and damage during the construction period. Replace damaged piping, valves, fixtures, or other appurtenances without additional cost. After installation, all pipe and valve ends shall be capped securely to prevent entry of duct, dirt, and moisture in the pipe.

e. Vent pipes in roof spaces shall be run as close as possible to the underside of the roof, with horizontal piping pitched down to stacks without forming traps in pipes, using fittings as required. Plumber shall extend vent stacks approximately 2' above roof. All flashing of vents will be done by Roofing Contractor.

    (1) No union connection shall be allowed on waste or vents.

f. No plumbing fixture, equipment, or pipe connection shall be installed which will provide a cross connection or interconnection between a potable water supply and any source of non-potable water.

3.2 STORM DRAINAGE SYSTEMS

3.2.1 Scope:

a. Complete storm drainage system starting at building storm drains, connecting same to the existing site roof storm sewer systems including all horizontal storm lines, vertical leaders, and final connections to all roof and canopy drains.

b. Storm water drainage piping unless otherwise noted shall be service weight cast iron pipe and fittings. No-hub cast iron or galvanized standard weight steel pipe with galvanized threaded cast iron drainage fittings. All piping below floor slab to be of service heavy cast iron.

c. Piping beyond 10 ft outside of the building wall may be extra strength vitrified clay bell and spigot sewer pipe and fittings extra strength or reinforced concrete bell and spigot or tongue and groove sewer pipe and fittings or asbestos cement sewer pipe and fittings. The governing code shall be checked.

3.2.2 General:

a. The Roofing Contractor to do all the flashing. Rain water leaders shall be furnished, installed, and connected to roof drains. Horizontal roof leaders shall be installed close to roof steel and given a pitch as indicated on Drawings. Vertical roof leaders shall be installed close to walls and columns and provided with cleanout at base. Run horizontal building storm drain mains below floor slab at grade 1/8" per foot and all branch lines at grade 1/4" per foot, where possible.

b. Roof drain: Furnish and install roof drains and marquee and canopy drains in sizes and locations indicated on Drawings. For description and location see plans and schedule on Drawings.

c. Plumbing Contractor shall be responsible for cutting holes in sump pans before setting drains.

3.3 COLD AND HOT WATER DISTRIBUTION SYSTEMS

3.3.1 Scope:

a. The Plumbing Contractor shall extend the domestic water service lines from 5' outside building into building, including all fittings, valves, strainers, piping, hangers, anchors, guides, drain valves, air vents, vacuum breakers, backflow preventers, water hammer arresters, etc., for complete hot and cold water distribution systems and final connections to all plumbing fixtures and equipment. Valved water service shall be provided for heating and air conditioning equipment as shown on Drawings, specified and as required. Furnish water meter as detailed on Drawings. Water meter shall be installed by Water Company. (Check with the plumbing inspector.) Plumbing Contractor shall pay for water meter.

3.3.2 General:

a. All hot and cold water piping above slab to be Type L with standard wrought copper fittings using 95-5 tin antimony solder and non-corrosive flux. Piping below slab and underground water services to be Type K in sizes 2" and smaller.

(1) Water service lines 3" and larger to be cast iron water pipe class 250 cement lines and mechanical or flanged joints.

(2) Fittings shall be cement lined and rated for 250#.

b. Run piping free of traps wherever possible and grade and valve for completion control and drainage of system with drain cocks at low points and at base of valved riser.

c. Install gate valves on all supply mains and branch lines.

d. Each separate fixture and equipment is to have a valved shut-off for both hot and cold water connections as specified or required so that repairs can be made without disturbing any other fixtures.

e. Furnish and install air vents or approved equal at each high point of piping runs and risers which may become air bound.

f. Vacuum breakers shall be installed on supply lines to beauty salon sink, water heaters, faucets with hose connections, including wall hydrants. Backflow preventers shall be installed in supply lines to mechanical equipment and fire protection piping system.

g. Provide insulating bushings, couplings, or unions where brass or copper piping connects to steel piping.

h. Furnish and install balancing valves in the recirculating piping and balance systems so there will be an even temperature drop throughout.

i. Pressure reducing valves shall be installed by the Plumbing Contractor where indicated on the Drawings or when the main pressure entering building exceeds 80 psi.

j. Branch lines from service or main lines may be taken off the top or bottom of main, using such crossover fittings as may be required by structural or installation conditions. All service pipes, fittings, and valves shall be kept a sufficient distance from other work to permit finished covering to be not less than 1-1/2" from other work and not less than 1/2" between finished coverings on the different services.

k. Pipes shall be run parallel and graded evenly to the drainage points, there shall be 1/2" drain valve provided for each low point in the piping, so that all parts of systems can be drawn off. Provide suitable means of thermal expansion for all hot water piping using swing joints, expansion compensators, expansion loops, and long turn off-sets as required. Piping connections to equipment shall be provided with unions to permit alterations and repairs.

l. All underground copper water piping shall be protected in a PVC casing pipe 2" greater than the largest diameter supply line. Piping shall be coated with asphaltum or bitumastic.

m. Access doors: Where cleanouts, valves, or other equipment requiring service occur in furred or chased walls and ceilings, access doors of an acceptable size and type shall be installed. Doors shall consist of 14-gauge steel frame and door with invisible hinge, and cam lock fastenings. For plaster walls or ceilings, frames shall have a 2" wide lath plaster bond. For masonry walls, the frame shall be set flush with masonry with provisions in the jamb for anchoring. Door shall be solid flush steel for bearing tile or for bearing plaster as required to match adjacent areas. Doors shall be furnished with a coat of gray metal primer.

n. Gauges: Gauges shall be furnished and installed with equipment as indicated on the Drawings and as specified, and shall be complete with all shut-off cocks and extensions necessary to clear insulation and maintain visibility.

Gauges shall have a black case and shall be 4-1/2" nominal diameter with phosphor bronze Bourdon tubes (beryllium copper bellows), 1/4" npt male connection, stainless steel rack and pinion movement, micro-adjustment for calibration white dials and black figures and threaded ring case. Gauges shall have a guaranteed accuracy of at least one percent of scale range. Gauges shall be installed with brass nipples not shorter than 2" and shall have a test cock, female outlets, and shut-off cock. The gauge shall be mounted directly in the outlet of the tee-bearing test cock.

o. Dissimilar Pipe Materials: Connections to water heaters and connections between ferrous and nonferrous metallic pipe shall be made with dielectric fittings. Connecting joints between plastic and metallic pipe shall be made with transition fitting for the specific purpose.

p. Waterproofing: Waterproofing at floor mounted water closets shall be accomplished by forming a flashing guard from soft tempered cooper.

q. Use of Plumbing Fixtures: The use of new permanent water closets and other new plumbing fixtures during the progress of the work is prohibited.

r. Angle stops, straight stops, stops integral with the faucets, or concealed type of lock shield, and loose key pattern stops shall be furnished and installed with fixtures. Exposed traps and supply pipes for fixtures and equipment shall be connected to the rough piping systems, unless otherwise specified under the item. All exposed fixtures and equipment supply and piping, fittings, valves, stops, traps, escutcheons, washers, nuts, etc., shall be chromium plated with polished finish. Drain lines and hot water lines of fixtures for handicapped personnel usage shall be insulated. Grout areas where fixture surfaces rest against wall or floor surfaces with clean white plaster or silicone.

s. Fixture Connections: Where space limitations prohibit the use of standard fittings, special short radius fittings shall be provided.

t. Flush Valves: Flush valves shall be secured by anchoring to wall adjacent to valve with an approved metal bracket to prevent movement.

u. Shower Bath Outfits: The area around the water supply piping to the mixing valves shall be made watertight.

v. Fixture Supports: Fixture supports for off the floor lavatories, urinals, water closets, and other fixtures of similar size, design, and use, shall be of the chair carrier type and shall conform to ANSI A112.6.

3.4 DISINFECTION OF DOMESTIC WATER SYSTEM PIPING

3.4.1 The Contractor shall disinfect water piping before it is placed in service.

The Contractor shall furnish all equipment and materials necessary to do the work of disinfecting, and shall perform the work in accordance with the procedure outlined in the AWWA Standard for Disinfecting Water Mains, Designation C601-68. Chlorination is detailed in AWWA Standard M20.

The dosage shall be such as to produce a chlorine residual of not less than 10 ppm after a contact period of no less than 24 hours. After treatment, the piping shall be flushed with clean water until the residual chlorine content does not exceed 0.2 ppm.

During the disinfection period, care shall be exercised to prevent contamination of water in the street main.

3.5 GAS PIPING SYSTEM

3.5.1 The Plumbing Contractor shall provide all labor and materials for the installation of a complete system of gas piping from Gas Company's meter service to roof top heating equipment, gas-fired water heater, and emergency generator.

a. The Plumbing Contractor shall provide a complete propane gas piping system including gas cylinders and regulating valve.

3.5.2 Gas piping throughout the building shall be Schedule 40 black steel with malleable iron 150# beaded pattern screwed fittings or welded fittings in sizes approved by the local authorities. Install interior piping in ventilated chases.

a. All piping on roof shall be welded steel. Wrap all buried piping in accordance with the local code and utility company's requirements.

3.5.3 Gas service lines from exterior gas meter or meter bank to various consumers shall be furnished and installed by the Plumbing Contractor. Occupant to make application to the Gas Company for meter service to be installed by the Gas Company.

3.5.4 Gas service lines from meter banks to consumers' spaces. These lines shall be terminated at a plug valve approximately 6" from inside face of exterior walls or partition walls and shall be located above the finished floor.

3.5.5 The gas distribution systems installations shall include all pipe, fittings, valves, and all accessories and incidentals to conform with the code and Gas Company requirements.

a. All horizontal lines shall grade to risers and from risers to the equipment or appliances. Provide drips in any point in the line where condensate may collect, and are to be installed in such locations that they will be readily accessible. All risers and drops shall be provided with drips at the base consisting of full size tees with 6" nipples and caps for drainage. After piping had been checked, all piping receiving gas shall be fully purged.

3.5.6 No gas pipe shall be installed in ceiling spaces which are used as return air plenums for air conditioning equipment. Gas piping in these areas must be installed on roof or a system of pipe within a pipe and supported.

3.5.7 The Plumbing Contractor shall make all final connections between each piece of equipment requiring gas and the gas distribution systems.

3.5.8 Furnish and install gas shut-off valves at each piece of equipment, appliance, at base of each riser and where indicated on the Drawings.

3.6 TESTING

3.6.1 General:

a. Labor, materials, instruments, and power required for testing shall be furnished by the Plumbing Contractor. All tests shall be performed in the presence and to the satisfaction of the Plumbing Inspector and such other parties as may have legal jurisdiction. No piping in any location shall be closed up, furred in, or covered before testing.

The plumbing system shall be tested in accordance with NAPHCC National Standard Plumbing Code.

b. Test Procedures:

(1) The Contractor shall furnish detailed test procedures for testing the system.

c. Make tests in stages so ordered by the Engineer to facilitate the work of other trades.

d. Repair, or if required by the Architect, replace, defective work with new work without extra charge to the Owner. Repeat tests as directed, until all work is proven satisfactory.

e. Test all fixtures for soundness, stability, or support and satisfactory operation.

f. Notify the Engineer and Inspectors having jurisdiction at least 48 hours in advance of making the required tests, so that arrangements may be made for their presence to witness the tests.

g. The Plumbing Contractor shall obtain certificates of approval from the State and all other local governmental agencies. Each certificate of approval shall be delivered to the Engineer before final acceptance.

3.6.2 Storm and Sanitary System:

a. Before the installation of fixtures, equipment, tanks and insulation, the entire storm, sanitary, and acid waste drainage piping systems including all vents shall have all necessary openings plugged to permit the entire system to be filled with water to the level of the highest vent stack of each system above the roof where practical. The systems shall hold this water for four hours without a drop in water level. Where a portion of the systems is to be tested, the test shall be conducted in the same manner as described for the entire system, except that a vertical stack 10' above the highest horizontal line to be tested may be installed and filled with water to supply the required pressure. The pressure shall be maintained for a minimum of four hours.

3.6.3 Acid Waste System:

a. Same test as storm and sanitary systems.

3.6.4 Cold Water and Hot Water Systems:

a. Upon completion of the roughing-in and before setting fixtures and final connections to all equipment, all water piping systems shall be tested not less than the hydrostatic pressures specified herein and proved tight at these pressures for not less than four hours in order to permit inspection of all joints. Where a portion of the water piping systems is to be concealed before completion, this portion shall be tested separately in a manner described for the entire system.

(1) Systems operating on city pressure shall be tested to a pressure of 150 psi gauge but not less then 1.5 times the working pressure.

(2) All boosted water piping systems shall be tested to a pressure of 250 psi gauge but not less then 1.5 times the working pressure.

3.6.5 Gas Piping System:

a. Testing of gas piping shall conform to the requirement of the utility company and authorities having jurisdiction.

3.6.6 Compressed Air System:

a. The compressed air piping systems shall be tested at a hydrostatic pressure of 300 psi gauge and proved tight at this pressure for not less than 2 hours. If the pressure drops, the Plumbing Contractor shall make all repairs and alterations in the system necessary to meet a retest. Retest to pressures and duration specified herein.

3.6.7 Air & Oil Piping Systems:

a. Air and oil piping systems shall be tested with 200 psi air pressure for a period of 30 minutes.

3.6.8 Performance Test Reports:

a. Upon completion and testing of the installed system, test reports shall be submitted showing all field tests performed to prove compliance with the specified performance criteria.

3.7 CONNECTIONS TO EQUIPMENT NOT FURNISHED UNDER THIS SECTION OF THE WORK

3.7.1 Mechanical Equipment:

a. The Plumbing Contractor shall provide water services to all heating and air conditioning equipment requiring same and shall make all final connections in strict accordance with equipment manufacturer requirements and per State Plumbing Codes. Where final connections are made by the HVAC Contractor, piping shall be terminated at a gate valve approximately 36" from the equipment.

3.7.2 Kitchen & Cafeteria Equipment:

a. All equipment will be furnished and set in place by the Food Service Equipment Contractor. Plumbing Contractor shall furnish all labor and material for the installation of all roughing as shown on Drawings and in accordance with Food Service Equipment Manufacturer's shop drawings. Plumbing Contractor shall be present when equipment is being set in place and make all final plumbing connections as required and in such manner as not to interfere with progress of the job.

3.7.3 Beauty Salon Equipment:

a. The following listed equipment will be furnished by the beauty salon and shall be installed by the Plumbing Contractor.

(1) Shampoo sinks

(2) Washer (Automatic cloth washer)

(3) Flo-Trol model 39 pressure valves

b. The Plumbing Contractor shall receive the equipment at the job site and shall immediately check the equipment for breakage, missing, or defective parts and shall notify Owner or supplier of any such deficiencies. All equipment and fixtures shall be stored in a safe place and installed as soon as possible. All roughing for waste and water supplied shall be done in accordance with shop drawings furnished by the beauty salon operator.

c. Plumbing Contractor shall furnish and install hair intercepting traps for shampoo sinks if required by code.

3.8 WASTE OIL TANK & PIPING

3.8.1 Furnish and install a 550 gallon, 42" diameter x 84" long UL approved underground waste oil tank complete with fittings and piping as shown on the Drawings. Tank to have 1 coat of black asphaltum and shall be tested tight at 5 pound air pressure. Size is for example only.

3.8.2 All underground waste oil piping shall be standard weight galvanized Yoloy with galvanized 150# screwed malleable iron fittings and above floor piping to be Schedule 40 galvanized steel piping with 150# galvanized screwed malleable fittings.

3.8.3 3" extractor line cap shall be Buckeye No. A0517 or approved equal as manufactured by Moore & Kling Company.

3.8.4 Grade box shall be Buckeye No. A0717 or approved equal as manufactured by Moore & Kling Company.

3.9 AIR PIPING

3.9.1 Plumbing Contractor shall furnish and install all air piping from compressors to equipment and fixtures requiring compressed air as shown on Drawings and shall make all piping connections to equipment. All air underground piping to lifts shall be standard weight galvanized Yoloy with 300# galvanized malleable iron fittings.

a. All above ground air piping shall be standard weight galvanized steel pipe with 300# galvanized malleable iron fittings.

3.10 REMOVALS, REPLACEMENTS & ADJUSTMENTS

3.10.1 The Plumbing Contractor shall remove, relocate, replace, adjust, or adapt all existing piping, fixtures, and other plumbing equipment or apparatus as required by the Drawings and Specifications, and also as required when such plumbing work is uncovered or when found to interfere in any way with carrying out and completing the work of this Contractor or other Contractors, including Contractors for General Construction, Heating and Ventilating, Sprinkler, and Electrical Work. The work shall include the furnishing of all materials, all necessary extensions, connections, cutting, core drilling, repairing, adapting, and other work incidental thereto, together with such temporary connections as may be required to maintain service pending completion of the permanent work, and shall be left in good working order and in a condition equal to the adjacent new or existing work.

3.10.2 When existing fixtures which are to remain, are disconnected from present systems, or piping that are to be removed, these fixtures shall be reconnected to new plumbing and drainage system as shown on plans or as required.

3.10.3 New work shall be installed in sections and made ready for a quick tie-in to existing fixtures and equipment before removing old lines, to obtain the shortest period of non-service possible for each case. The outage times shall coincide with the staging of the work for specific areas of the building as closely as possible. Where new lines are shown running in the same locations previously occupied by existing lines which are to be removed, minimize the time of substitution and reactivation.

Index

H

I

K

L

M

P

Other Titles Offered by BNP

Refrigeration Licenses Unlimited, Second Edition
A/C Cutters Ready Reference
How to Solve Your Refrigeration and A/C Service Problems
Blueprint Reading Made Easy
Starting in Heating and Air Conditioning Service
Legionnaires' Disease: Prevention & Control
Schematic Wiring, A Step-By-Step Guide
Sales Engineering
Valve Selection and Service Guide
Good Piping Practice
Hydronics
Refrigeration Fundamentals
Humidity Control
How to Design Heating-Cooling Comfort Systems
Industrial Refrigeration
Modern Soldering and Brazing
Inventing from Start to Finish
Indoor Air Quality in the Building Environment
Electronic HVAC Controls Simplified
4000 Q & A for Licensing Exams
Heat Pump Systems and Service
Ice Machine Service
Troubleshooting and Servicing A/C Equipment
Stationary Engineering
Medium and High Efficiency Gas Furnaces
Water Treatment Specifications Manual, Second Edition
The Four R's: Recovery, Recycling, Reclaiming, Regulation
SI Units for the HVAC/R Professional

BUSINESS NEWS
PUBLISHING COMPANY
Troy, Michigan
USA